Dissertation

Generalized Coorbit Theory and Applications to Shearlets

Lukas Sawatzki

2019

Bibliographic information published by the Deutsche Nationalbibliothek

The Deutsche Nationalbibliothek lists this publication in the Deutsche Nationalbibliografie; detailed bi graphic data are available in the Internet at http://dnb.d-nb.de .

ISBN 978-3-8325-5106-3

Logos Verlag Berlin GmbH

Comeniushof, Gubener Str. 47,

10243 Berlin

Tel.: +49 030 42 85 10 90

Fax: +49 030 42 85 10 92

INTERNET: http://www.logos-verlag.de

Generalized Coorbit Theory
and Applications to Shearlets

Dissertation

zur

Erlangung des akademischen Grades

Doktor der Naturwissenschaften

(Dr. rer. nat.)

vorgelegt

dem Fachbereich Mathematik und Informatik

der

Philipps–Universität Marburg

von

Lukas Sawatzki

geboren am 17. Februar 1992

in Hamburg

Am Fachbereich Mathematik und Informatik

der Philipps-Universität Marburg (Hochschulkennziffer: 1180)

als Dissertation eingereicht am: 2. Oktober 2019

Erstgutachter: **Prof. Dr. Stephan Dahlke**, Philipps-Universität Marburg

Zweitgutachterin: **Prof. Dr. Gabriele Steidl**, Technische Universität Kaiserslautern

Drittgutachter: **Prof. Dr. Filippo De Mari**, Universität Genua

Tag der mündlichen Prüfung: 24. Januar 2020

THE ROAD TO WISDOM
by Piet Hein (1905–1996)

The road to wisdom?—Well, it's
plain and simple to express:
Err
and err
and err again,
but less
and less
and less.

Für Franzi.

Contents

Acknowledgement

There are numerous people who supported me throughout the last three years and whom I would like to mention here. First and foremost I thank my advisor Stephan Dahlke for accepting me as a Ph.D. student and introducing me to research. I am grateful for your guidance not only in mathematical matters, but also concerning scientific culture. You gave me the chance to participate in various conferences—which I enjoyed very much—as well as the chance to collaborate with other researchers. I have learned a great deal in the past years because of the opportunities you have opened up for me. Thank you for your trust.

I would also like to thank Filippo De Mari, Ernesto De Vito, Gabriele Steidl, Gerd Teschke and Felix Voigtlaender for the collaboration; it was an enlightening experience working with you.

Furthermore, my thanks go out to my colleagues in the Workgroup Numerics, for the necessary daily coffee breaks and vivid discussions on topics far, far away from mathematics.

A special thank you to Deutsche Bahn for taking me (almost always) reliably through the republic. And to my friends and family I am indebted for your support and distraction, something invaluable after tackling unintelligible problems. I thank my parents for encouraging me to take this path and last but definitely not least I thank my fiancée Franzi for the constant encouragement, support and patience.

Introduction

This thesis deals with specific problems arising in the context of signal analysis. In general the main goal in signal analysis is the efficient extraction of information from a given signal. For this purpose the signal—usually modeled in suitable function spaces—needs to be preprocessed, denoised, compressed, etc. There are two main steps necessary for this. The first step is to apply a suitable transformation specifically designed to extract the desired properties, for example the wavelet transform, Gabor transform or shearlet transform. Which transform to choose obviously depends on what type of information one wants to extract. Many interesting transformations are connected to representations of certain locally compact groups. For instance, the wavelet transform is associated with the affine group whereas the Gabor transform is related to the Weyl-Heisenberg group. After transforming, the second step is to decompose the signal into universal building blocks that match the application. Similarly, this process needs to be reversible, thus a reconstruction technique is necessary to regain signals from their discrete decomposition. A stable discretization of signals is inevitable for applications, such as numerics. In order to perform computations involving the signal, a digital representation of the signal is needed. And for this the first step is a uniform discretization of families of signals, that is, the decomposition and reconstruction is performed according to a fixed technique.

Coorbit theory is designed to find function spaces related to transformations of functions that are related to representations of locally compact groups and to describe uniform discretization techniques for these spaces. This theory was originally developed by Feichtinger and Gröchenig [44–46, 67, 68] in the 1980s. The main idea is to measure the smoothness of a function via properties of the transform of the signal. To be more precise, one asks whether the transform is contained in certain function spaces on the index set of the transform, which usually is the underlying group, and this allows to define spaces of functions associated to the transform. Moreover, coorbit theory provides a uniform approach to discretize these function spaces and to characterize the associated discrete sequence spaces, where the sequence spaces directly depend on the group structure. This way the existence of Banach frames and atomic decompositions is naturally ensured. By an application of this theory classical homogeneous Besov-Triebel-Lizorkin spaces [97, 98, 100] can be identified as coorbit spaces [101] via the wavelet transform. Similarly, modulation spaces are related to the Gabor transform [41, 69], Bergman spaces can be treated as well [49] and there are applications of the classical coorbit theory to various shearlet transforms [25].

Classical Coorbit Theory

The original coorbit theory as developed by Feichtinger and Gröchenig [44–46, 67, 68] relies on a locally compact topological group G and a representation π of the group on a Hilbert space $\mathcal{H}$. Associated to an admissible vector $\psi \in \mathcal{H}$ the voice transform V_ψ is given as the map

$$V_\psi \colon \mathcal{H} \to L_2(G), \quad V_\psi f(x) = \langle f, \pi(x)\psi \rangle_{\mathcal{H}}.$$

The admissibility of ψ is just the well-definedness of the map above, we also say the map is square-integrable. This voice transform takes the role of the afore mentioned transform related to the group G and it lays the foundation for the coorbit spaces. One very important property of the voice transform is the fact that the convolution from the right with the element $V_\psi \psi$ is the identity. By denoting with $K_\psi = V_\psi \psi$ the so-called reproducing kernel, then this means $V_\psi f * K_\psi = V_\psi f$ for all $f \in \mathcal{H}$. For some fixed weight w on G we can now define the spaces $\mathcal{H}_{1,w} = \{f \in \mathcal{H} : V_\psi f \in L_{1,w}(G)\}$. The motivation behind these spaces is that we are interested in their dual space $\mathcal{H}'_{1,w}$. This set of distributions will then be the reservoir from which we choose the elements of the coorbit spaces. In order to do so we first extend the voice transform to the dual space $\mathcal{H}'_{1,w}$ via a Gelfand triple setting and denote the extension with $V_{e,\psi}$. Then, we are interested in those functionals in $\mathcal{H}'_{1,w}$, whose extended voice transform decays rapidly enough. In other words, they need to be contained in a certain function space Y, which characterizes the decay of transforms:

$$\text{Co}(Y) = \{T \in \mathcal{H}'_{1,w} : V_{e,\psi}T \in Y\}.$$

Examples for the space Y include L_p-spaces and their weighted counterparts. Under certain assumptions these coorbit spaces are well-defined Banach spaces. Another very important and much used property of the coorbit spaces is that they are isometrically isomorphic to the reproducing kernel Banach space $\{f \in Y : f * K_\psi = f\}$. This means it suffices to prove almost all properties for the latter space and then transfer them to the coorbit space. In other words, the reproducing kernel space already contains all structural properties. Finally, the coorbit theory provides a unified approach to discretize coorbit spaces. To be more precise, the existence of atomic decompositions and Banach frames is ensured. This is a very powerful tool to analyze the structure and other properties of coorbit spaces.

Coorbit Theory with Non-Integrable Kernel

The theory above is based on several fundamental assumptions, one of them states that the representation π is not only square-integrable but also integrable. This means the reproducing kernel K_ψ is an element of $L_{1,w}(G)$. Unfortunately, this condition is quite restrictive and even in the example of Paley-Wiener spaces not fulfilled. To work around this Dahlke et al. [18] have developed a coorbit theory for those spaces where the kernel is contained in a general Fréchet space of functions. This includes the setting where K_ψ is not contained in $L_{1,w}(G)$ but in all weighted Lebesgue spaces $L_{p,w}(G)$ for $1 < p < \infty$. We also say the kernel is non-integrable. In fact, it has been shown that also in this setting associated coorbit spaces are well-defined; the theory, however, needs certain adjustments. We need, for example, a substitute for the space $\mathcal{H}_{1,w}$ and accordingly the voice transform needs to be extended differently.

Yet, one question was left unanswered in [18]. It was unclear how one can obtain discretization results for these spaces as in the classical case, including atomic decompositions and Banach frames. The answer to this question is a part of this thesis and the results have been published in [19]. The first surprising result is the fact that the methods from the classical setting do not carry over to the Fréchet setting. Instead we present a different result that does not require the integrability of the kernel and therefore uses convolution inequalities other than Young's inequality. More precisely, Theorem 3.2.17 includes the following statement concerning reconstruction operators: for sequences $d \in \ell_{q,m}(I)$ with countable index set I in certain weighted sequence spaces we have

$$T = \sum_{i \in I} d_i \pi(x_i)\psi \in \text{Co}(L_{r,m}) \quad \text{for } q < r.$$

Here, $(x_i)_{i \in I}$ are certain points in G that are sufficiently close together. The price we pay for this result is that the integrability parameters of the discrete norm of the coefficient space and of the coorbit norm are different. This may appear strange at first but is a direct consequence of the altered convolution inequalities we employ. Conversely, we also obtain decompositions of functions, but only approximative. This means for every $\varepsilon > 0$ there is a finite sequence $c \in \ell_{r,m}(J)$ such that

$$\left\| T - \sum_{j \in J} c_j \pi(x_j) \psi \right\|_{\mathrm{Co}(L_{r,m})} \leqslant \varepsilon.$$

Both the reconstruction and the decomposition of functions is continuous, but not uniform.

The decomposition techniques described above are based on one assumption, namely we assume the convolution with the reproducing kernel K_ψ is a bounded operator on $L_{r,m}(G)$. This is not clear for non-integrable kernel and is therefore an additional condition. In Theorem 3.3.1 we show, however, that this condition is not only sufficient for a first discretization but also necessary for atomic decompositions and Banach frames for coorbit spaces. Therefore the assumption is inevitable and it seems like we have reached the limit of what is possible.

Still, the results above are not optimal. To obtain proper atomic decompositions and Banach frames similar to the classical coorbit theory we introduce the additional assumption that there is a second kernel W on G satisfying, among others, the following conditions:

(i) $W \in L_{1,w}(G)$,

(ii) $W * K_\psi = K_\psi$ or $K_\psi * W = K_\psi$, respectively.

With these properties at hand, Theorem 3.4.8 shows that the family $(\pi(x_i)\psi)_{i \in I}$ constitutes a Banach frame for suitably chosen discrete points $(x_i)_{i \in I} \subset G$. Analogously, Theorem 3.4.15 shows that under very similar conditions the same family of functions paves the way for atomic decompositions. It is therefore possible to describe exact conditions under which the same discretization results hold for coorbit spaces with non-integrable kernel as in the classical setting.

Generalized Coorbit Theory

The extension of coorbit theory described above is not the only generalization developed. In the last 15 years the coorbit theory has been rediscovered and generalized in many forms. For example, the classical theory only admitted Banach spaces of functions Y as target spaces and this was later extended to quasi-Banach spaces [90]. In [79] this has been further expanded to the setting of quasi-Banach spaces with variable smoothness and integrability. Moreover it was recognized that many interesting examples do not fit the group setting described above and the theory was generalized by Dahlke et al. to also fit the setting of homogeneous spaces, that is, quotients of groups via subgroups G/H [21, 26, 27]. This allowed to view modulation spaces on spheres from the standpoint of coorbit theory, as well as α-modulation spaces. The latter spaces were originally introduced by Feichtinger and Gröbner and can be seen as a merging of modulation spaces and Besov spaces [43]. Another approach is to exploit symmetry properties of functions, where a group is considered modulo a symmetry group, like radial symmetry [88, 89].

The approach we will discuss in the following is based on the realization that group theory is not needed at all to develop a coorbit theory [53]. This was originally proposed by Fornasier and Rauhut and later extended [76, 102]. The main idea is, instead of having a group, to take an arbitrary measure space (X, μ) as the index set of our transform. Without a group structure available the notion of representations of said groups on Hilbert spaces makes no sense. Instead, we replace both by continuous frames [1]. Assume we have a family of functions $\mathfrak{F} = (\psi_x)_{x \in X}$

indexed by the measure space X, which constitutes a continuous frame for the Hilbert space $\mathcal{H}$. Then the frame elements replace the functions $\pi(x)\psi$ above and therefore the voice transform associated to $\mathfrak{F}$ is defined via

$$V_{\mathfrak{F}} \colon \mathcal{H} \to L_2(X, \mu), \quad V_{\mathfrak{F}} f(x) = \langle f, \psi_x \rangle_{\mathcal{H}}.$$

Contrary to the setting before, we have no convolution available on arbitrary measure spaces. Instead, we assign a kernel operator to functions $K \colon X \times X \to \mathbb{C}$ via the integral

$$Kf(x) = \int_X K(x, y) f(y) \, d\mu(y).$$

If we now look at the kernel operator associated to the kernel $K_{\mathfrak{F}}(x, y) = V_{\mathfrak{F}} \psi_y(x)$, then the operator fulfills a similar reproducing identity as above given by $K_{\mathfrak{F}} V_{\mathfrak{F}} f = V_{\mathfrak{F}} f$ for all $f \in \mathcal{H}$. In this sense the kernel operator is a substitute, or generalization, of convolution. To measure the integrability of the reproducing kernel we introduce weighted kernel spaces $\mathcal{A}_{p,w}$ for weights w on $X \times X$ and integrability parameters $1 \leqslant p \leqslant \infty$. Obviously, we need substitutes for Young's inequality for the kernel spaces, which we provide. Then, by assuming $K_{\mathfrak{F}} \in \mathcal{A}_{1,w}$, in a certain sense the integrability of the kernel, we can define the test spaces $\mathcal{H}_{1,v} = \{ f \in \mathcal{H} : V_{\mathfrak{F}} f \in L_{1,v} \}$ similar to the classical case. Again, the voice transform can be extended by $V_{e,\mathfrak{F}}$ to the dual space $\mathcal{H}'_{1,v}$, which serves as a reservoir for the coorbit spaces. For suitable function spaces Y measuring the decay rate of the voice transform, the corresponding coorbit spaces are given by

$$\mathrm{Co}(Y) = \{ T \in \mathcal{H}'_{1,v} : V_{e,\mathfrak{F}} T \in Y \}.$$

These spaces have the same properties as above, that is, they are Banach spaces and they are isometrically isomorphic to the reproducing kernel space $\{ f \in Y : K_{\mathfrak{F}} f = f \}$. Likewise, we can pose conditions under which the existence of both atomic decompositions and Banach frames is ensured.

Generalized Coorbit Theory with Non-Integrable Kernel

As in the group setting, the theory described above relies on several fundamental assumptions. One assumption is named above, namely the integrability of the reproducing kernel with respect to weighted kernel spaces: $K_{\mathfrak{F}} \in \mathcal{A}_{1,w}$. This assumption is restrictive and not always fulfilled. We therefore develop a new theory that allows frames indexed by arbitrary measure spaces but also includes $K_{\mathfrak{F}} \notin \mathcal{A}_{1,w}$. Parts of the results have been published in [50]. The main idea is to assume $K_{\mathfrak{F}} \in \mathcal{A}_{p,w}$ for all $1 < p < \infty$, which is a weaker assumption than the integrability above. This setting is similar to coorbit theory with non-integrable kernel and group structure. Also in this case we can find substitutes for the test space $\mathcal{H}_{1,v}$ and it is possible to properly define meaningful coorbit spaces.

Yet again we are faced with the challenge of finding discretizations for these new coorbit spaces. For this we apply similar ideas as the ones for coorbit spaces with non-integrable kernel in the group case and obtain comparable results. To be more precise, we show the following reconstruction result in Theorem 5.2.17: for a countable sequence $d \in \ell_{q,m}(I)$ in a certain weighted sequence space we have

$$T = \sum_{i \in I} d_i \psi_{x_i} \in \mathrm{Co}(L_{r,m}) \quad \text{for } q < r,$$

where the discrete points $(x_i)_{i \in I} \subset X$ are chosen sufficiently close together, which is achieved using suitable coverings of the index set. Again, the downside is the different integrability parameters of the discrete norm of the coefficient space and the coorbit norm. Conversely, we can decompose the elements approximately in the following fashion. Take $\varepsilon > 0$, then there is a finite sequence $c \in \ell_{r,m}(J)$ such that

$$\left\| T - \sum_{j \in J} c_j \psi_{x_j} \right\|_{\mathrm{Co}(L_{r,m})} \leqslant \varepsilon.$$

These results require an additional assumption on the reproducing kernel $K_{\mathfrak{F}}$, that is, we assume the corresponding kernel operator is bounded as an operator on weighted Lebesgue spaces $L_{p,m}(X,\mu)$. This assumption needs to be checked individually for applications. Furthermore, we show in Theorem 5.3.1 that this assumption is also necessary for atomic decompositions and Banach frames to exist. So again, the boundedness of the kernel operator is inevitable and it seems like we have reached the limit of what is possible here as well.

As before, however, the results can still be improved under the following assumption. We assume there exists a second kernel W which, among others, fulfills the following conditions:

(i) $W \in \mathcal{A}_{1,w}$,

(ii) $W \circ K_{\mathfrak{F}} = K_{\mathfrak{F}}$ or $K_{\mathfrak{F}} \circ W = K_{\mathfrak{F}}$, respectively.

The operation $\circ$ denotes the multiplication of two kernels. This additional kernel W allows us to prove in Theorem 5.4.7 that the family $(\psi_{x_i})_{i \in I}$ constitutes a Banach frame for $\mathrm{Co}(L_{r,m})$ for suitably chosen points $(x_i)_{i \in I}$. And similarly Theorem 5.4.11 shows that under very similar assumptions and using the same family of functions the existence of atomic decompositions is also ensured.

Applications

There are numerous applications of coorbit theory, some were mentioned above. The following two additional applications provide good examples of the developed theory. The Paley-Wiener spaces illustrate how the coorbit theory for non-integrable kernels in the group setting can be applied to find proper discretization results. And we apply the ideas of generalized coorbit spaces with non-integrable kernel to certain shearlet frames to define new function spaces.

Paley-Wiener Spaces

The Paley-Wiener space B_{Ω}^p is the collection of functions in the Lebesgue space $L_p(\mathbb{R})$, where the Fourier transform is supported in a fixed subset $\Omega \subset \mathbb{R}$. In [18] it was shown that these spaces can be interpreted as coorbit spaces with non-integrable kernel and the underlying group is the additive group $\mathbb{R}$. The reproducing kernel is then given by the sinc-function, which is clearly not an element of $L_1(\mathbb{R})$. The developed theory is therefore applicable and we show that for different subsets Ω, the Paley-Wiener spaces represent both positive and negative examples for the discretization ideas. The negative examples include compact subsets, for which the convolution operator associated with the reproducing kernel is not bounded on Lebesgue spaces. Hence, the discretization techniques are not applicable. There are, however, also positive examples where the contrary is true and we can indeed show the existence of Banach frames and atomic decompositions. This includes the symmetric intervals $\Omega = [-\omega, \omega]$ and provides an alternative proof for the Whittaker-Kotelnikov-Shannon sampling theorem [63, 96].

Shearlets

As mentioned above, the classical coorbit theory can be applied to wavelets, yielding homogeneous Besov spaces. While wavelets are especially suitable in signal analysis for identifying isolated singularities, they are less efficient when dealing with signals with anisotropic features due to their isotropic nature. Since the identification of anisotropic features of signals, such as directional information, is of great importance in practice, other directional representation systems have been developed like curvelets [10, 12, 94], ridgelets [11], contourlets [34, 35] or shearlets.

Shearlets were designed as an extension to wavelets in multiple dimensions. While wavelet systems only consist of isotropically dilated and translated versions of a mother function, shearlet systems consist of anisotropically dilated, translated and sheared copies of the function. The additional shearing parameter allows changing the direction of the function, which is supported by the anisotropy. This makes them especially well-suited to deal with localized directional features in a signal. Indeed, it was shown in [80] that the shearlet transform can be used to resolve the wavefront set of a signal and in [72] that the shearlets yield an optimal N-term approximation error for cartoon-like functions. For further insights on shearlet algorithms we refer to [81].

Another great advantage of shearlets, which sets them apart from the other systems mentioned above, is the fact that the continuous shearlet transform, introduced and investigated in [24, 25, 28, 71], stems from the action of a square-integrable representation of a topological group, the so-called full shearlet group $\mathbb{S} = \mathbb{R}^* \times \mathbb{R}^{d-1} \times \mathbb{R}^d$. This makes it possible to apply the classical coorbit theory described above, which was investigated by Dahlke et al. in a series of papers [22, 29]. Since the shearlets used in the construction of these shearlet coorbit spaces are required to have vanishing moments, any polynomial part in a signal is ignored by the transform. This leads to the resulting spaces being homogeneous spaces, in the sense that, intuitively speaking, the shearlet transform possesses a "blind spot" in the Fourier domain. In practice, however, the smoothness spaces being used, for example to analyze the regularity of the solution space of an operator equation, are not homogeneous. This calls for inhomogeneous smoothness spaces related to the shearlet transform. We note there are already approaches, though not based on coorbit theory, to develop inhmogeneous shearlet smoothness spaces. In [82] Labate et al. used the notion of decomposition spaces [42, 43], while in [103, 104] Vera applied the framework of the φ-transform, introduced by Frazier and Jawerth [54], for this purpose.

In our approach we use the ideas of the inhomogeneous wavelet transform [53] and transport them to the shearlet group. This means we define the index space

$$X = \big(\{\infty\} \times \mathbb{R}^{d-1} \times \mathbb{R}^d\big) \cup \big([-1, 1]\backslash\{0\} \times \mathbb{R}^{d-1} \times \mathbb{R}^d\big),$$

where the left part is designed to analyze the low frequency part of a signal, which is ignored in the homogeneous setting. Using this index space we define a corresponding frame $\mathfrak{F} = (\psi_x)_{x \in X}$ via

$$\psi_{(\infty, s, t)} = \Phi(S_s^{-1}(\cdot - t)) \quad \text{and} \quad \psi_{(a, s, t)} = |\det A_a|^{-1/2} \Psi(A_a^{-1} S_s^{-1}(\cdot - t)),$$

where the functions Φ and Ψ can be chosen such that $\mathfrak{F}$ is a continuous frame for $L_2(\mathbb{R}^d)$, the so-called inhomogeneous shearlet frame. The corresponding reproducing kernel is in fact contained in all weighted kernel space $\mathcal{A}_{p,w}$ for $p > 1$ such that the generalized coorbit theory with non-integrable kernel is applicable. This way we can introduce new inhomogeneous shearlet coorbit spaces and give a first reconstruction result for these spaces.

Outline

This thesis starts in Chapter 1 with some preliminaries. This includes standard concepts as well as new results, including new convolution inequalities for weighted Lebesgue spaces similar to Young's inequality in Subsection 1.3.2 and new weighted kernel spaces for arbitrary measure spaces including norm inequalities for kernel operators in Section 1.8. In Chapter 2 we recall the classical coorbit theory and their discretization techniques, as well as two important applications of the theory, namely homogeneous Besov spaces and shearlet coorbit spaces. Then, in Chapter 3 we recall the coorbit theory with non-integrable kernel and show new discretization results. We add several applications of the Paley-Wiener spaces to motivate the theory. In Chapter 4 we recall the generalized coorbit theory as well as their discretizations and include the example of inhomogeneous Besov spaces. Finally, in Chapter 5 we present the new generalized coorbit theory with non-integrable kernel and prove the well-definedness of coorbit spaces and some discretization results. This is followed by the application to the newly defined inhomogeneous shearlet coorbit spaces. In the final Chapter 6 we include some conclusions and discuss proceeding problems and ideas that might invigorate further research.

Chapter 1

Preliminaries

The aim of this chapter is to recall and to introduce some mathematical concepts and notations forming the foundation of this dissertation. While some of these concepts are standard knowledge in applied and harmonic analysis, others are not very common and in parts altered or complemented for our setting.

We start by introducing notational conventions in Section 1.1 followed by brief insights in well-known areas of group theory, function space theory and representation theory for groups in Sections 1.2, 1.3 and 1.4, respectively. In Section 1.5 we introduce coverings of index spaces, which are used in Section 1.6 to define sequence spaces associated with certain spaces of functions. Then, in Section 1.7 we recall the theory of continuous frames followed by a definition of kernel spaces and a collection of important properties in Section 1.8. Finally, in Section 1.9 we recall two fundamental discretization techniques for function spaces, namely atomic decompositions and Banach frames.

1.1 Notations and Conventions

Here, we fix standard notations used throughout this thesis. More specific notations are introduced in the text and not listed here.

- The letters $\mathbb{Z}$, $\mathbb{Q}$, $\mathbb{R}$ and $\mathbb{C}$ have their usual meaning. We denote the natural numbers by $\mathbb{N} = \{n \in \mathbb{Z} : n \geq 1\}$ and $\mathbb{N}_0 = \mathbb{N} \cup \{0\}$. We also use the conventions $\mathbb{R}^* = \mathbb{R} \backslash \{0\}$ and $\mathbb{R}_+$ for all positive real numbers and $\mathbb{R}_{\geq 0}$ for all non-negative real numbers.

- For a dimension $d \in \mathbb{N}$ we denote with $\mathbb{R}^d$ the Euclidean d-space and for two elements $x, y \in \mathbb{R}^d$ we use the canonical inner product

$$\langle x, y \rangle = \sum_{i=1}^{d} x_i y_i.$$

  Any other inner product will be provided with a subscript.

- If $d \geq 2$ we write $x = (x_1, \tilde{x})$, where $\tilde{x} = (x_2, \ldots, x_d) \in \mathbb{R}^{d-1}$

- We write $I_d \in \mathbb{R}^{d \times d}$ for the d-dimensional identity matrix and $0_d \in \mathbb{R}^d$ for a vector containing only zeros.

- For $x \in \mathbb{R}^d$ we use $\|x\|_2$ for the Euclidean norm, $\|x\|_1$ for the 1-norm and $\|x\|_\infty$ for the maximum norm of x.

- If $A \in \mathbb{R}^{d \times d}$ is a matrix, $\|A\|_{2 \to 2}$ denotes the spectral norm of A.

- For two sets X, Y we write $X \subset Y$ if X is a proper subset of Y with $X \neq Y$ and we write $X \subseteq Y$ if we allow $X = Y$.

- For a set X and any subset $M \subset X$ the indicator function $\chi_M : X \to \{0, 1\}$ is defined as

$$\chi_M(x) = \begin{cases} 1, & \text{if } x \in M, \\ 0, & \text{if } x \notin M. \end{cases}$$

- If X is a topological space and $M \subset X$ a subset, we denote with $\overline{M}$ the closure of M and with $\mathring{M} = \operatorname{int} M$ the interior of M.

- The Lebesgue measure on $\mathbb{R}^d$ is denoted by $\mathrm{d}x$.

- For a function $f : \mathbb{R} \to \mathbb{C}$ and $n \in \mathbb{N}$ the n-th derivative is denoted by $\frac{\mathrm{d}^n f}{\mathrm{d}x^n}$.

- For a function $g : \mathbb{R}^d \to \mathbb{C}$ and a multiindex $\alpha \in \mathbb{N}_0^d$ the partial derivative is denoted by $\frac{\partial^\alpha g}{\partial x^\alpha}$. If $\alpha \in \mathbb{N}_0^d$ is the i-th unit vector then the i-th partial derivative is given by $\frac{\partial g}{\partial x_i}$.

- We denote with $\mathcal{C}^k(\mathbb{R}^d)$, $k \in \mathbb{N}_0$, the set of functions $f : \mathbb{R}^d \to \mathbb{C}$ for which all partial derivatives $\frac{\partial^\alpha f}{\partial x^\alpha}$, $|\alpha| = \alpha_1 + \ldots + \alpha_d \leqslant k$, exist and are continuous.

- The space $\mathcal{C}_0^\infty(\mathbb{R}^d)$ denotes the space of smooth functions on $\mathbb{R}^d$ with compact support.

- We use $\mathcal{S}(\mathbb{R}^d)$ for the space of Schwartz functions on $\mathbb{R}^d$.

- If X is a topological space then $\mathcal{C}_c(X)$ denotes the space of compactly supported and continuous functions on X and $L_0(X)$ denotes the space of all equivalence classes of measurable functions on X.

- We write L_p, $0 < p \leqslant \infty$, for the Lebesgue spaces on a suitable measure space. For an integrability parameter $1 \leqslant p \leqslant \infty$ we denote with $p' = \frac{p}{p-1}$ the Hölder-dual of p.

- For topological spaces X, Y we write $X \hookrightarrow Y$ for the embedding of X in Y, i.e., $X \subseteq Y$ and the identity map is continuous.

- Concerning the Fourier transform of a function $f \in L_1(\mathbb{R}^d)$ we write $\hat{f} = \mathcal{F}f$ using the convention

$$\mathcal{F}f(\omega) = \int_{\mathbb{R}^d} f(x)e^{-2\pi i \langle \omega, x \rangle} \, \mathrm{d}x.$$

The inverse Fourier transform is likewise given by

$$\mathcal{F}^{-1}f(x) = \int_{\mathbb{R}^d} f(\omega)e^{2\pi i \langle \omega, x \rangle} \, \mathrm{d}x.$$

We use the same symbols for the unitary automorphism on $L_2(\mathbb{R}^d)$.

- For quantities a and b we write $a \lesssim b$ if there exists a finite constant $C > 0$ such that $a \leqslant C \cdot b$, with C being independent of all relevant parameters. If $a \lesssim b$ and $b \lesssim a$ we write $a \asymp b$.

1.2   Group Theory

One of the main ingredients of coorbit theory, at least in the classical setup, are locally compact topological groups. We give some basic definitions and results in this section, a more thorough exposition can be found in [51].

Definition 1.2.1. A *topological group* G is a group G endowed with a topology such that the group operations

$$(g, h) \mapsto gh, \quad \text{and} \quad g \mapsto g^{-1}$$

are continuous as maps $G \times G \to G$ and $G \to G$, respectively. We shall denote the unit element of G with $e_G = e$. We say G is *locally compact* if every point has a compact neighborhood. Furthermore we say G is *Hausdorff* if for every $x, y \in G$ there exist open sets $U_x, U_y \subset G$ with $U_x \cap U_y = \varnothing$.

The left translation $\lambda(x)$ on G by x is denoted by

$$\lambda(x)y = x^{-1}y, \quad y \in G,$$

and similarly the right translation $\rho(x)$ on G by x is denoted by

$$\rho(x)y = yx, \quad y \in G.$$

We use the same notation for left and right translations of a function f on G by x:

$$\lambda(x)f(y) = f(\lambda(x)y) = f(x^{-1}y), \qquad \rho(x)f(y) = f(\rho(x)y) = f(yx),$$

for $y \in G$.

Definition 1.2.2. A Borel measure μ on the σ-algebra generated by the open sets of the topological space X is called a *Radon measure*, if

(i) it is finite on compact sets;

(ii) it is *outer regular* on ther Borel sets, i.e., for every Borel set E we have

$$\mu(E) = \inf\{\mu(U) : U \supset E, \ U \text{ open}\};$$

(iii) it is *inner regular* on the open sets, i.e., for every open set U we have

$$\mu(U) = \sup\{\mu(K) : K \subset E, \ K \text{ compact}\}.$$

Definition 1.2.3. A *left Haar measure* μ on the topological group G is a nonzero Radon measure μ on G that satisfies $\mu(\lambda(x)E) = \lambda(E)$ for every Borel set $E \subset G$ and every $x \in G$. Similarly, a *right Haar measure* ϱ on G is a nonzero Radon measure ϱ on G that satisfies $\varrho(\rho(x)E) = \lambda(E)$ for every Borel set $E \subset G$ and every $x \in G$.

We are mainly interested in left Haar measures and whenever the left Haar measure μ is fixed and no confusion is possible we write dx instead of $d\mu(x)$.

Theorem 1.2.4 (Theorem 2.10 and Theorem 2.20 of [51]). *Every locally compact group G has a left Haar measure μ, which is unique up to a factor in the sense that if η is a second left Haar measure on G, then there exists a constant $c > 0$ such that $\mu = c \cdot \eta$.*

In general a left Haar measure is not a right Haar measure on G. In the abelian setting they obviously coincide, but for non-commutative groups the discrepancy between left and right translations is quantified in the modular function denoted by Δ. For any left Haar measure μ and $x \in G$ we set

$$\mu_x(E) = \mu(\rho(x)E) = \mu(Ex),$$

which is again a left Haar measure on G. By Theorem 1.2.4 there exists a positive constant $c =: \Delta(x)$ such that

$$\mu_x = \Delta(x)\mu.$$

The value $\Delta(x)$ is independent of the choice of μ and therefore the function $\Delta : G \to \mathbb{R}_+$ is well-defined. We call Δ the *modular function* on G. If $\Delta \equiv 1$ we say that G is *unimodular*, that is, the left and right Haar measure coincide. Unimodular groups have very useful properties, but in this thesis most groups of interest are not unimodular.

Proposition 1.2.5 (Proposition 2.24 of [51]). *Let G be a locally compact group. The modular function $\Delta : G \to \mathbb{R}_+$ is a continuous homomorphism from G into the multiplicative group $\mathbb{R}_+$, i.e., $\Delta(e_G) = 1$ and $\Delta(x)^{-1} = \Delta(x^{-1})$ for all $x \in G$. Furthermore, for every measureable function f on G with $\int_G |f(x)| \, d\mu(x) < \infty$ and $y \in G$ we have*

$$\int_G \rho(y)f(x) \, dx = \Delta(y)^{-1} \int_G f(x) \, dx. \tag{1.2.1}$$

We give two elementary examples.

Example 1.2.6. Let $G = \mathbb{R}^d$ with the group action $(x, y) \mapsto x + y$. Obviously, the inverse of $x \in \mathbb{R}^d$ under the group action is given by $-x$ and the group is abelian. Endowed with the standard topology this space is a locally compact Hausdorff topological group. A Haar measure on $\mathbb{R}^d$ is now given by the Lebesgue measure dx.

Example 1.2.7. Let $G = \mathbb{R}^*$ be endowed with the group action $(x, y) \mapsto x \cdot y$ and the inverse of $x \in \mathbb{R}^*$ is given by $\frac{1}{x}$. This group is abelian as well and endowed with the standard topology this is a locally compact Hausdorff topological group. A Haar measure on $\mathbb{R}^*$ is given by $dx/|x|$, where dx is the Lebesgue measure.

For non-abelian examples we refer to Section 2.3, where the *full affine group* and *full shearlet group* are discussed in detail. In both cases the following proposition provides a way to compute the left Haar measure on the group.

Proposition 1.2.8 (Proposition 2.21 of [51]). *Suppose that G is a manifold which is an open subset of $\mathbb{R}^N$ for some $N \in \mathbb{N}$ and the left translations are given by affine maps*

$$xy = A(x)y + b(x), \quad x, y \in G,$$

where $A(x) \in \mathbb{R}^{N \times N}$ and $b(x) \in \mathbb{R}^N$. Then $|\det A(x)|^{-1}dx$ is a left Haar measure on G, where dx denotes the Lebesgue measure on $\mathbb{R}^N$.

1.3 Function Space Theory

In this section we give the definitions and recall some basic properties of spaces of functions. In Subsection 1.3.1 we discuss Banach, Hilbert and Fréchet spaces of functions on arbitrary measure spaces. Then, in Subsection 1.3.2, we look at spaces of functions on groups and their specific properties. A particular focus lies on the convolution of functions on groups.

1.3.1   Banach, Hilbert and Fréchet Spaces

We start by giving the definitions of Banach and Hilbert spaces.

Definition 1.3.1. A linear space $\mathcal{B}$ is called *Banach space*, if $\mathcal{B}$ is equipped with a *norm* $\|\cdot\|_{\mathcal{B}} : \mathcal{B} \to \mathbb{R}_{\geqslant 0}$ and if $\mathcal{B}$ is complete. If $\|\cdot\|_{\mathcal{B}}$ is a quasi-norm, i.e., the triangle inequality holds true up to a positive constant, then $\mathcal{B}$ is called *quasi-Banach space*.

Definition 1.3.2. If the norm $\|\cdot\|_{\mathcal{H}}$ of a Banach space $\mathcal{H}$ is induced by a scalar product $\langle\cdot,\cdot\rangle_{\mathcal{H}} : \mathcal{H} \times \mathcal{H} \to \mathbb{C}$ on $\mathcal{H}$, that is $\|\cdot\|_{\mathcal{H}} = \langle\cdot,\cdot\rangle_{\mathcal{H}}^{1/2}$, then $\mathcal{H}$ is called *Hilbert space*. The scalar product is linear in the first factor and conjugate linear in the second factor.

We are mostly interested in Banach spaces of functions, that is, Banach spaces where the elements are measurable functions $f : X \to \mathbb{C}$ on a measure space X. By the usual abuse of notation two functions are identified if they coincide almost everywhere on X. A Banach space of functions is called *solid*, if the condition $|f(x)| \leqslant |g(x)|$ for almost all $x \in X$ implies $\|f\| \leqslant \|g\|$ is fulfilled.

We give two examples.

Example 1.3.3. The most prominent and important examples of function spaces are *Lebesgue spaces*. Let X be a measure space endowed with a Radon measure μ and $0 < p \leqslant \infty$, then the spaces

$$L_p(X, \mu) = \{f \colon X \to \mathbb{C} \text{ measurable} : \|f\|_{L_p} < \infty\}$$

endowed with the (quasi-)norm

$$\|f\|_{L_p} = \left(\int_X |f(x)|^p \, d\mu(x)\right)^{1/p}, \quad p < \infty,$$

$$\|f\|_{L_\infty} = \operatorname*{ess\,sup}_{x \in X} |f(x)|,$$

are quasi-Banach spaces for $0 < p \leqslant \infty$ and Banach spaces for $p \geqslant 1$. For $p = 2$ the space $L_2(X, \mu)$ is a Hilbert space where the norm is induced by the scalar product

$$\langle f, g\rangle_{L_2} = \int_X f(x)\overline{g(x)} \, d\mu(x), \quad f, g \in L_2(X, \mu).$$

By taking a positive measurable weight function $w : X \to \mathbb{R}_{\geqslant 0}$ we can define the *weighted Lebesgue spaces* via

$$L_{p,w}(X, \mu) = \{f \colon X \to \mathbb{C} \text{ measurable} : \|f\|_{L_{p,w}} = \|f \cdot w\|_{L_p} < \infty\}.$$

These spaces have the same properties as the unweighted ones.

Example 1.3.4. If we choose a countable set I and endow it with the counting measure, then the corresponding Lebesgue spaces are the $\ell_p(I)$-spaces, where the norm of a sequence $c = (c_i)_{i \in I} \in \mathbb{C}^I$ is given by

$$\|c\|_{\ell_p} = \left(\sum_{i \in I} |c_i|^p\right)^{1/p}, \quad p < \infty,$$

$$\|c\|_{\ell_\infty} = \sup_{i \in I} |c_i|.$$

Again, there are weighted versions of these spaces denoted with $\ell_{p,w}(I)$ for suitable weights $w = (w_i)_{i \in I}$.

Next, we define Fréchet spaces.

Definition 1.3.5. A topological linear space F is called *Fréchet space*, if

(i) it is Hausdorff;

(ii) its topology may be induced by a countable family of semi-norms $\|\cdot\|_k$, $k \in \mathbb{N}$. This means that a subset $U \subset F$ is open if and only if for every $u \in U$ there exists $N \geqslant 0$ and $\varepsilon > 0$ such that $\{v : \|u - v\|_k < \varepsilon, \ k \leqslant N\} \subseteq U$;

(iii) it is complete with respect to the family of semi-norms.

A sequence $(u_n)_{n \in \mathbb{N}}$ in a Fréchet space F converges to an element $u \in F$ if and only if $\lim_{n \to \infty} \|u - u_n\|_k = 0$ for all semi-norms $\|\cdot\|_k$. Obviously, every Banach space is a Fréchet space.

Again we are mainly interested in Fréchet spaces of functions, that is, Fréchet spaces F where the elements are measurable functions $f : X \to \mathbb{C}$ on a measure space X.

Let us give two examples.

Example 1.3.6. Let X be a measure space endowed with a Radon measure μ, then the space

$$\mathcal{T} := \bigcap_{1 < p < \infty} L_p(X, \mu)$$

is a Fréchet space of functions where the topology is induced by the family of norms $(\|\cdot\|_p)_{1 < p < \infty}$. This means that a sequence $(f_n)_{n \in \mathbb{N}}$ converges in $\mathcal{T}$ to f if and only if $\lim_{n \to \infty} \|f - f_n\|_p$ for all $1 < p < \infty$. This space plays a vital role in Chapter 3.

Example 1.3.7. The *Schwartz space* $\mathcal{S}(\mathbb{R}^d)$, or space of rapidly decreasing functions, defined via

$$\mathcal{S}(\mathbb{R}^d) = \left\{ f \in \mathcal{C}^\infty(\mathbb{R}^d) : \|f\|_{\alpha, \beta} < \infty \text{ for all } \alpha, \beta \in \mathbb{N}^d \right\},$$
$$\|f\|_{\alpha, \beta} = \sup_{x \in \mathbb{R}^d} \left| x^\alpha D^\beta f(x) \right|, \quad \alpha, \beta \in \mathbb{N}^d,$$

is a Fréchet space where the topology is induced by the family of semi-norms $(\|\cdot\|_{\alpha, \beta})_{\alpha, \beta \in \mathbb{N}^d}$.

Before closing this subsection let us clarify the notation of different dual spaces. For a space of functions Y we denote with Y' the classical *(anti-)dual space* of Y, that is, the space of (anti-)linear continuous functionals $f' : Y \to \mathbb{C}$. When convenient, we write $f'(f) = \langle f', f \rangle_{Y' \times Y}$ for the dual pairing of $f' \in Y'$ and $f \in Y$. In contrast, we write $Y^\#$ for the *Köthe dual space* of Y defined as

$$Y^\# = \{g \text{ measurable} : gf \in L_1 \text{ for all } f \in Y\}.$$

In general Y' and $Y^\#$ cannot be identified, however $Y^\#$ is a subset of Y'.

1.3.2 Function Spaces on Groups

Let G be a locally compact topological group equipped with a left Haar measure μ and modular function Δ. When the measure is fixed and no confusion is possible we will write dx instead of $d\mu(x)$. Denote with $L_p(G)$ the space of L_p-integrable functions on the group for $0 < p < \infty$, and $L_{p,w}(G)$ for the weighted Lebesgue spaces on the group. As before, $\langle \cdot, \cdot \rangle_{L_2}$ is the inner product on $L_2(G)$, for a function f on G we write $\check{f}(x) = f(x^{-1})$ and the *involution* f^* is given by $\overline{f^*} = \check{f}$. Now the *convolution* of two functions $f, g \in L_1(G)$ is defined via

$$f * g(x) = \int_G f(y) g(y^{-1} x) \, d\mu(y) = \langle f, [\lambda(x)g]^* \rangle_{L_2}, \quad x \in G. \tag{1.3.1}$$

By Fubini's theorem this is well-defined and it can be easily verified that convolution and translation have the following behaviour:

$$\lambda(z)(f * g) = (\lambda(z)f) * g, \quad \rho(z)(f * g) = f * (\rho(z)g), \quad z \in G. \tag{1.3.2}$$

The convolution of two functions on non-abelian groups is in general not commutative, since we have

$$g * f(x) = \int_G f(y)g(xy^{-1})\Delta(y^{-1})\, dy = \langle f, \lambda(x^{-1})g^* \cdot \Delta^{-1} \rangle_{L_2}.$$

The next step is to extend the convolution to $L_p(G)$.

Proposition 1.3.8 (Young's inequality for convolutions, Proposition 2.40 of [51]). *Assume* $1 \leqslant p \leqslant \infty$, $f \in L_1(G)$ or $f \in L_{1,\Delta^{-1}}(G)$, respectively, and $g \in L_p(G)$. Then,

(i) $f * g \in L_p(G)$ with $\|f * g\|_{L_p} \leqslant \|f\|_{L_1} \cdot \|g\|_{L_p}$;

(ii) $g * f \in L_p(G)$ with $\|g * f\|_{L_p} \leqslant \max\{\|f\|_{L_1}, \|f\|_{L_{1,\Delta^{-1}}}\} \cdot \|g\|_{L_p}$.

A similar result holds for weighted Lebesgue spaces. For this we let $w\colon G \to \mathbb{R}_+$ be a weight satisfying

$$w(xy) \leqslant w(x)w(y), \quad w(x^{-1}) = w(x), \quad x, y \in G.$$

Proposition 1.3.9 (Generalized Young's inequality, Proposition 3.8 of [20]). *Let* $m\colon G \to \mathbb{R}_+$ *be a weight fulfilling*

$$m(xyz) \leqslant w(x)m(y)w(z) \quad \text{for all } x, y, z \in G,$$

and $1 \leqslant p \leqslant \infty$. Then it follows for $f \in L_{1,w}(G)$ or $L_{1,w\Delta^{-1}}(G)$, respectively, and $g \in L_{p,m}(G)$, that

(i) $f * g \in L_{p,m}(G)$ with $\|f * g\|_{L_{p,m}} \leqslant \|f\|_{L_{1,w}} \cdot \|g\|_{L_{p,m}}$;

(ii) $g * f \in L_{p,m}(G)$ with $\|g * f\|_{L_{p,m}} \leqslant \max\{\|f\|_{L_{1,w}}, \|f\|_{L_{1,w\Delta^{-1}}}\} \cdot \|g\|_{L_{p,m}}$.

To extend this result we first need a weighted version of Schur's test.

Lemma 1.3.10 (Schur's test). *Let* $K\colon G \times G \to \mathbb{C}$ *be measurable, let* $w > 0$ *denote a weight on* G, *and let* $1 \leqslant p, q, r \leqslant \infty$ *with* $1 + 1/p = 1/q + 1/r$ *and the convention* $\frac{1}{\infty} = 0$. *Assume there is a constant* $C_K > 0$ *such that*

$$\left\| K(x, \cdot)\frac{w(x)}{w} \right\|_{L_r} \leqslant C_K \quad \text{for a.e. } x \in G, \tag{1.3.3}$$

$$\left\| K(\cdot, y)\frac{w}{w(y)} \right\|_{L_r} \leqslant C_K \quad \text{for a.e. } y \in G. \tag{1.3.4}$$

If $f \in L_{q,w}(G)$, then the integral

$$I_f(x) = \int_G K(x, y)f(y)\, dy$$

converges for almost every $x \in G$. The function I_f is in $L_{p,w}(G)$ and fulfills

$$\|I_f\|_{L_{p,w}} \leqslant C_K \|f\|_{L_{q,w}}.$$

Proof. It suffices to assume $f \geqslant 0$ and $K \geqslant 0$. Indeed, temporarily writing $I_{K,f}$ instead of I_f to emphasize the role of the function K, we have $|I_{K,f}| \leqslant I_{|K|,|f|}$; furthermore, if (1.3.3) and (1.3.4) hold for K, then they also hold for $|K|$, with the same constants, and we have $\|f\|_{L_{q,w}} = \| |f| \|_{L_{q,w}}$. Hence, if the claim holds for $K, f \geqslant 0$, then also

$$\|I_{K,f}\|_{L_{p,w}} \leqslant \|I_{|K|,|f|}\|_{L_{p,w}} \leqslant C_{|K|} \cdot \| |f| \|_{L_{q,w}} = C_K \cdot \|f\|_{L_{q,w}}.$$

Thus, we will assume in the following that $K, f \geqslant 0$. Hence also $I_f \geqslant 0$, so that [52, Theorem 6.14] shows

$$\|I_f\|_{L_{p,w}} = \sup_{0 \leqslant h \in L_{p',w^{-1}}(G) \setminus \{0\}} \frac{\langle I_f, h \rangle_{L_2}}{\|h\|_{L_{p',w^{-1}}}}. \tag{1.3.5}$$

We denote with $d(x,y)$ the product measure on $G \times G$. Furthermore, for brevity we set $M_x(y) := \frac{w(x)}{w(y)} \cdot K(x,y)$ and observe $\|M_x\|_{L_r} \leqslant C_K$ for almost all $x \in G$, thanks to (1.3.3). Likewise, (1.3.4) shows $\|M^{(y)}\|_{L_r} \leqslant C_K$ for almost all $y \in G$, where $M^{(y)}(x) := \frac{w(x)}{w(y)} \cdot K(x,y)$.

We first consider a number of special cases, so that we can then concentrate on the case where $1 < p, q, r < \infty$.

Case 1: At least one of p, q, r is infinite. In case of $p < \infty$, we have $1 < 1 + p^{-1} = q^{-1} + r^{-1}$. But if $q = \infty$, then the right-hand side of this inequality is $r^{-1} \leqslant 1$, which leads to a contradiction. Similarly, we see that $r = \infty$ leads to a contradiction. Therefore, we necessarily have $p = \infty$ in the present case.

Because of $1 = 1 + p^{-1} = q^{-1} + r^{-1}$, this implies $q = r'$, and hence

$$w(x) \cdot I_f(x) = \int_G M_x(y) \cdot w(y) \cdot f(y) \, dy \leqslant \|M_x\|_{L_r} \|f\|_{L_{r',w}} \leqslant C_K \cdot \|f\|_{L_{q,w}} < \infty$$

for almost all $x \in G$, proving the claim in Case 1, since $p = \infty$.

Case 2: We have $p, q, r < \infty$, but at least one of p, q, r is equal to one. This leaves three subcases:

Case 2-A: We have $p = 1$, and hence $2 = 1 + p^{-1} = q^{-1} + r^{-1} \leqslant 2$, which implies $q = r = 1$. Hence, by Fubini's theorem,

$$\|I_f\|_{L_{p,w}} = \int_G w(x) \int_G K(x,y) \cdot f(y) \, dy \, dx = \int_G w(y) \cdot f(y) \cdot \int_G M_x(y) \, dx \, dy$$
$$\leqslant C_K \cdot \|f\|_{L_{1,w}} = C_K \cdot \|f\|_{L_{q,w}},$$

which proves the claim in Case 2-A.

Case 2-B: We have $1 < p < \infty$, but $r = 1$. Since $1 + p^{-1} = q^{-1} + r^{-1} = 1 + q^{-1}$, this implies $p = q \in (1, \infty)$. Hence, for each nonnegative $h \in L_{p',w^{-1}}(G) \setminus \{0\}$, Fubini's theorem and Hölder's inequality show

$$\langle I_f, h \rangle_2 = \int_G h(x) \int_G K(x,y) \, f(y) \, dy \, dx$$
$$= \int_{G \times G} \frac{h(x)}{w(x)} \cdot [M^{(y)}(x)]^{\frac{1}{p'}} [M^{(y)}(x)]^{\frac{1}{p}} \cdot w(y) \, f(y) \, d(x,y)$$
$$\leqslant \left(\int_G \left(\frac{h(x)}{w(x)} \right)^{p'} \int_G M^{(y)}(x) \, dy \, dx \right)^{1/p'}$$
$$\cdot \left(\int_G (w(y) \, f(y))^p \int_G M^{(y)}(x) \, dx \, dy \right)^{1/p}$$
$$\leqslant C_K \cdot \|h\|_{L_{p',w^{-1}}} \cdot \|f\|_{L_{p,w}}.$$

In view of (1.3.5) and because of $p = q$, this proves the claim in Case 2-B.

Case 2-C: We have $1 < p, r < \infty$, but $q = 1$. This implies $p = r \in (1, \infty)$, since we have $1 + p^{-1} = q^{-1} + r^{-1} = 1 + r^{-1}$, For nonnegative $h \in L_{p', w^{-1}}(G) = L_{r', w^{-1}}(G)$, we thus have

$$
\begin{aligned}
\langle I_f, h \rangle_{L_2} &= \int_G w(y) \cdot f(y) \int_G M_x(y) \cdot \frac{h(x)}{w(x)} \, dx \, dy \\
&\leqslant \int_G w(y) \cdot f(y) \cdot \|M_x\|_{L_p} \cdot \|h\|_{L_{p', w^{-1}}} \, dy \\
&\leqslant C_K \cdot \|h\|_{L_{p', w^{-1}}} \cdot \|f\|_{L_{1,w}} = C_K \cdot \|h\|_{L_{p', w^{-1}}} \cdot \|f\|_{L_{q,w}}.
\end{aligned}
$$

In view of (1.3.5), this proves the claim in Case 2-C.

Case 3: Finally, we handle the case $1 < p, q, r < \infty$. By elementary calculations one can show $r/p + r/q' = q/p + q/r' = p'/q' + p'/r' = 1$, where all occurring numbers $\frac{r}{p}, \frac{r}{q'}$ and so on are elements of the interval $(0, 1)$. Thus, for any $0 \leqslant h \in L_{p', w^{-1}}(G)$, it follows from Hölder's inequality and Fubini's theorem that

$$
\begin{aligned}
\langle I_f, h \rangle_{L_2} &= \int_{G \times G} K(x, y) \frac{w(x)}{w(y)} \cdot f(y) w(y) \cdot h(x) w(x)^{-1} \, d(x, y) \\
&= \int_{G \times G} \left(M^{(y)}(x) \right)^{r/p} \cdot \left(f(y) w(y) \right)^{q/p} \cdot \left(M_x(y) \right)^{r/q'} \\
&\qquad \cdot \left(h(x) w(x)^{-1} \right)^{p'/q'} \cdot \left(f(y) w(y) \right)^{q/r'} \cdot \left(h(x) w(x)^{-1} \right)^{p'/r'} \, d(x, y) \\
&\leqslant \left(\int_G |f(y) w(y)|^q \int_G \left(M^{(y)}(x) \right)^r \, dx \, dy \right)^{1/p} \\
&\qquad \cdot \left(\int_G |h(x) w(x)^{-1}|^{p'} \int_G \left(M_x(y) \right)^r \, dy \, dx \right)^{1/q'} \\
&\qquad \cdot \left(\int_{G \times G} |f(y) w(y)|^q |h(x) w(x)^{-1}|^{p'} \, d(x, y) \right)^{1/r'} \\
&\leqslant C_K \cdot \|f\|_{L_{q,w}} \cdot \|h\|_{L_{p', w^{-1}}} < \infty,
\end{aligned}
$$

where we used $\frac{1}{p} + \frac{1}{q'} + \frac{1}{r'} = \frac{1}{p} + 1 - \frac{1}{q} + 1 - \frac{1}{r} = 1$. In view of (1.3.5), this proves the claim for the case $p, q, r \in (1, \infty)$. $\qquad\square$

Proposition 1.3.9 can now be extended in the following fashion.

Proposition 1.3.11 (Extended Young's inequality). *Let $m \colon G \to \mathbb{R}_+$ be a weight fulfilling*

$$
m(xyz) \leqslant w(x) m(y) w(z) \quad \text{for all } x, y, z \in G,
$$

and let $1 \leqslant p, q, r \leqslant \infty$ such that $1 + 1/p = 1/q + 1/r$ with the convention $\frac{1}{\infty} = 0$. Then it follows for $f \in L_{q,m}(G)$ and $g \in L_{r,w}(G) \cap L_{r, w\Delta^{-1/r}}(G)$ that $f * g \in L_{p,m}(G)$ and

$$
\|f * g\|_{L_{p,m}} \leqslant w(e) \cdot \max\{\|g\|_{L_{r,w}}, \|g\|_{L_{r, w\Delta^{-1/r}}}\} \cdot \|f\|_{L_{q,m}}. \tag{1.3.6}
$$

If, instead of $g \in L_{r,w}(G) \cap L_{r, w\Delta^{-1/r}}(G)$, we have $g \in L_{r,w}(G)$ and $|g(x)| = |g(x^{-1})|$ as well as $w(x) = w(x^{-1})$ for all $x \in G$, or if $g \in L_{r,w}(G)$ and G is unimodular, then

$$
\|f * g\|_{L_{p,m}} \leqslant w(e) \cdot \|g\|_{L_{r,w}} \cdot \|f\|_{L_{q,m}}. \tag{1.3.7}
$$

Proof. We apply Lemma 1.3.10 for the case $K(x,y) = g(y^{-1}x)$ and the weight m. It suffices to show that there exists a constant C_K that fulfills (1.3.3) and (1.3.4). We first consider the case $r < \infty$ and use the left invariance of the Haar measure to conclude

$$\int_G |g(y^{-1}x)|^r \cdot \frac{m(x)^r}{m(y)^r} \, dx = \int_G |g(z)|^r \cdot \frac{m(yz)^r}{m(y)^r} \, dz$$

$$\leqslant w(e)^r \cdot \int_G |g(z)|^r \cdot \frac{m(y)^r w(z)^r}{m(y)^r} \, dz$$

$$= w(e)^r \cdot \int_G |g(z)|^r \cdot w(z)^r \, dz = w(e)^r \cdot \|g\|_{L_{r,w}}^r$$

for almost all $y \in G$. Now, using the change of variables $z = x^{-1}y$, and recalling the formula $d\varrho(x) = \Delta(x^{-1})dx$, see (1.2.1), for the right Haar measure ϱ given by $\varrho(M) = \mu(M^{-1})$, we see

$$\int_G |g(y^{-1}x)|^r \cdot \frac{m(x)^r}{m(y)^r} \, dy = \int_G |g(z^{-1})|^r \cdot \frac{m(x)^r}{m(xz)^r} \, dz$$

$$\leqslant w(e)^{-r} \cdot \int_G |g(z^{-1})|^r \cdot [w(z^{-1})]^r \, dz$$

$$= w(e)^r \cdot \int_G |g(y)|^r \cdot [w(y)]^r \cdot \Delta(y)^{-1} \, dy = w(e)^r \cdot \|g\|_{L_{r,w\Delta^{-1/r}}}^r$$

for almost all $x \in G$. By setting $C_K = w(e) \cdot \max\{\|g\|_{L_{r,w}}, \|g\|_{L_{r,w\Delta^{-1/r}}}\} < \infty$, Lemma 1.3.10 yields

$$\|f * g\|_{L_{p,m}} \leqslant C_K \cdot \|f\|_{L_{q,m}} \qquad \text{for all } f \in L_{q,m}(G) \, ,$$

which proves (1.3.6).

Finally, for the case $r = \infty$, observe $m(x) = m(yy^{-1}x) \leqslant m(y) \cdot w(y^{-1}x)$, so that we get

$$|g(y^{-1}x)| \cdot \frac{m(x)}{m(y)} \leqslant |g(y^{-1}x)| \cdot w(y^{-1}x) \leqslant \|g\|_{L_{\infty,w}}$$

for almost every $x \in G$ and almost every $y \in G$, which establishes (1.3.3) and (1.3.4).

It remains to prove (1.3.7). If we assume $|g(x)| = |g(x^{-1})|$ and $w(x) = w(x^{-1})$, the formula $d\varrho(x) = \Delta(x^{-1}) \, dx$ from above yields for $r < \infty$ that

$$\|g\|_{L_{r,w\Delta^{-1/r}}}^r = \int_G |g(y)|^r \cdot [w(y)]^r \cdot \Delta(y^{-1}) \, dy = \int_G |g(z^{-1})|^r \cdot [w(z^{-1})]^r \, dz$$

$$= \int_G |g(z)|^r \cdot [w(z)]^r \, dz = \|g\|_{L_{r,w}}^r \, .$$

This identity trivially holds if G is unimodular, so that $\Delta \equiv 1$. For $r = \infty$, we always have $\|g\|_{L_{r,w\Delta^{-1/r}}} = \|g\|_{L_{r,w}}$. In all of these cases (1.3.7) is a direct consequence of (1.3.6). $\qquad\square$

For Hölder-dual exponents the convolution also yields a continuous function.

Corollary 1.3.12 (Theorem 20.16 of [74]). *Let $1 \leqslant p,q \leqslant \infty$ such that $\frac{1}{p} + \frac{1}{q} = 1$ with the convention $\frac{1}{\infty} = 0$. For $f \in L_p(G)$ and $\check{g} \in L_q(G)$ the function $f * g$ is bounded and continuous with $\|f * g\|_{L_\infty} \leqslant \|f\|_{L_p} \cdot \|\check{g}\|_{L_q}$.*

Remark 1.3.13. The results above, especially Proposition 1.3.9 and Proposition 1.3.11, can be reformulated as follows. Let $K \in L_{1,w}(G)$ and denote with RC_K the *right convolution operator* associated with K, that is, $RC_K(f) = f * K$ for any function f on G, assuming the convolution is explained. Then, Proposition 1.3.9 states that the operator $RC_K \colon L_{p,m}(G) \to L_{p,m}(G)$ for suitabe weights m and w is a bounded operator. We can also see that $L_{1,w}(G)$ is an Banach-$*$-algebra of functions under the convolution product and the involution $f^*(x) = \Delta(x^{-1})\overline{f(x^{-1})}$.

Similarly, if we assume $K \in L_{r,w}(G)$ for all $1 < r < \infty$, then for any $1 \leqslant p < q \leqslant \infty$ the operator $RC_K \colon L_{q,m}(G) \to L_{p,m}(G)$ is bounded for suitable weights m and w by Proposition 1.3.11. The function K is also called *kernel* and the operator RC_K *kernel operator* associated with K. This idea is later extended to measure spaces in Section 1.8.

Let us now assume that F is a Fréchet space of functions on G. The following result shows that the convolution of elements of F and its Köthe dual $F^{\#}$ are well-defined.

Proposition 1.3.14 (Proposition 2.2 of [18]). *Let F be a Fréchet space and assume $g \in F^{\#}$, then*

(i) for all $f \in \mathcal{F}$, the convolution $f * g$ exists everywhere and is a continuous function;

(ii) the map $f \mapsto f * g$ is a continuous map from F to $\mathcal{C}(G)$, when the latter space is endowed with the topology of compact convergence.

1.4 Representation Theory

In this section we recall some basic definitions and results from representation theory of groups. Representations of groups play a vital role in Chapters 2 and 3 and lay the foundation for coorbit theory. In the following, let G be a locally compact topological group. We recall that for some Hilbert space $\mathcal{H}$ the space of unitary operators is denoted by $\mathcal{U}(\mathcal{H})$ and as unitary operators we understand bijective isometries on $\mathcal{H}$. Obviously, $\mathcal{U}(\mathcal{H})$ forms a group with group law $AB = A \circ B$ for $A, B \in \mathcal{U}(\mathcal{H})$.

Definition 1.4.1. A *unitary representation* of G is a homomorphism $\pi \colon G \to \mathcal{U}(\mathcal{H})$ into the group $\mathcal{U}(\mathcal{H})$ of unitary operators on some nonzero Hilbert space $\mathcal{H}$ that is continuous with respect to the strong operator topology. This means that π satisfies

(i) $\pi(xy) = \pi(x)\pi(y)$, for $x, y \in G$;

(ii) $\pi(x^{-1}) = \pi(x)^*$, for $x \in G$;

(iii) the map $x \mapsto \pi(x)u$ is continuous from G to $\mathcal{H}$ for any $u \in \mathcal{H}$.

Example 1.4.2. As a basic example we choose $\mathcal{H} = L_2(G)$. Then, with slight abuse of notation, define the *left regular representation* via

$$\lambda \colon G \to \mathcal{U}(L_2(G)), \quad \lambda(x)f(y) = f(\lambda(x)y) = f(x^{-1}y).$$

This is a unitary representation as one can easily check.

Similarly, we define the *right regular representation* as

$$\rho \colon G \to \mathcal{U}(L_2(G)), \quad \rho(x)f(y) = \Delta(x)^{1/2}f(\rho(x)y)) = \Delta(x)^{1/2}f(yx).$$

The modular function is necessary for the representation to be unitary with respect to the left Haar measure.

Definition 1.4.3. Suppose $\mathcal{M}$ is a closed subspace of $\mathcal{H}$ and π a unitary representation of G on $\mathcal{H}$. $\mathcal{M}$ is called an *invariant subspace* for π if $\pi(x)\mathcal{M} \subset \mathcal{M}$ for all $x \in G$. If π admits an invariant subspace $\mathcal{M}$ that is nontrivial, i.e., $\mathcal{M} \neq \{0\}$ or $\mathcal{M} \neq \mathcal{H}$, then π is called *reducible*, otherwise π is *irreducible*

A possible way to check if a unitary representation is irreducible is by looking at the so-called coefficients.

Definition 1.4.4. Let π be a unitary representation of G on $\mathcal{H}$ and let $u, v \in \mathcal{H}$. The map $x \mapsto \langle u, \pi(x)v \rangle_{\mathcal{H}}$ is called *coefficient* of π relative to (u, v). If $u = v$, it is called *diagonal coefficient*.

Proposition 1.4.5 (Proposition 2.47 of [20])**.** *The following two conditions for a unitary representation π of G on $\mathcal{H}$ are equivalent:*

(i) *π is irreducible;*

(ii) *if $u, v \in \mathcal{H}$ are non-zero, then the coefficient $x \mapsto \langle u, \pi(x)v \rangle_{\mathcal{H}}$ is nonzero as a map.*

We close the section with the following definition.

Definition 1.4.6. Let π be a unitary representation of G on $\mathcal{H}$ and $u \in \mathcal{H}$. The closed linear subspace $\mathcal{M}_u = \text{span}\{\pi(x)u : x \in G\} \subset \mathcal{H}$ is called *cyclic subspace* generated by u. If $\mathcal{M}_u = \mathcal{H}$, the vector u is called *cyclic vector* for π.

1.5 Coverings

A main ingredient for the discretization techniques in the following chapters are coverings. This means we cover an arbitrary measure space (X, μ) using only a particular family $\mathcal{U} = (U_i)_{i \in I}$ of subsets of X for some countable index set I. Naturally, we are interested in those coverings of X that have structural properties we can exploit. In this section we present definitions and properties for coverings of groups and arbitrary measure spaces.

The Group Setting

We first assume G to be a locally compact topological group with Haar measure dx. This is the setting we need for Chapters 2 and 3. We call a set $U \subset G$ *unit neighborhood* if $e_G \in U$.

Definition 1.5.1. Let $\mathcal{X} = (x_i)_{i \in I}$ be a family in G.

(i) $\mathcal{X}$ is *U-dense* in G, for a unit neighborhood $U \subset G$, if $G = \bigcup_{i \in I} x_i U$.

(ii) $\mathcal{X}$ is *U-separated* in G, for a unit neighborhood $U \subset G$, if the family $(x_i U)_{i \in I}$ is pairwise disjoint.

(iii) $\mathcal{X}$ is *relatively separated* if for every compact unit neighborhood $Q \subset G$ there is a constant $N \in \mathbb{N}$ with

$$\sum_{i \in I} \chi_{x_i Q}(x) \leqslant N \quad \text{for all } x \in G.$$

(iv) $\mathcal{X}$ is *U-well-spread*, for a unit neighborhood $U \subset G$, if $\mathcal{X}$ is relatively separated and U-dense.

Remark 1.5.2. (i) If we additionally assume the underlying group G to be second countable, G is in particular σ-compact. Therefore, [105, Lemma 2.3.10] shows that the index set of every relatively separated family in G is countable. If G is a connected Lie group, which is usually the case, it is always second countable.

(ii) Usually, $\mathcal{X}$ is called relatively separated if $\mathcal{X}$ is a finite union of U-separated sets, for some compact unit neighborhood U. The two definitions are shown to be equivalent in in [43, Lemma 2.9] and [105, Lemma 2.3.11].

We see in Definition 1.5.1 that a covering $(x_i U)_{i \in I}$ of a group consists of two parts: some mother unit neighborhood $U \subset G$ and a family $\mathcal{X} = (x_i)_{i \in I}$ that defines the spreading of the covering. Moreover, by the left-invariance of the Haar measure, we have $|x_i U| = |U|$ for all $i \in I$.

Given a U-well-spread family $\mathcal{X} = (x_i)_{i \in I}$, one often wants to decompose a given function f into building blocks f_i which are supported in the sets $(x_i V)_{i \in I}$. This can be done using suitable partitions of unity.

Definition 1.5.3 (Definition 3.6 of [45]). Let $U \subset G$ be a compact unit neighborhood. A family $\Phi = (\varphi_i)_{i \in I}$ is called a U-BUPU (bounded uniform partition of unity) with localizing family $\mathcal{X} = (x_i)_{i \in I}$ if the following holds:

(i) Each $\varphi_i \colon G \to [0, 1]$ is a measurable function.

(ii) $\mathcal{X}$ is relatively separated and $\operatorname{supp} \varphi_i \subset x_i U$ for all $i \in I$.

(iii) We have $\sum_{i \in I} \varphi_i \equiv 1$ on G.

The following lemma shows that one can always find a U-BUPU for any compact unit neighborhood U.

Lemma 1.5.4 (Theorem 2 of [40] and Lemma 2.3.12 of [105]). *Let $U \subset G$ be a compact unit neighborhood. Then there exists a U-BUPU $\Phi = (\varphi_i)_{i \in I}$ with $\varphi_i \in \mathcal{C}_c(G)$ for all $i \in I$.*

The General Setting

The next step is to generalize the idea of coverings to locally compact Hausdorff spaces X endowed with a positive Radon measure μ.

Definition 1.5.5. A family $\mathcal{U} = (U_i)_{i \in I}$ of subsets of X is called a *(discrete) admissible covering* of X, if the following conditions are satisfied.

(i) Each set U_i, $i \in I$, is relatively compact and has non-void interior.

(ii) It holds that $X = \bigcup_{i \in I} U_i$.

(iii) There is a constant $N \in \mathbb{N}$ such that

$$\sum_{i \in I} \chi_{U_i}(x) \leqslant N \quad \text{for all } x \in X.$$

Furthermore, we say that an admissible covering $\mathcal{U} = (U_i)_{i \in I}$ is *moderate* if it fulfills the following additional conditions.

(iv) There is a constant $C > 0$ such that $\mu(U_i) \geqslant C$ for all $i \in I$.

(v) There is a constant $D > 0$ such that

$$\mu(U_i) \leqslant D \, \mu(U_j) \quad \text{for all } i, j \in I \text{ with } U_i \cap U_j \neq \varnothing.$$

Remark 1.5.6. If, in addition, X is σ-compact, the index set I is also countable.

Obviously, Definition 1.5.5 is a generalization of Definition 1.5.1 in the following sense. If we choose a compact unit neighborhood $U \subset G$ and $\mathcal{X} = (x_i)_{i \in I}$ relatively separated and U-well-spread in the sense of Definition 1.5.1, then $\mathcal{U} = (x_i U)_{i \in I}$ is a moderate admissible covering of G in the sense of Definition 1.5.5.

Note that for moderate coverings the definition above does not require $\mu(U_i)$ to be bounded from above, by part (v) it is only required that neighboring sets are of similar size and part (iv) assumes a lower bound for all $\mu(U_i)$. Usually, one tries to construct coverings where the sets are of equal size, in which case every admissible covering is moderate.

If the sets U_i do not overlap, any admissible covering fulfills condition (v). For a more general discussion on how to construct admissible coverings that fulfill property (v) and relevant examples we refer to [42].

Again we are interested in partitions of unity subordinate to coverings. The definition is similar to Definition 1.5.3.

Definition 1.5.7. Let $\mathcal{U} = \{U_i\}_{i \in I}$ be a moderate admissible covering of X. A family $\Phi = (\varphi_i)_{i \in I}$ is called $\mathcal{U}$-BAPU (bounded admissible partition of unity) if the following hold:

  (i) Each $\varphi_i \colon X \to [0, 1]$ is a measurable function.

 (ii) We have $\operatorname{supp} \varphi_i \subset U_i$ for all $i \in I$.

(iii) We have $\sum_{i \in I} \varphi_i \equiv 1$ on X.

Again we can always find such a $\mathcal{U}$-BAPU for moderate admissible coverings $\mathcal{U}$.

Lemma 1.5.8 (Proposition 4.41 and Exercise 57 b of [52])**.** *Let $\mathcal{U}$ be a moderate admissible covering of a σ-compact and locally compact Hausdorff space X. Then there exists a $\mathcal{U}$-BAPU $\Phi = (\varphi_i)_{i \in I}$ with $\varphi_i \in \mathcal{C}_c(X)$ for all $i \in I$.*

1.6   Sequence Spaces

In this section we introduce some sequence spaces associated to weighted Lebesgue spaces on groups and arbitrary measure spaces. In order to do so we first define sequence spaces associated to Banach spaces of functions. Then we apply this concept to weighted Lebesgue spaces and prove some basic properties. As in the preceding section, this section is divided into two parts. First, we look at the group case and then we generalize the ideas to certain measure spaces.

The Group Setting

Let G be a locally compact topological group with Haar measure dx. Again, this is the setting we need for Chapters 2 and 3.

Definition 1.6.1. Let $\mathcal{X} = (x_i)_{i \in I}$ be a family in G and $U \subset G$ a unit neighborhood. Furthermore, let $(Y, \| \cdot \|_Y)$ be a solid, translation invariant Banach space of functions on G, then the associated sequence space $Y_s(\mathcal{X}, I, U)$ is defined as

$$Y_s(\mathcal{X}, I, U) := \left\{ c = (c_i)_{i \in I} \in \mathbb{C}^I \ : \ \|c\|_{Y_s(\mathcal{X}, I, U)} := \left\| \sum_{i \in I} |c_i| \chi_{x_i U} \right\|_Y < \infty \right\}.$$

We will usually omit the indication of the dependence on $\mathcal{X}$ and U and simply write $Y_s(\mathcal{X}, I, U) = Y_s(I)$ or just Y_s whenever no confusion is possible.

Let us collect some properties of the sequence spaces.

Lemma 1.6.2 (Lemma 3.5 of [45]). *Let Y, $\mathcal{X}$ and U as in Definition 1.6.1. Then the following holds true.*

(i) *The space Y_s is a Banach space of sequences.*

(ii) *If the functions with compact support are dense in Y, the finite sequences form a dense subspace of Y_s.*

(iii) *Let $\mathcal{X}$ and $\mathcal{X}'$ be two U-well-spread families over the same index set I such there is a compact set $Q \subset G$ with $x_i^{-1} x_i' \in Q$ for all $i \in I$, then $Y_s(\mathcal{X}, I, U) = Y_s(\mathcal{X}', I, U)$ and the corresponding norms are equivalent.*

If we consider $Y = L_{p,w}(G)$, that is weighted Lebesgue spaces on G, the associated sequence space $(L_{p,w}(G))_s$ is a weighted ℓ_p-space, as the following lemma shows.

Lemma 1.6.3 (Lemma 3.5 of [45]). *Given a weight $w: G \to \mathbb{R}$ and $1 \leqslant p \leqslant \infty$ as well as a family $\mathcal{X} = (x_i)_{i \in I}$ in G, the associated sequence space $(L_{p,w}(G))_s(\mathcal{X}, I, U) = \ell_{p,w}(I)$ is given by*

$$\ell_{p,w}(I) := \left\{ c = (c_i)_{i \in I} \in \mathbb{C}^I \ : \ \|c\|_{\ell_{p,w}}^p = \sum_{i \in I} |c_i|^p w(x_i)^p < \infty \right\}.$$

Remark 1.6.4. Note that the weight w is evaluated at the points x_i for $i \in I$. If we choose a second U-well-spread family $\mathcal{X}'$ in G over the same index set with $x_i^{-1} x_i' \in Q$ for all $i \in I$ and a fixed compact set $Q \subset G$, then Lemma 1.6.2 (iii) shows that the spaces $\ell_{p,w}(I)$ coincide up to equivalent norms. In the following, sequence spaces $\ell_{p,w}(I)$ are usually defined for a fixed U-well-spread family $\mathcal{X}$. Since no confusion is possible, we will implicitly assume this family to define the weight of the sequence space.

The General Setting

Assume X is a locally compact Hausdorff space endowed with a positive Radon measure μ. This is the setting that appears in Chapter 4 and Chapter 5.

Definition 1.6.5. Let $\mathcal{U} = (U_i)_{i \in I}$ be a countable admissible covering of X and $(Y, \|\cdot\|_Y)$ be a solid Banach space of functions on X, then the associated sequence space $Y_s(I, \mathcal{U})$ is defined as

$$Y_s(I, \mathcal{U}) := \left\{ c = (c_i)_{i \in I} \in \mathbb{C}^I \ : \ \|c\|_{Y_s(I, \mathcal{U})} := \left\| \sum_{i \in I} |c_i| \chi_{U_i} \right\|_Y < \infty \right\}.$$

Again, we will only write Y_s when it is obvious which covering $\mathcal{U}$ is intended. It is also possible to define sequence spaces which have a different norm, namely an additional factor $\mu(U_i)$ in each summand above. We omit this second definition.

Next, we can show the following.

Lemma 1.6.6 (Theorem 4 of [53]). *Let $\mathcal{U}$ and Y as in Definition 1.6.5. Then the following holds true.*

(i) *The space Y_s is a Banach space of sequences.*

(ii) *If the functions with compact support are dense in Y, the finite sequences are dense in Y_s.*

Again, we are particularly interested in the sequence space associated to weighted Lebesgue spaces $L_{p,w}(X, \mu)$ and again these spaces coincide with $\ell_{p,m}(I)$-spaces.

Lemma 1.6.7 (Theorem 4 of [53]). *Suppose $w\colon X \to \mathbb{R}_+$ is a weight on X, $1 \leqslant p \leqslant \infty$ and let $\mathcal{U}$ be a countable admissible covering of X and set $w = (w_i)_{i \in I} = (\sup_{x \in U_i} w(x) \cdot \mu(U_i)^{1/p})_{i \in I}$; then the associated sequence space $(L_{p,w}(X, \mu))_s(I, \mathcal{U}) = \ell_{p,w}(I, \mathcal{U})$ is given by*

$$\ell_{p,w}(I, \mathcal{U}) := \left\{ c = (c_i)_{i \in I} \in \mathbb{C}^I \ : \ \|c\|_{\ell_{p,w}}^p = \sum_{i \in I} |c_i|^p \mu(U_i) \sup_{x \in U_i} w(x)^p < \infty \right\}, \tag{1.6.1}$$

with the usual norm for $p = \infty$. We simply write $\ell_{p,w}(I)$ when no confusion is possible.

This characterization is different to the one in the group setting, since on the one hand the size of the covering may vary in this general setting, and on the other hand the local supremum of the weight w enters the equation. In practice, this is no big deal since we are interested in sufficiently well-behaving weights, in the sense that they do not vary too much locally. This is made precise in the following lemma.

Lemma 1.6.8. *Let $w\colon X \to \mathbb{R}_{\geqslant 0}$ be a positive weight on X and $1 \leqslant p \leqslant \infty$.*

(i) Suppose $\mathcal{U} = (U_i)_{i \in I}$ and $\mathcal{V} = (V_i)_{i \in I}$ are two moderate admissible coverings of X over the same index set I. Assume further there exist constants $C_1, C_2, C_3 > 0$ such that

$$C_1 \cdot \mu(U_i) \leqslant \mu(V_i) \leqslant C_2 \cdot \mu(U_i) \quad \text{for all } i \in I$$

and a weight $v\colon X \times X \to \mathbb{R}_{\geqslant 0}$ with

$$w(x) \leqslant v(x, y) \cdot w(y) \quad \text{for all } x, y \in X$$

as well as

$$\sup_{x \in U_i} \sup_{y \in V_i} v(x, y) \leqslant C_3 \quad \text{for all } i \in I.$$

Then the sequence spaces $\ell_{p,w}(I, \mathcal{U})$ and $\ell_{p,w}(I, \mathcal{V})$ coincide with equivalence of norms.

(ii) Let $\mathcal{U} = (U_i)_{i \in I}$ be a moderate admissible covering of X and assume there is a constant $C > 0$ such that

$$w(x) \leqslant C \cdot w(y) \quad \text{for all } x, y \in U_i, i \in I. \tag{1.6.2}$$

Then the norm in (1.6.1) is equivalent to

$$\sum_{i \in I} |c_i|^p \mu(U_i) w(x_i)^p$$

for any fixed set of points $(x_i)_{i \in I} \subset X$ with $x_i \in U_i$ for all $i \in I$.

Proof. Part (i) is proven in Lemma 6 of [53] and part (ii) is a simple application of (1.6.2). $\square$

1.7 Continuous Frames

In this section we define the notion of continuous frames in Hilbert spaces. This idea is a generalization of the concept of *discrete frames* in Hilbert spaces [15], which was originally introduced by Duffin and Schaeffer [37]. In the wavelet era this concept has been picked up by Daubechies and her coauthors [31, 32] and was made popular. The main idea behind discrete frames is to find a countable family of functions in a Hilbert space that allows stable but possibly redundant decompositions of arbitrary elements into expansions of the frame element. While this idea is similar to finding a basis of the Hilbert space, frames admit a larger freedom in choosing the frame elements due to the redundancy they induce. A detailed review of frames can be found in [15].

Later, this concept has been generalized by Antoine et al., Kaiser and others [1, 2, 62, 78] to families of functions in a Hilbert space that are indexed by elements of a measure space.

Definition 1.7.1. Let $\mathcal{H}$ be a complex Hilbert space and X a measure space with positive measure μ. A *continuous frame* for $\mathcal{H}$ is a family of vectors $\mathfrak{F} = (f_x)_{x \in X}$ with the following properties.

(i) For all $f \in \mathcal{H}$ the map $x \mapsto \langle f, f_x \rangle_\mathcal{H}$ is a measurable function on X;

(ii) There exist constants $A, B > 0$ such that

$$A \cdot \|f\|_\mathcal{H}^2 \leqslant \int_X |\langle f, f_x \rangle_\mathcal{H}|^2 \, d\mu(x) \leqslant B \cdot \|f\|_\mathcal{H}^2 \quad \text{for all } f \in \mathcal{H}. \tag{1.7.1}$$

If $A = B$ then the frame is called *tight* and if $A = B = 1$ it is called *Parseval frame*.

Associated to a frame $\mathfrak{F}$ is the *frame operator* $S = S_{\mathfrak{F}}$, which we define in the following.

Lemma 1.7.2. *Let $\mathfrak{F} = (f_x)_{x \in X}$ be a frame, then the frame operator $S : \mathcal{H} \to \mathcal{H}$ defined in the weak sense by*

$$Sf := \int_X \langle f, f_x \rangle_\mathcal{H} f_x \, d\mu(x) \tag{1.7.2}$$

is well-defined, bounded, positive, self-adjoint and boundedly invertible. If $\mathfrak{F}$ is tight, then S is a multiple of the identity and if $\mathfrak{F}$ is a Parseval frame, then S is the identity.

Proof. Fix $f \in \mathcal{H}$, then the frame expansion

$$g \mapsto \int_X \langle f, f_x \rangle_\mathcal{H} \langle f_x, g \rangle_\mathcal{H} \, d\mu(x),$$

which is a functional on $\mathcal{H}$, is a conjugated linear map. Moreover it is bounded, since by Cauchy-Schwarz' inequality we have

$$\left| \int_X \langle f, f_x \rangle_\mathcal{H} \langle f_x, g \rangle_\mathcal{H} \, d\mu(x) \right|^2 \leqslant \int_X |\langle f, f_x \rangle_\mathcal{H}|^2 \, d\mu(x) \cdot \int_X |\langle f_x, g \rangle_\mathcal{H}|^2 \, d\mu(x)$$
$$\leqslant B^2 \cdot \|f\|_\mathcal{H}^2 \cdot \|g\|_\mathcal{H}^2. \tag{1.7.3}$$

Therefore by Riesz' representation theorem there exists a unique element in $\mathcal{H}$ that we call $\int_X \langle f, f_x \rangle_\mathcal{H} f_x \, d\mu(x)$ such that

$$\left\langle \int_X \langle f, f_x \rangle_\mathcal{H} f_x \, d\mu(x), g \right\rangle_\mathcal{H} = \int_X \langle f, f_x \rangle_\mathcal{H} \langle f_x, g \rangle_\mathcal{H} \, d\mu(x) \quad \text{for all } g \in \mathcal{H}.$$

Thereby we have shown that S is well-defined. It is easy to check that S is linear and the boundedness follows from (1.7.3) by writing

$$\|Sf\|_\mathcal{H} = \sup_{\|g\|_\mathcal{H}=1} |\langle Sf, g \rangle_\mathcal{H}| \leqslant B \cdot \|f\|_\mathcal{H}$$

for all $f \in \mathcal{H}$, hence $\|S\|_{\mathcal{H} \to \mathcal{H}} \leqslant B$. Obviously, for any $f \in \mathcal{H}$ we have

$$\langle Sf, f \rangle_{\mathcal{H}} = \left\langle \int_X \langle f, f_x \rangle_{\mathcal{H}} f_x \, d\mu(x), f \right\rangle_{\mathcal{H}} = \int_X |\langle f, f_x \rangle_{\mathcal{H}}|^2 \, d\mu(x),$$

which implies

$$A \cdot \|f\|_{\mathcal{H}}^2 \leqslant \langle Sf, f \rangle_{\mathcal{H}} \leqslant B \cdot \|f\|_{\mathcal{H}}^2 \qquad (1.7.4)$$

by the frame property (1.7.1). This shows that S is positive.

Let us define the operator

$$V : \mathcal{H} \to L_2(X, \mu), \quad Vf(x) := \langle f, f_x \rangle_{\mathcal{H}}, \qquad (1.7.5)$$

which is obviously bounded. Since

$$\langle Vf, G \rangle_{L_2} = \int_X \langle f, f_x \rangle_{\mathcal{H}} \overline{G(x)} \, d\mu(x) = \left\langle f, \int_X G(x) f_x \, d\mu(x) \right\rangle_{\mathcal{H}} \quad \text{for } f \in \mathcal{H}, \ G \in L_2(X, \mu),$$

the adjoint of V is given weakly by

$$V^* : L_2(X, \mu) \to \mathcal{H}, \quad V^*F := \int_X F(y) f_y \, d\mu(y). \qquad (1.7.6)$$

Then, $S = V^*V$ and $S^* = (V^*V)^* = V^*V = S$, so S is self-adjoint.

To show that S is boundedly invertible we use a Neumann series argument, see for example [85, Theorem 4.15]. Using (1.7.4) we have

$$\begin{aligned} \|\operatorname{id} - B^{-1}S\|_{\mathcal{H} \to \mathcal{H}} &= \sup_{\|f\|_{\mathcal{H}}=1} |\langle (\operatorname{id} - B^{-1}S)f, f \rangle_{\mathcal{H}}| \\ &= \sup_{\|f\|_{\mathcal{H}}=1} |\langle f, f \rangle_{\mathcal{H}} - B^{-1}\langle Sf, f \rangle_{\mathcal{H}}| \\ &= 1 - \frac{A}{B} < 1, \end{aligned}$$

so S is boundedly invertible.

Finally, (1.7.4) shows that if $A = B$, then $S = A \cdot \operatorname{id}_{\mathcal{H} \to \mathcal{H}}$. $\qquad \square$

Since S is invertible and self-adjoint, we can write

$$f = SS^{-1}f = \int_X \langle f, S^{-1}f_x \rangle_{\mathcal{H}} f_x \, d\mu(x), \qquad (1.7.7)$$

$$f = S^{-1}Sf = \int_X \langle f, f_x \rangle_{\mathcal{H}} S^{-1}f_x \, d\mu(x) \qquad (1.7.8)$$

for all $f \in \mathcal{H}$ in the weak sense. By taking the inner product with f_x in (1.7.8) we obtain

$$Vf(x) = \langle S^{-1}Sf, f_x \rangle_{\mathcal{H}} = \int_X Vf(y) \langle f_y, S^{-1}f_x \rangle_{\mathcal{H}} \, d\mu(y) \qquad (1.7.9)$$

This formula reveals that the function

$$K : X \times X \to \mathbb{C}, \quad K(x, y) := K_{\mathfrak{F}}(x, y) := \langle f_y, S^{-1}f_x \rangle_{\mathcal{H}} \qquad (1.7.10)$$

plays a special role—it reproduces Vf. Functions of this type are called *kernels* and we will take a closer look at spaces of these functions in the following section.

1.8   Kernel Spaces

The aim of this section is to define spaces of kernels that suffice certain decay properties. For this let (X, μ) be a σ-finite measure space throughout this section. Let us first define admissible weights on X.

Definition 1.8.1. A positive function $w \colon X \times X \to \mathbb{R}_+$ is called *admissible weight function*, if w is continuous and it satisfies

$$w(x, y) \leqslant w(x, z) \cdot w(z, y), \tag{1.8.1}$$
$$w(x, y) = w(y, x), \tag{1.8.2}$$
$$w(x, x) \leqslant C < \infty \tag{1.8.3}$$

for all $x, y, z \in X$.

This definition implies $w(x, y) \geqslant 1$ for all $x, y \in X$ as follows. By (1.8.1) and (1.8.2) we have $w(x, y) \leqslant w(x, y) \cdot w(y, y) \leqslant w(x, y)^3$, therefore $w(x, y) \geqslant 1$.

For an admissible w we now define weighted kernel spaces $\mathcal{A}_{p,w}$.

Definition 1.8.2. Let $1 \leqslant p \leqslant \infty$, then the kernel spaces $\mathcal{A}_p$ are defined via

$$\mathcal{A}_p := \left\{ K \colon X \times X \to \mathbb{C} \,:\, K \text{ is measurable}, \|K\|_{\mathcal{A}_p} < \infty \right\}, \tag{1.8.4}$$

where

$$\|K\|_{\mathcal{A}_p} := \max \left\{ \operatorname*{ess\,sup}_{x \in X} \left(\int_X |K(x,y)|^p \, d\mu(y) \right)^{1/p}, \, \operatorname*{ess\,sup}_{y \in X} \left(\int_X |K(x,y)|^p \, d\mu(x) \right)^{1/p} \right\} \tag{1.8.5}$$

is its norm with the usual definition for $p = \infty$. For an admissible weight w the weighted kernel spaces $\mathcal{A}_{p,w}$ are given by

$$\mathcal{A}_{p,w} := \{ K \colon X \times X \to \mathbb{C} \,:\, K \cdot w \in \mathcal{A}_p \} \tag{1.8.6}$$

with the natural norm $\|K\|_{\mathcal{A}_{p,w}} = \|K \cdot w\|_{\mathcal{A}_p}$.

For a given kernel function K, we also denote the kernel operator associated to the kernel function acting on a function f on X through

$$Kf(x) := \int_X K(x,y) f(y) \, d\mu(y), \quad \text{for } x \in X. \tag{1.8.7}$$

Remark 1.8.3.   (i) The spaces $\mathcal{A}_{p,w}$ are Banach spaces.

(ii) The space $\mathcal{A}_{1,w}$ with $p = 1$ is a Banach $*$-algebra, where the multiplication is given by

$$K_1 \circ K_2(x, y) := \int_X K_1(x, z) K_2(z, y) \, d\mu(z).$$

The well-definedness of the multiplication is ensured by property (1.8.1) of the weight w and the involution is given by $K^*(x, y) = \overline{K(y, x)}$. It is then straightforward that all properties of a Banach $*$-algebra are fulfilled. In view of the corresponding kernel operators defined in (1.8.7), the multiplication of two kernels implies $(K_1 \circ K_2)f = K_1(K_2 f)$. And interpretating the kernel operator for some $K \in \mathcal{A}_{1,w}$ as an operator on $L_2(X, \mu)$, the adjoint operator is in fact given by the involution K^*.

Remark 1.8.4. The application of a kernel operator associated to a kernel K can be seen as a substitute for the convolution on groups. Let $X = G$ be a group with left Haar measure and k a function on G satisfying $|k(x^{-1})| = |k(x)|$ for all $x \in X$. Then, by setting $K(x,y) = \lambda(y)k(x) = k(y^{-1}x)$ the associated kernel operator is given by

$$Kf(x) = \int_G k(y^{-1}x) f(y) \, d\mu(x) = (f * k)(x).$$

Furthermore, $\|K\|_{\mathcal{A}_p} = \|k\|_{L_p}$ for all $1 \leqslant p \leqslant \infty$. So the kernels can be seen as generalizations of convolutions to arbitrary measure spaces.

A very important property of the spaces $\mathcal{A}_{1,w}$ is that the associated kernel operators are bounded on certain weighted Lebesgue spaces. For this we first recall the classical Schur's test.

Lemma 1.8.5 (Schur's test, Theorem 6.18 of [52]). *Let $K \in \mathcal{A}_1$, then the associated kernel operator defined by (1.8.7) is bounded on $L_p(X, \mu)$ for all $1 \leqslant p \leqslant \infty$, that is*

$$\|Kf\|_{L_p} \leqslant \|K\|_{\mathcal{A}_p} \cdot \|f\|_{L_p} \quad \text{for all } f \in L_p(X, \mu).$$

Using this we can now easily prove a weighted version of Schur's test. We include the proof for the reader's convenience.

Corollary 1.8.6 (Weighted version of Schur's test). *Let $m \colon X \times X \to \mathbb{R}_+$ be a weight satisfying $m(x) \leqslant w(x,y) \cdot m(y)$ for all $x, y \in X$ and let $K \in \mathcal{A}_{1,w}$. Then, the associated kernel operator is bounded on $L_{p,m}(X, \mu)$ for all $1 \leqslant p \leqslant \infty$, that is*

$$\|Kf\|_{L_{p,m}} \leqslant C \cdot \|f\|_{L_{p,m}} \quad \text{for all } f \in L_{p,m}(X, \mu).$$

Proof. Let $f \in L_{p,m}(X, \mu)$ and $1 \leqslant p < \infty$, then by the definition of the weight m we have

$$\|Kf\|_{L_{p,m}}^p = \int_X \left(\int_X K(x,y) f(y) \, d\mu(y) \right)^p m(x)^p \, d\mu(x)$$
$$\leqslant \int_X \left(\int_X K(x,y) w(x,y) f(y) m(y) \, d\mu(y) \right)^p d\mu(x) = \|(K \cdot w)(f \cdot m)\|_{L_p}^p.$$

By the unweighted Schur's test above and since $K \cdot w \in \mathcal{A}_1$ this implies

$$\|Kf\|_{L_{p,m}} \leqslant C \cdot \|f \cdot m\|_{L_p} = C \cdot \|f\|_{L_{p,m}}.$$

Analogously, for $f \in L_{\infty,m}(X, \mu)$ it is $\|Kf\|_{L_{\infty,m}} = \|(K \cdot w)(f \cdot m)\|_{L_\infty} \leqslant C \cdot \|f\|_{L_{\infty,m}}$ and we are done. $\qquad \square$

For kernels in $\mathcal{A}_{p,w}$ with $p \neq 1$ a similar statement to Corollary 1.8.6 is possible, with the significant difference that the kernel operator does not map into the same space. To be more precise, the integrability parameter of the Lebesgue space increases as we apply the kernel operator.

Lemma 1.8.7 (Generalized weighted version of Schur's test). *Let $m \colon X \times X \to \mathbb{R}_+$ be a weight satisfying $m(x) \leqslant w(x,y) \cdot m(y)$ for all $x, y \in X$ and suppose $K \in \mathcal{A}_{q,w}$ for some $1 < q \leqslant \infty$. Then the corresponding kernel operator is well-defined and bounded as an operator $K \colon L_{p,m}(X, \mu) \to L_{r,m}(X, \mu)$ for all $1 \leqslant p < r \leqslant \infty$ satisfying $1 - 1/q = 1/p - 1/r$.*

Proof. For fixed $1 < p < r < \infty$ and $g \in L_{p,m}(X,\mu)$ with $\|g\|_{L_{p,m}} \leqslant 1$ arbitrary one has

$$
\|Kg\|_{L_{r,m}} = \sup_{\|h\|_{L_{r',m-1}} \leqslant 1} |\langle Kg, h\rangle_{L_2}|
$$

$$
\leqslant \sup_{\|h\|_{L_{r',m-1}} \leqslant 1} \int_X \int_X |K(x,y)g(y)h(x)| \, d\mu(x) \, d\mu(y)
$$

$$
=: \sup_{\|h\|_{L_{r',m-1}} \leqslant 1} I_{K,g,h},
$$

where r' denotes the Hölder-dual of r satisfying $1/r + 1/r' = 1$. We now set $\alpha := r > 0$, $\beta := p' > 0$, $1/\gamma := 1/p - 1/r > 0$, $a := q/r$, $b := p/r$, $c := q/p'$, $d := r'/p'$, $e := 1 - p/r$, $f := r'/p - r'/r$. These choices suffice the following relations:

$$
1/\alpha + 1/\beta + 1/\gamma = 1, \quad a + c = 1, \quad b\alpha = p, \quad d\beta = r', \quad a\alpha = q,
$$
$$
b + e = 1, \quad e\gamma = p, \quad f\gamma = r', \quad c\beta = q,
$$
$$
d + f = 1.
$$

By applying Young's inequality for products we obtain

$$
I_{K,g,h} \leqslant \int_X \int_X |K(x,y)w(x,y)|^a |g(y)m(y)|^b \cdot |K(x,y)w(x,y)|^c |h(x)m(x)^{-1}|^d
$$
$$
\cdot |g(y)m(y)|^e |h(x)m(x)^{-1}|^f \, d\mu(x) \, d\mu(y)
$$
$$
\leqslant \frac{1}{\alpha} \int_X \int_X |K(x,y)w(x,y)|^{a\alpha} |g(y)m(y)|^p \, d\mu(x) \, d\mu(y)
$$
$$
+ \frac{1}{\beta} \int_X \int_X |K(x,y)w(x,y)|^{c\beta} |h(x)m(x)^{-1}|^{r'} \, d\mu(x) \, d\mu(y)
$$
$$
+ \frac{1}{\gamma} \int_X \int_X |g(y)m(y)|^p |h(x)m(x)^{-1}|^{r'} \, d\mu(x) \, d\mu(y).
$$

For the first summand we deduce the estimate

$$
\int_X \int_X |K(x,y)w(x,y)|^{a\alpha} |g(y)m(y)|^p \, d\mu(x) \, d\mu(y)
$$
$$
\leqslant \left(\operatorname*{ess\,sup}_{y \in X} \int_X |K(x,y)|^{a\alpha} |w(x,y)|^{a\alpha} \, d\mu(x) \right) \int_X |g(y)|^p |m(y)|^p \, d\mu(y)
$$
$$
\leqslant \|K\|_{\mathcal{A}_{a\alpha,w}}^{a\alpha} \cdot \|g\|_{L_{p,m}}^p = \|K\|_{\mathcal{A}_{q,w}}^q \cdot \|g\|_{L_{p,m}}^p
$$

and the other two summands can be treated analogously. Thus we obtain

$$
I_{K,g,h} \leqslant \frac{1}{\alpha} \|K\|_{\mathcal{A}_{a\alpha,w}}^{a\alpha} \cdot \|g\|_{L_{p,m}}^p + \frac{1}{\beta} \|K\|_{\mathcal{A}_{c\beta,w}}^{c\beta} \cdot \|h\|_{L_{r',m-1}}^{r'} + \frac{1}{\gamma} \|g\|_{L_{p,m}}^p \cdot \|h\|_{L_{r',m-1}}^{r'}
$$
$$
\leqslant \max \left\{ 1, \|K\|_{\mathcal{A}_{q,w}}^q, \|K\|_{\mathcal{A}_{q,w}}^q \right\} =: C_{K,p,r,m}
$$

for all g, h. This implies $\|K\|_{L_{p,m} \to L_{r,m}} \leqslant C_{K,p,r,m}$.

If $1 = p < r < \infty$ — which implies $q = r$ — and $g \in L_{1,m}(X,\mu)$ arbitrary with $\|g\|_{L_{1,m}} \leqslant 1$ we follow the lines of the proof above and set $1/\beta = c = d = 0$ and $a = f = 1$ as well as $\alpha = r$

and $\gamma = r'$. Then, for any $h \in L_{r',m^{-1}}(X,\mu)$ with $\|h\|_{L_{r',m^{-1}}} \leqslant 1$, it follows

$$
\begin{aligned}
I_{K,g,h} &\leqslant \frac{1}{\alpha} \int_X \int_X |K(x,y)w(x,y)|^{a\alpha} |g(y)v(y)| \, d\mu(x) \, d\mu(y) \\
&\quad + \frac{1}{\gamma} \int_X \int_X |g(y)m(y)| |h(x)m(x)^{-1}|^{f\gamma} \, d\mu(x) \, d\mu(y) \\
&\leqslant \frac{1}{r} \|K\|_{\mathcal{A}_{r,w}}^r \cdot \|g\|_{L_{1,m}} + \frac{1}{r'} \|g\|_{L_{1,m}} \cdot \|h\|_{L_{r',m^{-1}}}^{r'} \\
&\leqslant \max\{1, \|K\|_{\mathcal{A}_{r,w}}^r\}.
\end{aligned}
$$

Therefore, $K\colon L_{1,m}(X,\mu) \to L_{q,m}(X,\mu)$ is well-defined and bounded.

If $1 \leqslant p < r = \infty$ — which implies $p = q'$ — and $g \in L_{p,m}(X,\mu)$ arbitrary, it follows with Hölder's inequality that

$$
\begin{aligned}
\|K(g)\|_{L_{\infty,m}} &\leqslant \operatorname*{ess\,sup}_{x \in X} \int_X |K(x,y)w(x,y)| \cdot |g(y)m(y)| \, d\mu(y) \\
&\leqslant \|K\|_{\mathcal{A}_{p',w}} \cdot \|g\|_{L_{p,m}},
\end{aligned}
$$

which proves that $K\colon L_{q',m}(X,\mu) \to L_{\infty,m}(X,\mu)$ is well-defined and bounded. □

Remark 1.8.8. Actually, Lemma 1.8.7 is a generalization of the weighted version of Schur's test in Corollary 1.8.6, which can be seen as follows. By setting $q = 1$, i.e., by assuming $K \in \mathcal{A}_{1,w}$, the condition $1 - 1/q = 1/p - 1/r$ implies $p = r$. Then, we set $1/\gamma = 0$ and $e = f = 0$ and only apply Young's inequality with two factors in the proof. The rest of the proof remains identical and we are in the setting of a weighted version of Schur's test.

The following auxiliary corollary is an easy consequence of the previous lemma. The only difference is that we require the kernel to be in all kernel spaces $\mathcal{A}_{q,w}$, $1 < q \leqslant \infty$, which gives us a broader family of bounded kernel operators.

Corollary 1.8.9. *Let $m\colon X \times X \to \mathbb{R}_+$ be a weight satisfying $m(x) \leqslant w(x,y) \cdot m(y)$ for all $x, y \in X$ and $K \in \mathcal{A}_{q,w}$ for all $1 < q \leqslant \infty$. Then the corresponding kernel operator is well-defined and bounded as an operator $K\colon L_{p,m}(X,\mu) \to L_{r,m}(X,\mu)$ for all $1 \leqslant p < r \leqslant \infty$.*

Proof. Fix $1 \leqslant p < r \leqslant \infty$ and let $1 < q \leqslant \infty$ such that $1 - 1/q = 1/p + 1/r$. Then, Lemma 1.8.7 yields the assertion. □

1.9　Discretization of Function Spaces

The concept of discrete frames in Hilbert spaces gives rise to families of functions which allow possibly redundant representations as convergent series, where the norm can be estimated form above and below—up to a constant—by a discrete norm of the coefficients [15]. Similar expansions in the context of Banach spaces are referred to as *atomic decompositions*. Another conceptpt, the *Banach frames* were introduced by Gröchenig [68] and provide a refined tool very similar to discrete frames. Both discretization techniques are useful for describing and examining Banach spaces of functions.

We start by defining atomic decompositions for Banach spaces of functions.

Definition 1.9.1. A family $(f_i)_{i \in I}$ in a Banach space of functions $(B, \|\cdot\|_B)$ is called an *atomic decomposition* for B, if there exists an associated *coefficient space* $(B^\sharp(I), \|\cdot\|_{B^\sharp(I)})$ for some index set I and a *coefficient operator* $C\colon B \to B^\sharp(I)$ such that

(i) C is bounded;

(ii) the *synthesis operator*

$$S\colon B^{\sharp}(I) \to B, \quad (c_i)_{i\in I} \mapsto \sum_{i\in I} c_i \cdot f_i$$

is well-defined and bounded, with unconditional convergence of the defining series in some suitable topology;

(iii) $S \circ C = \mathrm{id}_B$, i.e., $h = \sum_{i\in I} C(h)_i \cdot f_i$ for all $h \in B$.

Dual to the concept of atomic decompositions we define Banach frames.

Definition 1.9.2. Let $(B, \|\cdot\|_B)$ be a Banach space of functions. A family $(g_i)_{i\in I}$ in the dual space B' is called a *Banach frame* for B , if there exists an associated *coefficient space* $(B^{\flat}(I), \|\cdot\|_{B^{\flat}(I)})$ for some index set I and a linear, bounded reconstruction operator $R\colon B^{\flat}(I) \to B$ such that

(i) for all $h \in B$ we have $(g_i(h))_{i\in I} \in B^{\flat}(I)$, and there exist constants $0 < C_1, C_2 < \infty$ independent of h such that

$$C_1 \cdot \|h\|_B \leqslant \|(g_i(h))_{i\in I}\|_{B^{\flat}(I)} \leqslant C_2 \cdot \|h\|_B \,;$$

(ii) $R[(g_i(h))_{i\in I}] = h$ for all $h \in B$.

Since all the applications we are interested in will rely on the weighted Lebesgue spaces $L_{p,m}$ over some suitable measure space, the associated coefficient space will be the sequence spaces $\ell_{p,m}(I)$ with countable index set.

Chapter 2

Classic Coorbit Theory

In this chapter we recall the coorbit theory as originally developed by Feichtinger and Gröchenig in a series of papers [44–46]. We call it the "classic" coorbit theory not only because it was historically first, but also because the coorbit theories described in Chapter 3, 4 and 5 are generalizations of different kinds of the theory presented in this chapter. Since the theory is standard and well-known we omit the proofs.

This chapter is organized as follows. In Section 2.1 we recall the theory that allows us to define coorbit spaces. Then, in Section 2.2 we discuss the two major discretization techniques, namely atomic decompositions and Banach frames, as well as the necessary conditions. In Section 2.3 two interesting and reoccurring applications are presented. First we apply the coorbit theory to the full affine group to obtain homogeneous Besov spaces, and then we define coorbit spaces associated to the so-called full shearlet group.

2.1   Coorbit Spaces

Let G be a locally compact topological group with Haar measure dx and modulation function Δ. As in Section 1.2 we denote with $\lambda(x)$ and $\rho(x)$ the left- and right translation operators acting on a function f on G by $x \in G$. Furthermore, let π be a unitary representation of G in a Hilbert space $\mathcal{H}$. We now define the voice transform and kernel related to π to be the coefficients of the representation.

Definition 2.1.1. Let $\psi \in \mathcal{H}$ be an arbitrary vector. The *voice transform* V_ψ associated with ψ is defined as the mapping

$$V_\psi \colon \mathcal{H} \to L_\infty(G), \quad V_\psi f(x) := \langle f, \pi(x)\psi\rangle_{\mathcal{H}}. \tag{2.1.1}$$

We denote with

$$K_\psi(x) := V_\psi \psi(x) = \langle \psi, \pi(x)\psi\rangle_{\mathcal{H}} \tag{2.1.2}$$

the *kernel function* associated with ψ.

By the Cauchy-Schwarz inequality the voice-transform V_ψ is well-defined. Note that for a unitary representation π we have $\|\pi(x)\psi\|_{\mathcal{H}} = \|\psi\|_{\mathcal{H}}$ for all $x \in G$.

We collect some basic facts of the voice transform in the following lemma.

Lemma 2.1.2 (Proposition 2.56 of [20])**.** *Let π be a unitary representation of G on $\mathcal{H}$ and let $\psi \in \mathcal{H}$ be a fixed vector. Then:*

(i) the transforms $V_\psi f$ are bounded continuous functions on G, for all $f \in \mathcal{H}$;

(ii) the voice transform satisfies $V_\psi \circ \pi(x) = \lambda(x) \circ V_\psi$ for every $x \in G$;

(iii) ψ is a cyclic vector for π if and only if V_ψ is an injective map of $\mathcal{H}$ into $\mathcal{C}(G) \cap L_\infty(G)$;

(iv) the kernel satisfies $\overline{K_\psi} = \widecheck{K_\psi}$.

Note that part (i) implies that K_ψ is a continuous function on G. Additionally, the kernel K_ψ is positive, see Section 3.3 of [51], i.e., $\sum_{i,j=1}^{n} c_i \overline{c_j} K(x_i^{-1} x_j) \geqslant 0$ for all $c_1, \dots, c_n \in \mathbb{C}$ and $x_1, \dots, x_n \in G$. This, together with part (iv) above, uniquely defines the representation π up to a unitary equivalence, see Theorem 3.20 and Proposition 3.35 of [51].

We now define the notion of admissibility.

Definition 2.1.3. Let π be a unitary irreducible representation of G on $\mathcal{H}$. If there exists a vector $\psi \in \mathcal{H}$ such that $V_\psi \psi = K_\psi \in L_2(G)$, then we say that π is *square integrable* and ψ is an *admissible vector* for π.

In general, π does not need to be irreducible as assumed in the definition above. In fact, many important examples in analysis are connected with reducible representations. However, we will only be concerned with irreducible representations and shall assume this from now on.

The following classical result is due to Duflo and Moore [38], later reviewed with a slightly different argument in [70].

Theorem 2.1.4 ([38]). *Let π be a square integrable unitary representation of G on $\mathcal{H}$. Then there exists a unique, positive, densely defined self-adjoint operator A on $\mathcal{H}$ such that:*

(i) If ψ_1, ψ_2 are in the domain of A, then for all $f_1, f_2 \in \mathcal{H}$ we have

$$\int_G \langle f_1, \pi(x)\psi_1 \rangle_\mathcal{H} \langle \pi(x)\psi_2, f_2 \rangle_\mathcal{H} \, dx = \int_G V_{\psi_1} f_1(x) \overline{V_{\psi_2} f_2(x)} \, dx = \langle A\psi_2, A\psi_1 \rangle_\mathcal{H} \langle f_1, f_2 \rangle_\mathcal{H}.$$
$$(2.1.3)$$

In particular, the set of admissible vectors coincides with the subset of the domain of A consisting of those vectors ψ for which $\|A\psi\|_\mathcal{H} < \infty$. We shall assume the normalization $\|A\psi\|_\mathcal{H} = 1$.

(ii) The operator A satisfies the semi-invariance relation

$$\pi(x)A\pi(x)^* = \Delta(x)^{1/2} A \qquad (2.1.4)$$

for every $x \in G$.

The operator A is usually referred to as the *Duflo-Moore operator* of π, whereas (2.1.3) is known as the *orthogonality relation*.

As a consequence of Theorem 2.1.4 we obtain that the voice transform is an isometry.

Proposition 2.1.5. *Let π be a square integrable unitary representation of G on $\mathcal{H}$ and ψ be an admissible vector for π. Then the voice transform V_ψ is an isometry, that is*

$$\|V_\psi f\|_{L_2(G)} = \|f\|_\mathcal{H}, \quad f \in \mathcal{H}, \qquad (2.1.5)$$

and the reproducing identity

$$V_\psi f * K_\psi = V_\psi f, \quad f \in \mathcal{H}, \qquad (2.1.6)$$

holds true.

Proof. By Theorem 2.1.4 (i) it holds that $\|A\psi\|_{\mathcal{H}} = 1$ where A is the Duflo-Moore operator of π. Applying (2.1.3) to the special case $\psi_1 = \psi_2 = \psi$ and $f_1 = f_2 = f$ for some $f \in \mathcal{H}$ we have

$$\|V_\psi f\|_{\mathcal{H}} = \|A\psi\|_{\mathcal{H}} \cdot \|f\|_{\mathcal{H}} = \|f\|_{\mathcal{H}},$$

thus V_ψ is an isometry. Similarly, setting $\psi_1 = \psi_2 = \psi$ and $f_1 = f$, $f_2 = \pi(y)\psi$ for some $y \in G$, we obtain

$$V_\psi f * K_\psi(y) = \int_G V_{\psi_1} f_1(x) \overline{V_{\psi_2} f_2(x)} \, dx = \|A\psi\|_{\mathcal{H}}^2 \cdot \langle f, \pi(y)\psi \rangle_{\mathcal{H}} = V_\psi f(y),$$

proving the reproducing identity. □

Obviously, by Definition 1.7.1, for an admissible vector $\psi \in \mathcal{H}$ the family $(\pi(x)\psi)_{x \in X}$ constitutes a tight frame with frame constant C_ψ. This observation is later generalized in Chapters 4 and 5.

By the polarization identity, the isometry property (2.1.5) is equivalent to

$$\langle V_\psi f, V_\psi g \rangle_{L_2(G)} = \langle f, g \rangle_{\mathcal{H}}, \quad \text{for all } f, g \in \mathcal{H}. \tag{2.1.7}$$

This implies that if ψ is an admissible vector for π, then it is a cyclic vector for π as well. This is because if V_ψ is an isometry, then it is injective on $\mathcal{H}$ and (iii) of Lemma 2.1.2 applies.

The following proposition contains the important reproducing formula associated to the voice transform and characterizes its adjoint.

Proposition 2.1.6 (Proposition 2.58 of [20]). *Let π be a unitary representation of G on $\mathcal{H}$ and ψ an admissible vector of π. Then the* reproducing formula

$$f = \int_G V_\psi f(x) \pi(x)\psi \, dx = \int_G \langle f, \pi(x)\psi \rangle_{\mathcal{H}} \pi(x)\psi \, dx \tag{2.1.8}$$

holds for every $f \in \mathcal{H}$, where the right hand side is interpreted as a weak integral. The adjoint of the voice transform is given as weak integral by the formula

$$V_\psi^* F = \int_G F(x) \pi(x)\psi \, dx, \quad F \in L_2(G), \tag{2.1.9}$$

and $V_\psi^* V_\psi = \mathrm{id}_{\mathcal{H}}$.

Proposition 2.1.7 (Proposition 2.61 of [20]). *Let π be a unitary representation of G on $\mathcal{H}$ and ψ an admissible vector of π. Then the projection onto the range of the voice transform is given by*

$$V_\psi V_\psi^* F = F * K_\psi, \quad F \in L_2(G). \tag{2.1.10}$$

Let us introduce a fixed *weight* w on G, which is a continuous function $w \colon G \to \mathbb{R}_+$ assumed to be submultiplicative, that is

$$w(xy) \leqslant w(x)w(y) \quad \text{for all } x, y \in G. \tag{2.1.11}$$

Further we assume throughout this chapter that

$$w(x) = w(x^{-1}) \quad \text{for all } x \in G, \tag{2.1.12}$$

which implies $w(x) \geqslant 1$ for all $x \in G$. The symmetry (2.1.12) can always be satisfied by replacing w with $w + \breve{w}$, where the latter weight is easily seen to still satisfy the submultiplicativity condition (2.1.11).

From now on we will always assume that π is a square integrable unitary representation of G on $\mathcal{H}$ and we denote with ψ a corresponding admissible vector for π.

Definition 2.1.8. An admissible function $\psi \in \mathcal{H}$ is called an *analyzing vector* if $K_\psi \in L_{1,w}(G)$, i.e., it is contained in the set

$$\mathcal{A}_w := \{\psi \in \mathcal{H} : K_\psi = \langle \psi, \pi(\cdot)\psi \rangle_\mathcal{H} \in L_{1,w}(G)\}. \tag{2.1.13}$$

For an analyzing vector $\psi \in \mathcal{A}_w$ we consider the space

$$\mathcal{H}_{1,w} := \{f \in \mathcal{H} : V_\psi f = \langle f, \pi(\cdot)\psi \rangle_\mathcal{H} \in L_{1,w}(G)\} \tag{2.1.14}$$

with the natural norm $\|f\|_{\mathcal{H}_{1,w}} := \|V_\psi f\|_{L_{1,w}}$. In [44, Lemma 4.2] it was shown that $\mathcal{A}_w$ and $\mathcal{H}_{1,w}$ coincide as sets.

Remark 2.1.9. (i) It is not always the case that such an analyzing vector exists. In fact, since K_ψ is a bounded function on G by Lemma 2.1.2 (i), the integrability of K_ψ with respect to $L_{1,w}(G)$ is a strong assumption. This assumption can be weakend by assuming $K_\psi \in L_{q,w}(G)$ for all $q > 1$, as will be done in Chapter 3. We postpone a more thorough discussion on this matter to Subsection 3.1.2.

(ii) The integrability of K_ψ with respect to $L_{1,w}(G)$ allows us to view the kernel as a convolution operator on a variety of function spaces on G, which is continuous on weighted Lebesgue spaces by virtue of the generalized Young's inequality, see Proposition 1.3.9. This is a crucial property which is exploited numerous times in this theory.

The next proposition collects some basic properties of $\mathcal{H}_{1,w}$.

Proposition 2.1.10 (Proposition 3.2 of [20])**.**

(i) The space $\mathcal{H}_{1,w}$ is a Banach space.

(ii) The space $\mathcal{H}_{1,w}$ is independent of the choice of $\psi \in \mathcal{A}_w$.

(iii) The spaces $\mathcal{H}$, $\mathcal{H}_{1,w}$ and its dual space $\mathcal{H}'_{1,w}$ (space of all continuous linear functionals on $\mathcal{H}_{1,w}$) form a Gelfand triple, i.e., we have the dense continuous embeddings

$$\mathcal{H}_{1,w} \hookrightarrow \mathcal{H} \hookrightarrow \mathcal{H}'_{1,w}. \tag{2.1.15}$$

(iv) The space $\mathcal{H}_{1,w}$ is π-invariant, i.e., $f \in \mathcal{H}_{1,w}$ implies $\pi(x)f \in \mathcal{H}_{1,w}$ for all $x \in G$.

The spaces $\mathcal{H}_{1,w}$ will serve as substitutes for the (Schwartz) spaces of test functions which are commonly used in the classical approaches to construct Besov, Sobolev, and Triebel-Lizorkin spaces. We refer to [100] for details. Similarly, the dual spaces $\mathcal{H}'_{1,w}$ serve as substitutes for the spaces of distributions.

By the Gelfand triple property in Proposition 2.1.10 (iii), the inner product of $\mathcal{H}$ can be extended to a sesquilinear form on $\mathcal{H}'_{1,w} \times \mathcal{H}_{1,w}$. Applying this to the voice transform gives us an extended voice transform.

Definition 2.1.11. For $\psi \in \mathcal{A}_w$, the *extended voice transform* $V_{e,\psi}$ is defined for $T \in \mathcal{H}'_{1,w}$ and $x \in G$ by

$$V_{e,\psi}T(x) := \langle T, \pi(x)\psi \rangle_{\mathcal{H}'_{1,w} \times \mathcal{H}_{1,w}}.$$

We now collect some properties of the extended voice transform in the following proposition. We place an emphasis on part (i), which states that the reproducing identity (2.1.6) carries over to the extended voice transform.

Proposition 2.1.12 (Proposition 3.4 of [20]).

(i) $V_{e,\psi}T * K_\psi = V_{e,\psi}T$ for all $T \in \mathcal{H}'_{1,w}$.

(ii) The map $V_{e,\psi} : \mathcal{H}'_{1,w} \to L_{\infty,w^{-1}}(G)$ is a bijection.

(iii) For every $F \in L_{\infty,w^{-1}}(G)$ satisfying the relation $F * K_\psi = F$ there exists a unique element $T \in \mathcal{H}'_{1,w}$ with $V_{e,\psi}T = F$.

Many classical function spaces such as Besov or Triebel-Lizorkin spaces are defined as subsets of spaces of tempered distributions. As mentioned above, this role is taken over by the dual space $\mathcal{H}'_{1,w}$, therefore we are interested in subspaces of these dual spaces. The subspace is determined by the decay properties of the extended voice transform $V_{e,\psi}$.

Definition 2.1.13. We call a function space Y on G a *target space*, if it fulfills the following conditions:

(i) $(Y, \|\cdot\|_Y)$ is a two-sided translation-invariant solid Banach function space, i.e., $f \in Y$ implies $\lambda(x)f, \rho(x)f \in Y$ for all $x \in G$;

(ii) $Y * \widetilde{L_{1,w}(G)} \subseteq Y$ and $\|f * g\|_Y \leqslant \|f\|_Y \cdot \|\breve{g}\|_{L_{1,w}}$ for all $f \in Y$, $g \in L_{1,w}(G)$.

It is also possible to relax property (i) and consider quasi-Banach spaces as target spaces using certain Wiener amalgam spaces [90, 91].

Now we are in the position to define the coorbit spaces.

Definition 2.1.14. The *coorbit space* $\mathrm{Co}(Y)$ associated with a target space Y is defined as

$$\mathrm{Co}(Y) := \{T \in \mathcal{H}'_{1,w} : V_{e,\psi}T \in Y\}. \tag{2.1.16}$$

We equip these spaces with the natural norm $\|T\|_{\mathrm{Co}(Y)} := \|V_{e,\psi}T\|_Y$.

In the following theorem we collect some basic properties of the coorbit spaces.

Theorem 2.1.15 (Section 5.2 of [44]).

(i) $\mathrm{Co}(Y)$ is a π-invariant Banach space with its natural norm.

(ii) $V_{e,\psi} \circ \pi(x) = \lambda(x) \circ V_{e,\psi}$ on $\mathrm{Co}(Y)$ for all $x \in G$.

(iii) The reproducing identity

$$V_{e,\psi}T * K_\psi = V_{e,\psi}T, \quad T \in \mathrm{Co}Y, \tag{2.1.17}$$

holds true and the convolution map

$$F \mapsto F * K_\psi, \quad F \in Y, \tag{2.1.18}$$

is a bounded projection onto the closed subspace $V_{e,\psi}(\mathrm{Co}(Y))$. It follows that $\mathrm{Co}(Y)$ is non-trivial.

(iv) $\mathrm{Co}(Y)$ is independent of the choice of the analyzing vector $\psi \in \mathcal{A}_w$.

It has been shown in [44], that the basic spaces $\mathcal{H}$, $\mathcal{A}_w$ and $\mathcal{H}'_{1,w}$ can be expressed by using coorbit spaces. More precisely, $\mathcal{H} = \mathrm{Co}(L_2(G))$, $\mathcal{H}'_{1,w} = \mathrm{Co}(L_{\infty,w^{-1}}(G))$ as well as $\mathcal{A}_w = \mathrm{Co}(L_{1,w^\#}(G))$, where $w^\# := w + \breve{w}\Delta^{-1}$.

We are interested in a specific family of target spaces. Given the weight w on G, a continuous, positive function $m\colon G \to \mathbb{R}_+$ is called a *w-moderate weight* on G if

$$m(xyz) \leqslant w(x)m(y)w(z), \quad \text{for all } x, y, z \in G. \tag{2.1.19}$$

The moderateness of m implies

$$m(xy) \leqslant C \cdot m(x)w(y) \quad \text{and} \quad m(xy) \leqslant C \cdot w(x)m(y) \tag{2.1.20}$$

with $C = w(e)$. It follows from the generalized Young's inequality, see Proposition 1.3.9, that the weighted Lebesgue spaces $L_{p,m}(G)$, $1 \leqslant p \leqslant \infty$, are target spaces. In the rest of the chapter we restrict ourselves to this kind of target spaces.

We briefly recall the definition of Wiener-type spaces. Let B and Y be solid Banach function spaces on G and $Q \subset G$ compact with non-void interior $\mathring{Q}$ and $e \in \mathring{Q}$, then

$$\mathcal{M}^R(B,Y) := \{F \in B : \widetilde{M}^R(F;B) \in Y\}, \quad \mathcal{M}^L(B,Y) := \{F \in B : \widetilde{M}^L(F;B) \in Y\}, \tag{2.1.21}$$

where the norms are given by

$$\|F\|_{\mathcal{M}^R(B,Y)} := \|\widetilde{M}^R(F;B)\|_Y, \quad \|F\|_{\mathcal{M}^L(B,Y)} := \|\widetilde{M}^L(F;B)\|_Y, \tag{2.1.22}$$

and

$$M^R(F;B)(x) := \|\rho(x)\chi_Q F\|_B, \quad M^L(F;B)(x) := \|\lambda(x)\chi_Q F\|_B. \tag{2.1.23}$$

Associated to the weighted Lebesgue spaces $L_{p,m}(G)$ we define the reproducing kernel subspaces

$$\mathcal{M}_{p,m} := \{F \in L_{p,m}(G) : F * K_\psi = F\}. \tag{2.1.24}$$

The following theorem states a fundamental correspondence principle between these reproducing kernel spaces and the coorbit spaces $\mathrm{Co}(L_{p,m})$.

Theorem 2.1.16 (Theorem 3.12 of [20]). *Let m be a w-moderate weight and ψ and analyzing vector with kernel $K_\psi \in \mathcal{M}^R(L_\infty, L_{1,w})$. Then the map $V_{e,\psi}$ induces an isomorphism, that is $V_{e,\psi}\colon \mathrm{Co}(L_{p,m}) \longleftrightarrow \mathcal{M}_{p,m}$.*

Remark 2.1.17. (i) The importance of Theorem 2.1.16 cannot be understated. The general approach to proving properties of coorbit spaces $\mathrm{Co}(L_{p,m})$ is to prove properties of the spaces $\mathcal{M}_{p,m}$ and then transfer the results to the coorbit spaces using the inverse map $V_{e,\psi}^{-1}$.

(ii) The spaces $\mathcal{M}_{p,m}$, $1 < p < \infty$, are reproducing kernel Banach Spaces (RKBS). This means that $\mathcal{M}_{p,m}$ is reflexive, the dual space $\mathcal{M}'_{p,m}$ is isometric to the Banach function space $\mathcal{M}_{p',m^{-1}}$, where p' denotes the Hölder-dual of p, and the point evaluation δ_x is continuous on both $\mathcal{M}_{p,m}$ and $\mathcal{M}_{p',m^{-1}}$.

We close this section with some properties of coorbit spaces $\mathrm{Co}(L_{p,m})$.

Lemma 2.1.18 (Section 5.7 of [44]). *Let m, m_1, m_2 be w-moderate weights on G. Then:*

(i) The spaces $\mathrm{Co}(L_{p,m})$ are monotonically increasing with $1 \leqslant p \leqslant \infty$.

(ii) Let $1 \leqslant p_2 < p_1 \leqslant \infty$ and assume the quotient m_1/m_2 belongs to $L_r(G)$ for some $1 \leqslant r < \infty$ fulfilling $r^{-1} \leqslant p_2^{-1} - p_1^{-1}$, then $\mathrm{Co}(L_{p_1,m_1}) \subset \mathrm{Co}(L_{p_2,m_2})$.

(iii) $\mathrm{Co}(L_{p_1,m_1}) = \mathrm{Co}(L_{p_2,m_2})$ if and only if $p_1 = p_2$ and there are two constants $0 < C_1 \leqslant C_2$ such that $C_1 m_1(x) \leqslant m_2(x) \leqslant C_2 m_1(x)$ for all $x \in G$.

2.2   Discretization

The content of this section is the presentation of well-known results on atomic decompositions and Banach frames for coorbit spaces constructed in Section 2.1. The proofs of the results can be found in [20]. For a general definition of atomic decompositions and Banach frames we refer to Section 1.9.

Let us first revisit some notions from Sections 1.5 and 1.6. In the following, let $U \subset G$ denote a compact set with non-void interior and $\mathcal{X} = (x_i)_{i \in I} \subset G$ a countable family. We will commonly assume $\mathcal{X}$ to be U-dense, that is $\bigcup_{i \in \mathcal{I}} x_i U = G$, and relatively seperated, that is, every $x \in \mathcal{X}$ is only contained in finitely many sets $x_i U$, $i \in I$. We define the U-oscillation of a function F on G to be given by

$$\operatorname{osc}_U F(x) = \sup_{y \in U} |F(yx) - F(x)| \quad \text{for } x \in G. \tag{2.2.1}$$

Finally the sequence spaces $\ell_{p,m}(I)$ associated with $\mathcal{X}$ are defined as

$$\ell_{p,m}(I) = \Big\{ (c_i)_{i \in I} \ : \ \|c\|_{\ell_{p,m}} = \sum_{i \in I} |c_i|^p m(x_i)^p < \infty \Big\}, \tag{2.2.2}$$

see also Section 1.6.

2.2.1   Atomic Decompositions

In this subsection we present atomic decomposition of the coorbit spaces introduced in Section 2.1. The main result is given in the next theorem.

Theorem 2.2.1 (Theorem 6.2 of [44])**.** *Let $1 \leqslant p < \infty$ and m be a w-moderate weight. Let ψ be an analyzing vector and choose a neighborhood of the identity U such that the U-oscillation of K_ψ fulfills $\operatorname{osc}_U K_\psi \in L_{1,w}(G) \cap L_{1,w\Delta^{-1}}(G)$ with*

$$\max\{\|\operatorname{osc}_U K_\psi\|_{L_{1,w}}, \|\operatorname{osc}_U K_\psi\|_{L_{1,w\Delta^{-1}}}\} < 1. \tag{2.2.3}$$

Furthermore let $\mathcal{X} = (x_i)_{i \in I}$ be a U-dense and relatively separated set.

Then the family $(\pi(x_i)\psi)_{i \in I}$ forms a family of atoms for the coorbit space $\operatorname{Co}(L_{p,m})$ with associated sequence space $\ell_{p,m}(I)$. This means:

(i) The synthesis operator

$$\operatorname{Synth} \colon \ell_{p,m}(I) \to \operatorname{Co}(L_{p,m}), \quad (c_i)_{i \in I} \mapsto \sum_{i \in I} c_i \cdot \pi(x_i)\psi$$

is well-defined and bounded, with unconditional convergence of the defining series. This even holds without assuming (2.2.3).

(ii) There is a bounded coefficient operator

$$\operatorname{Coeff} \colon \operatorname{Co}(L_{p,m}) \to \ell_{p,m}(I) \quad \text{with} \quad \operatorname{Synth} \circ \operatorname{Coeff} = \operatorname{id}_{\operatorname{Co}(L_{p,m})}.$$

The condition (2.2.3) may seem technical, but can be checked using the following lemma involving Wiener-type spaces.

Lemma 2.2.2 (Lemma 3.16 of [20]).

(i) *Let $F \in L_{1,w}(G) \cap \mathcal{M}^R(L_\infty, L_{1,w})$ then it holds that $\mathrm{osc}_U F \in L_{1,w}(G)$. If F is in addition continuous, then for every $\varepsilon > 0$ there exists a compact set $e \in U \subset G$ such that*

$$\|\mathrm{osc}_U F\|_{L_{1,w}} < \varepsilon, \tag{2.2.4}$$

or in short $\lim_{U \to \{e\}} \|\mathrm{osc}_U F\|_{L_{1,w}} = 0$.

(ii) *If $F \in L_{1,w}(G) \cap \mathcal{M}^R(L_\infty, L_{1,w})$, then $\widetilde{\mathrm{osc}}_U F \in L_{1,w\Delta^{-1}}(G)$. On the other hand if we assume $F \in L_{1,w}(G) \cap \mathcal{M}^R(L_\infty, L_{1,w\Delta^{-1}})$, then $\widetilde{\mathrm{osc}}_U F \in L_{1,w}(G)$. If F is in addition continuous, then (2.2.4) holds true for $\widetilde{\mathrm{osc}}_U F$ with respect to the $L_{1,w}(G)$-norm and the $L_{1,w\Delta^{-1}}(G)$-norm, respectively.*

The lemma above thus shows that if we assume $K_\psi \in \mathcal{M}^R(L_\infty, L_{1,w}) \cap \mathcal{M}^R(L_\infty, L_{1,w\Delta^{-1}})$, then (2.2.3) is fulfilled. Note that $K_\psi \in L_{1,w}(G)$ by the definition of the analyzing vector ψ, see Definition 2.1.3.

We will briefly discuss the idea of the proof of Theorem 2.2.1 and the implications. Firstly, in the proof of the boundedness of the coefficient operator a Neumann series argument is used for which the boundedness of the map $F \mapsto F * \mathrm{osc}_U K_\psi$ is needed. In light of Lemma 2.2.2 the condition $K_\psi \in L_{1,w}(G)$ is necessary. Secondly, to proof the boundedness of the synthesis operator we need the boundedness of the map $F \mapsto F * K_\psi$, which is also true using the integrability of the kernel K_ψ.

We conclude that to obtain atomic decompositions for coorbit spaces we heavily rely on the condition $K_\psi \in L_{1,w}(G)$. Therefore a different approach is necessary to obtain atomic decompositions for coorbit spaces with non-integrable kernel, i.e., kernels with $K_\psi \notin L_{1,w}(G)$, as we will discuss in Chapter 3.

2.2.2 Banach Frames

Analogously to the concept of atomic decompositions we can find Banach frames for coorbit spaces under very similar assumptions.

Theorem 2.2.3 (Theorem 3.19 of [20]). *Let $1 \leqslant p < \infty$ and m be a w-moderate weight. Let ψ be an analyzing vector and choose a neighborhood of the identity U such that the U-oscillation of K_ψ fulfills $\widetilde{\mathrm{osc}}_U K_\psi \in L_{1,w}(G) \cap L_{1,w\Delta^{-1}}(G)$ with*

$$\|K_\psi\|_{L_{1,w}} \cdot \max\{\|\widetilde{\mathrm{osc}}_U K_\psi\|_{L_{1,w}}, \|\widetilde{\mathrm{osc}}_U K_\psi\|_{L_{1,w\Delta^{-1}}}\} < 1. \tag{2.2.5}$$

Furthermore let $\mathcal{X} = (x_i)_{i \in I}$ be a U-dense and relatively separated set.

Then the family $(\pi(x_i)\psi)_{i \in I} \subset (\mathrm{Co}(L_{p,m})'$ forms a Banach frame for $\mathrm{Co}(L_{p,m})$ with associated coefficient space $\ell_{p,m}(I)$. This means:

(i) *The sampling operator*

$$\mathrm{Samp} \colon \mathrm{Co}(L_{p,m}) \to \ell_{p,m}(I), \quad T \mapsto (V_{e,\psi} T(x_i))_{i \in I} = (\langle T, \pi(x_i)\psi \rangle_{\mathcal{H}'_{1,w} \times \mathcal{H}_{1,w}})_{i \in I}$$

is well-defined and bounded.

(ii) *There exists a bounded linear reconstruction operator*

$$\mathrm{Reco} \colon \ell_{p,m}(I) \to \mathrm{Co}(L_{p,m}) \quad with \quad \mathrm{Reco} \circ \mathrm{Samp} = \mathrm{id}_{\mathrm{Co}(L_{p,m})}.$$

Note that condition (2.2.5) can be achieved using Lemma 2.2.2 above. If we again assume $K_\psi \in \mathcal{M}^R(L_\infty, L_{1,w}) \cap \mathcal{M}^R(L_\infty, L_{1,w\Delta^{-1}})$ then we can find U sufficiently small to fulfill the condition above. However this U may be different to the one in Theorem 2.2.1.

We will not go into the details of the proof of the preceding theorem, we just note that similarly to the atomic decompositions of coorbit spaces it is crucial to assume the integrability of the kernel K_ψ in order to obtain Banach frames.

2.3 Examples

We close this chapter with a section containing two examples where we apply the classic coorbit theory to specific settings. In Subsection 2.3.1 we show that the (homogeneous) Besov spaces $\dot{B}_{p,p}^s(\mathbb{R}^d)$ can be described by using the full affine group and in Subsection 2.3.2 we define coorbit spaces associated to the full shearlet group.

There are other examples in the literature; an application of the coorbit theory to the reduced Heisenberg group leads to modulation spaces [44], and an application to the Blaschke group to Bergman spaces [49], just to name a few.

2.3.1 Homogeneous Besov Spaces

In this subsection we will show that the (homogeneous) Besov spaces $\dot{B}_{p,p}^{s-1/2+1/p}(\mathbb{R})$ for an integrability parameter $1 \leqslant p < \infty$ and smoothness parameter $s \in \mathbb{R}$ can be described using coorbit theory. It is well-known that Besov spaces can be described using wavelets [107], see for example [54] or more generally [93, 99]. Since wavelets are constructed by using an underlying group structure, it is not surprising that the particular group we define in a moment is the foundation of the coorbit theory in this subsection.

Let us give a definition of homogeneous Besov spaces taken from [54], for a detailed discussion on Besov spaces we refer to the well-known books by Triebel [98–100].

Definition 2.3.1. Let $d \in \mathbb{N}$ and $\{\varphi_j\}_{j\in\mathbb{Z}}$ be a family of functions on $\mathbb{R}^d$ satisfying

$$\varphi_j \in \mathcal{S}(\mathbb{R}^d), \tag{2.3.1}$$

$$\operatorname{supp}\widehat{\varphi_j} \subseteq \left\{\xi \in \mathbb{R}^d : \frac{1}{2} \leqslant 2^{-j}\|\xi\|_2 \leqslant 2\right\}, \tag{2.3.2}$$

$$|\widehat{\varphi_j}(\xi)| \geqslant C > 0 \quad \text{if} \quad \frac{3}{5} \leqslant 2^{-j}\|\xi\|_2 \leqslant \frac{5}{3}, \tag{2.3.3}$$

$$\left|\frac{\partial^\beta \widehat{\varphi_j}}{\partial \xi^\beta}(\xi)\right| \leqslant C_\beta 2^{-j|\beta|} \quad \text{for every multi-index } \beta \in \mathbb{N}_0^d. \tag{2.3.4}$$

The *homogeneous Besov space* $\dot{B}_{p,q}^s(\mathbb{R}^d)$ for $s \in \mathbb{R}$ and $0 < p, q \leqslant \infty$ is defined as

$$\dot{B}_{p,q}^s(\mathbb{R}^d) := \left\{f \in \mathcal{S}'(\mathbb{R}^d)/\mathcal{P} : \|f\|_{\dot{B}_{p,q}^s}^q := \sum_{j\in\mathbb{Z}} 2^{jsq}\|\varphi_j * f\|_{L_p}^q < \infty\right\}, \tag{2.3.5}$$

with the usual change if $q = \infty$.

Here, $\mathcal{S}'/\mathcal{P}$ means distributions modulo polynomials, i.e., adding a polynomial to a distribution $f \in \dot{B}_{p,q}^s(\mathbb{R}^d)$ does not change the distribution. This is because of the fact that $\|p\|_{\dot{B}_{p,q}^s(\mathbb{R}^d)} = 0$ for all polynomials $p \in \mathcal{P}$. We call these spaces homogeneous, because of the inequality

$$\|f(\lambda\cdot)\|_{\dot{B}_{p,q}^s} \lesssim \lambda^{d-\frac{s}{p}} \cdot \|f\|_{\dot{B}_{p,q}^s}$$

for all $\lambda > 0$ and $f \in \dot{B}^s_{p,q}(\mathbb{R}^d)$, see [97, Section 5.1.3]. Later on, however, homogeneity refers to a wider range of function spaces developed in a similar pattern as the homogeneous Besov spaces.

Note that the Besov spaces are translation invariant (quasi-)Banach spaces of functions.

There are numerous equivalent characterizations of Besov spaces, we briefly name the following, which is a restatement of [97, Corollary 10] for the case $p = q$.

Corollary 2.3.2. *Let $0 < p \leq \infty$ and $s \in \mathbb{R}$. Let φ such that $\hat{\varphi}$ is an infinitely differentiable complex-valued function on $\mathbb{R}$ satisfying conditions (54), (90) and (91) in [97]. Then*

$$\|f\|_{\tilde{\dot{B}}^s_{p,p}} := \left(\int_{\mathbb{R}} \int_{\mathbb{R}*} |\langle \varphi(a^{-1}(\cdot - t)), f \rangle_{L_2}|^p |a|^{-sp} \frac{da}{|a|^2} \, dt \right)^{1/p} \tag{2.3.6}$$

is an equivalent (quasi-)norm on $\dot{B}^s_{p,p}(\mathbb{R})$.

The structure of the norm in (2.3.6) will be the blueprint for the coorbit spaces we construct in this section. The proofs of the next results presented can be found in [20, 44] and the references therein. For the sake of completeness we include some of the arguments.

The (Full) Affine Group

We consider $\mathbb{A} = \mathbb{R}^* \times \mathbb{R}$ as a manifold the "$ax + b$"-group, also called the *full affine group*. The group operation on $\mathbb{A}$ is given by

$$(a, b) \circ (a', b') = (aa', ab' + b), \quad (a, b), (a', b') \in \mathbb{A}. \tag{2.3.7}$$

It is easy to verify that this is indeed a locally compact topological group with neutral element $e_{\mathbb{A}} = (1, 0)$ and the inverse is given by $(a, b)^{-1} = (a^{-1}, -a^{-1}b)$. Clearly, $\mathbb{A}$ is not abelian.

We can also see the group action as the affine map

$$(a, b) \circ (a', b') = A_{(a,b)}(a', b')^T + b_{(a,b)}, \quad A_{(a,b)} = \begin{pmatrix} a & 0 \\ 0 & a \end{pmatrix}, \quad b_{(a,b)} = \begin{pmatrix} 0 \\ b \end{pmatrix}.$$

The Haar measure is therefore given by Proposition 1.2.8 and the modular function on $\mathbb{A}$ can be easily computed so that we have

$$d(a, b) = \frac{da}{|a|^2} \, db, \quad \Delta(a, b) = a^{-1}, \tag{2.3.8}$$

so $\mathbb{A}$ is clearly not unimodular.

The Wavelet Transform

As a Hilbert space we set $\mathcal{H} = L_2(\mathbb{R})$ and we consider the *translation* and *dilation* operators

$$T_b f(x) = f(x - b) \quad \text{and} \quad D_a f(x) = |a|^{-1/2} f\left(\frac{x}{a}\right),$$

for $(a, b) \in \mathbb{A}$ and $f \in L_2(\mathbb{R})$. Then the representation π of $\mathbb{A}$ on $L_2(\mathbb{R})$ is given by

$$\pi(a, b) f(x) = T_b D_a f(x) = |a|^{-1/2} f\left(\frac{x - b}{a}\right). \tag{2.3.9}$$

This representation is also called the *wavelet transform* of f. For $(a, b), (a', b') \in \mathbb{A}$ observe that

$$T_b T_{b'} = T_{b+b'}, \quad D_a D_{a'} = D_{aa'},$$

but additionally

$$D_a T_b = T_{ab} D_a,$$

so that in general the operators do not commute, i.e., $T_b D_a \neq D_a T_b$. We can use this to show that π is indeed a homomorphism.

Proposition 2.3.3. *The mapping* $\pi \colon \mathbb{A} \to \mathcal{U}(L_2(\mathbb{R}))$ *defined in (2.3.9) is a unitary continuous and irreducible representation of* $\mathbb{A}$.

Proof. For $(a, b), (a', b') \in \mathbb{A}$ we have

$$\pi(a,b)\pi(a',b') = (T_b D_a)(T_{b'} D_{a'}) = T_b T_{ab'} D_a D_{a'} = T_{b+ab'} D_{aa'} = \pi((a,b) \circ (a',b')),$$

so π is a homomorphism. To show that π is unitary take $(a, b) \in \mathbb{A}$ fixed, then an integral transformation yields

$$\langle \pi(a,b)f, \pi(a,b)g \rangle_{L_2} = \int_{\mathbb{R}} \pi(a,b)f(x) \cdot \overline{\pi(a,b)g(x)} \, \mathrm{d}x$$

$$= |a|^{-1} \int_{\mathbb{R}} f\left(\frac{x-b}{a}\right) \cdot \overline{g\left(\frac{x-b}{a}\right)} \, \mathrm{d}x$$

$$= \int_{\mathbb{R}} f(x) \cdot \overline{g(x)} \, \mathrm{d}x = \langle f, g \rangle_{L_2}$$

for all $f, g \in L_2(\mathbb{R})$, so π is a unitary representation.

Before we show the irreducibility of π we make some remarks. Using the Fourier transform $\mathcal{F}$ we can compute that

$$\mathcal{F}(\pi(a,b)f)(\xi) = |a|^{1/2} e^{-2\pi i b \xi} \cdot \hat{f}(a\xi), \quad (a,b) \in \mathbb{A}. \tag{2.3.10}$$

Moreover, the equality

$$\langle f, \pi(a,b)g \rangle_{L_2} = (f * \pi(a,0)g^*)(b) = |a|^{1/2} \mathcal{F}^{-1}(\hat{f}(\cdot)\overline{\hat{g}(a\cdot)})(b) \tag{2.3.11}$$

holds for all $f, g \in L_2(\mathbb{R})$, $(a, b) \in \mathbb{A}$. Now take two nonzero functions $f, g \in L_2(\mathbb{R})$, then by using Plancherel's formula twice and Fubini's theorem we have

$$\int_{\mathbb{A}} |\langle \pi(a,b)f, g \rangle_{L_2}|^2 \, d(a,b) = \int_{\mathbb{R}} \int_{\mathbb{R}} \left| (\mathcal{F}^{-1}(\hat{f}(a\cdot)\overline{\hat{g}(\cdot)}))(-b) \right|^2 \, \mathrm{d}b \, \frac{\mathrm{d}a}{|a|}$$

$$= \int_{\mathbb{R}} \left(\int_{\mathbb{R}} |\hat{f}(ab)|^2 \, \frac{\mathrm{d}a}{|a|} \right) |\hat{g}(b)|^2 \, \mathrm{d}b$$

$$= \left(\int_{\mathbb{R}} \frac{|\hat{f}(x)|^2}{|x|} \, \mathrm{d}x \right) \cdot \left(\int_{\mathbb{R}} |\hat{g}(y)|^2 \, \mathrm{d}y \right) \tag{2.3.12}$$

To use Proposition 1.4.5, we assume towards a contradiction that the map $(a, b) \mapsto \langle \pi(a,b)f, g \rangle_{L_2}$ is identical to zero. This implies that either $\hat{f}$ or $\hat{g}$ is identical to zero, which is the desired contradiction and therefore π is irreducible. The continuity of π is obvious. $\qquad \square$

Having shown the necessary properties of π, we search for admissible vectors. By (2.3.12) a function $\psi \in L_2(\mathbb{R}^d)$ is an admissible vector, if and only if the *Calderón equation*

$$\int_{\mathbb{R}} \frac{|\hat{\psi}(x)|^2}{|x|} \, \mathrm{d}x =: C_\psi < \infty \tag{2.3.13}$$

is fulfilled. An admissible vector ψ of the affine group is also called *wavelet*. One can find wavelets fulfilling (2.3.13) that are band-limited functions as well as compactly supported functions which are sufficiently smooth and satisfy vanishing moment conditions.

For any wavelet ψ we now define the corresponding voice transform $\mathcal{W}_\psi : L_2(\mathbb{R}) \to L_2(\mathbb{A})$ via

$$\mathcal{W}_\psi f(a, b) := V_\psi f(a, b) = \langle f, \pi(a, b)\psi \rangle_{L_2}, \quad f \in L_2(\mathbb{R}), \ (a, b) \in \mathbb{A}. \tag{2.3.14}$$

We call $\mathcal{W}_\psi f$ the *wavelet transform* of f; this transform is well-known in the theory of wavelets, see [31, 107]. We now intend to show that we can find analyzing vectors ψ, i.e., the integrability condition $K_\psi \in L_{1,w}(\mathbb{A})$ is fulfilled.

Wavelet Coorbit Spaces

From now on let $w_\rho(a, b) = w_\rho(a) := |a|^{-\rho} + |a|^\rho$ for $\rho \in \mathbb{R}$ be a fixed weight, which fulfills conditions (2.1.11) and (2.1.12). Further let $m_\sigma(a, b) = m_\sigma(a) = |a|^{-\sigma}$ for some $\sigma \in \mathbb{R}$ with $\sigma \leqslant \rho$, then elementary calculations show that m_σ is w_ρ-moderate in the sense of (2.1.19). The following theorem states that for smooth and band-limited wavelets the kernel K_ψ is contained in $L_{1,w_\rho}(\mathbb{A})$.

Theorem 2.3.4 (Theorem 3.24 of [20]). *Let $0 < a_0 < a_1 < \infty$ and $\psi \in \mathcal{S}(\mathbb{R})$ a Schwartz function satisfying supp $\hat{\psi} \subseteq [-a_1, -a_0] \cup [a_0, a_1]$. Then $K_\psi = \mathcal{W}_\psi \psi \in L_{1,w_\rho}(\mathbb{A})$ for all $\rho \in \mathbb{R}$.*

Proof. Let $a \in \mathbb{R}^*$ then $\mathcal{F}(\hat{\psi}\hat{\psi}(a\cdot)) = 0$ if and only if $\hat{\psi}\hat{\psi}(a\cdot) = 0$. This is true if we have supp $\hat{\psi} \cap$ supp $\hat{\psi}(a\cdot) = \varnothing$, or equivalently if $a \in [-a_1 a_0^{-1}, -a_0 a_1^{-1}] \cup [a_0 a_1^{-1}, a_1 a_0^{-1}] =: A$. Using (2.3.11) we then compute by applying Fubini's theorem that

$$\int_\mathbb{A} |\mathcal{W}_\psi \psi(a, b)| \cdot w_\rho(a) \, d(a, b) = \int_\mathbb{R} \int_\mathbb{R} |\mathcal{F}^{-1}(\hat{\psi}\overline{\hat{\psi}(a\cdot)})(b)| \cdot w_\rho(a) \frac{da}{|a|^{3/2}} \, db$$

$$= \int_A \|\mathcal{F}^{-1}(\hat{\psi}\overline{\hat{\psi}(a\cdot)})\|_{L_1(\mathbb{R})} \cdot w_\rho(a) \frac{da}{|a|^{3/2}}.$$

For all $a \in A$ and $\rho \in \mathbb{R}$ it holds that $\|\mathcal{F}^{-1}(\hat{\psi}\overline{\hat{\psi}(a\cdot)})\|_{L_1(\mathbb{R}^d)} \leqslant C_1$ and $w_\rho(a) \cdot |a|^{-1} \leqslant C_2$ for two constants $C_1, C_2 < \infty$, therefore $\|K_\psi\|_{L_{1,w}(G)} < \infty$. $\square$

Finally we define the coorbit spaces related to the affine group given an analyzing function ψ as

$$\mathcal{W}C_{p,\sigma}(\mathbb{R}) := \mathrm{Co}(L_{p,m_\sigma}) = \{T \in \mathcal{H}'_{1,w_\rho} : V_{e,\psi}T \in L_{p,m_\sigma}(\mathbb{A})\}, \quad 1 \leqslant p < \infty, \ \sigma \leqslant \rho.$$

By Theorem 2.1.15 these spaces are well-defined Banach spaces with respect to their natural norm.

The condition $V_{e,\psi}f \in L_{p,m_\sigma}(\mathbb{A}) < \infty$ is equivalent to $f \in \dot{B}^{s-1/2+1/p}_{p,p}(\mathbb{R})$ by virtue of (2.3.6), for a discussion on the technical issues see [101]. This shows that the homogeneous Besov spaces can indeed be described using coorbit theory, i.e., $\mathcal{W}C_{p,s}(\mathbb{R}) = \dot{B}^{s-1/2+1/p}_{p,p}(\mathbb{R})$.

Discretization and Properties

The next step is to construct atomic decompositions and Banach frames according to Section 2.2 for these spaces. For this we need to construct suitable U-dense and relatively separated sets.

Let $e_{\mathbb{A}} \in U \subset \mathbb{A}$ be a compact set with non-void interior. Let $\alpha > 1$ and $\tau > 0$ be defined such that

$$U_0 := [\alpha^{-1/2}, \alpha^{1/2}) \times \left[-\frac{\tau}{2}, \frac{\tau}{2}\right) \subseteq U.$$

It is easy to see that the set

$$\mathcal{X} := \{x_{\varepsilon,j,m} := (\varepsilon\alpha^j, \varepsilon\alpha^j\tau m) : \varepsilon \in \{-1, 1\}, j \in \mathbb{Z}, m \in \mathbb{Z}\}$$

is U-dense and relatively separated. Moreover it holds that $|U_0| = (\alpha^{\frac{1}{2}} - \alpha^{-\frac{1}{2}})\tau$ so U can be chosen arbitrarily small. Assume further that $\psi \in \mathcal{M}^R(L_\infty, L_{1,w})$, then Lemma 2.2.2 shows that condition (2.2.5) is fulfilled. Theorem 2.2.1 thus gives us that every $f \in \mathcal{WC}_{p,\sigma}(\mathbb{R})$ can be decomposed into

$$f = \sum_{\varepsilon \in \{-1,1\}} \sum_{j \in \mathbb{Z}} \sum_{m \in \mathbb{Z}} c_{j,m}(f) \cdot \pi(\varepsilon\alpha^j, \varepsilon\alpha^j\tau m)\psi,$$

for a sequence $c(f) \in \ell_{p,m_\sigma}(\{-1, 1\} \times \mathbb{Z} \times \mathbb{Z})$ depending on f and we have

$$\left(\sum_{\varepsilon \in \{-1,1\}} \sum_{j \in \mathbb{Z}} \sum_{m \in \mathbb{Z}} |c(f)_{\varepsilon,j,m}|^p |\alpha|^{j\sigma p}\right)^{1/p} = \|c(f)\|_{\ell_{p,m_\sigma}} \lesssim \|f\|_{\mathcal{W}_{p,\sigma}}. \tag{2.3.15}$$

Conversely, if $c \in \ell_{p,m_\sigma}(\{-1, 1\} \times \mathbb{Z} \times \mathbb{Z})$, then $f = \sum_{\varepsilon,j,m} c_{\varepsilon,j,m} \cdot \pi(\varepsilon\alpha^j, \varepsilon\alpha^j\tau m)\psi \in \mathcal{WC}_{p,\sigma}(\mathbb{R})$ and

$$\|f\|_{\mathcal{W}_{p,\sigma}} \lesssim \left(\sum_{\varepsilon \in \{-1,1\}} \sum_{j \in \mathbb{Z}} \sum_{m \in \mathbb{Z}} |c_{\varepsilon,j,m}|^p |\alpha|^{j\sigma p}\right)^{1/p}. \tag{2.3.16}$$

These results are similar to the discrete results in [54, Theorem 2.6].

2.3.2 Homogeneous Shearlet Coorbit Spaces

In this subsection we define smoothness spaces on $\mathbb{R}^d$ related to the shearlet group via coorbit theory which we will call homogeneous shearlet coorbit spaces. For this we first need to define the shearlet group in $\mathbb{R}^d$. Using the methods from Section 2.2 we discretize these function spaces.

The shearlet group arises when generalizing the wavelet approach to arbitrary space dimensions. This has been done in [6,55,86] and many more. More recently in [56,57,60] the approach has been broadened by looking at groups of the type $G = \mathbb{R}^d \times H$, where $H \subset \mathrm{GL}(\mathbb{R}^d)$ is a subgroup. By choosing a suitable subgroup H which we can identify with dilations and shearing we obtain the shearlet group. While the choice of H for $d = 2$ is straightforward [24,25], for arbitrary space dimension $d \geqslant 3$ one can choose different shearlet groups [17,28,30]. We restrict ourselves to one particular choice in this subsection.

Most of the proofs in this subsection can be found in [20], and are included for the sake of completeness.

The (Full) Shearlet Group

Fix $d \geqslant 2$ in this subsection, then for $a \in \mathbb{R}^*$ and $s = (s_1, \ldots, s_{d-1})^T \in \mathbb{R}^{d-1}$ we define the *parabolic dilation matrix*, or scaling matrix, A_a and the *shear matrix* S_s as follows:

$$A_a := \begin{pmatrix} a & 0_{d-1}^T \\ 0_{d-1} & \mathrm{sgn}(a)|a|^{\frac{1}{d}}I_{d-1} \end{pmatrix} \in \mathbb{R}^{d \times d} \quad \text{and} \quad S_s := \begin{pmatrix} 1 & s^T \\ 0_{d-1} & I_{d-1} \end{pmatrix} \in \mathbb{R}^{d \times d}. \tag{2.3.17}$$

Now consider the *(full) shearlet group* $\mathbb{S} := \mathbb{R}^* \times \mathbb{R}^{d-1} \times \mathbb{R}^d$ as a manifold endowed with the group operation

$$(a,s,t) \circ (a',s',t') := (aa', s + |a|^{1-\frac{1}{d}}s', t + S_s A_a t'), \quad (a,s,t),(a',s',t') \in \mathbb{S}. \tag{2.3.18}$$

To prove the group properties of $\mathbb{S}$ we first need the following auxiliary equality. Let $a,a' \in \mathbb{R}^*$ and $s,s' \in \mathbb{R}^{d-1}$, then the matrices A_a and S_s from (2.3.17) fulfill

$$
\begin{aligned}
S_s A_a S_{s'} A_{a'} &= \begin{pmatrix} 1 & s^T \\ 0_{d-1} & I_{d-1} \end{pmatrix} \begin{pmatrix} a & 0_{d-1}^T \\ 0_{d-1} & \operatorname{sgn}(a)|a|^{\frac{1}{d}} I_{d-1} \end{pmatrix} \begin{pmatrix} 1 & s'^T \\ 0_{d-1} & I_{d-1} \end{pmatrix} \begin{pmatrix} a' & 0_{d-1}^T \\ 0_{d-1} & \operatorname{sgn}(a')|a'|^{\frac{1}{d}} I_{d-1} \end{pmatrix} \\
&= \begin{pmatrix} a & \operatorname{sgn}(a)|a|^{\frac{1}{d}} s^T \\ 0_{d-1} & \operatorname{sgn}(a)|a|^{\frac{1}{d}} I_{d-1} \end{pmatrix} \begin{pmatrix} a' & \operatorname{sgn}(a')|a'|^{\frac{1}{d}} s'^T \\ 0_{d-1} & \operatorname{sgn}(a')|a'|^{\frac{1}{d}} I_{d-1} \end{pmatrix} \\
&= \begin{pmatrix} aa' & \operatorname{sgn}(aa')|aa'|^{\frac{1}{d}}(s^T + |a|^{1-\frac{1}{d}} s'^T) \\ 0_{d-1} & \operatorname{sgn}(aa')|aa'|^{\frac{1}{d}} I_{d-1} \end{pmatrix} \\
&= \begin{pmatrix} 1 & s+|a|^{1-\frac{1}{d}} s'^T \\ 0_{d-1} & I_{d-1} \end{pmatrix} \begin{pmatrix} aa' & 0_{d-1}^T \\ 0_{d-1} & \operatorname{sgn}(aa')|aa'|^{\frac{1}{d}} I_{d-1} \end{pmatrix} = S_{s+|a|^{1-\frac{1}{d}}s'} A_{aa'}. \tag{2.3.19}
\end{aligned}
$$

Furthermore it is easy to show

$$A_a^{-1} = A_{a^{-1}} = \begin{pmatrix} a^{-1} & 0_{d-1}^T \\ 0_{d-1} & \operatorname{sgn}(a)|a|^{-\frac{1}{d}} I_{d-1} \end{pmatrix} \quad \text{and} \quad S_s^{-1} = S_{-s} = \begin{pmatrix} 1 & -s^T \\ 0_{d-1} & I_{d-1} \end{pmatrix}. \tag{2.3.20}$$

Now we have all ingredients to show the group properties of $\mathbb{S}$.

Lemma 2.3.5. *The set $\mathbb{S} = \mathbb{R}^* \times \mathbb{R}^{d-1} \times \mathbb{R}^d$ endowed with the operation defined in (2.3.18) forms a non-abelian locally compact topological group.*

Proof. We first note that $\mathbb{S}$ is obviously closed under the group operation. We now prove the associativity of $\mathbb{S}$. Let $(a,s,t),(a',s',t'),(a'',s'',t'') \in \mathbb{S}$, then

$$
\begin{aligned}
(a,s,t) &\circ [(a',s',t') \circ (a'',s'',t'')] \\
&= (a,s,t) \circ (a'a'', s' + |a'|^{1-\frac{1}{d}}s'', t' + S_{s'}A_{a'}t'') \\
&= (aa'a'', s + |a|^{1-\frac{1}{d}}(s' + |a'|^{1-\frac{1}{d}}s''), t + S_s A_a(t' + S_{s'}A_{a'}t'')) \\
&= (aa'a'', s + |a|^{1-\frac{1}{d}}s' + |aa'|^{1-\frac{1}{d}}s'', t + S_s A_a t' + S_{s+|a|^{1-\frac{1}{d}}s'} A_{aa'}t'') \\
&= (aa', s + |a|^{1-\frac{1}{d}}s', t + S_s A_a t') \circ (a'',s'',t'') \\
&= [(a,s,t) \circ (a',s',t')] \circ (a'',s'',t'').
\end{aligned}
$$

The neutral element is given by $e = (1, 0_{d-1}, 0_d) \in \mathbb{S}$ since

$$
\begin{aligned}
(a,s,t) \circ (1,0_{d-1},0_d) &= (a \cdot 1, s + |a|^{\frac{1}{d}} 0_{d-1}, t + S_s A_a 0_d) = (a,s,t) \\
&= (1 \cdot a, 0_{d-1} + |1|^{\frac{1}{d}}s, 0_{d-1} + S_{0_{d-1}} A_1 t) = (1, 0_{d-1}, 0_d) \circ (a,s,t)
\end{aligned}
$$

for all $(a,s,t) \in \mathbb{S}$. The inverse element of $(a,s,t) \in \mathbb{S}$ is given by

$$(a,s,t)^{-1} = (a^{-1}, -|a|^{\frac{1}{d}-1}s, -A_a^{-1}S_s^{-1}t),$$

since by (2.3.19) we have

$$S_s A_a S_{-|a|^{\frac{1}{d}-1}s} A_{a^{-1}} = S_{s-|a|^{1-\frac{1}{d}}|a|^{\frac{1}{d}-1}s} A_{aa^{-1}} = I_d,$$

implying $(S_s A_a)^{-1} = S_{-|a|^{\frac{1}{d}-1}s} A_{a^{-1}}$, hence

$$
\begin{aligned}
(a,s,t) \circ (a^{-1}, -|a|^{\frac{1}{d}-1}s, -A_a^{-1}S_s^{-1}t) &= (aa^{-1}, s - |a|^{\frac{1}{d}-1}|a|^{1-\frac{1}{d}}s, t - S_s A_a A_a^{-1} S_s^{-1}t) \\
&= (1, 0_{d-1}, 0_d) \\
&= (a^{-1}a, -|a|^{\frac{1}{d}-1}s + |a^{-1}|^{\frac{1}{d}-1}s, -A_{a^{-1}}S_{-s}t + S_{-|a|^{\frac{1}{d}-1}s}A_{a^{-1}}t) \\
&= (a^{-1}, -|a|^{\frac{1}{d}-1}s, -A_a^{-1}S_s^{-1}t) \circ (a,s,t).
\end{aligned}
$$

Altogether $\mathbb{S}$ forms a group, which is clearly non-abelian. Viewing $\mathbb{S}$ as a real manifold it also inherits a topological structure that is locally compact.   □

We can also view $\mathbb{S}$ as a Lie group whose group structure is given by the affine map

$$
(a,s,t) \circ (a',s',t') = A_{(a,s,t)}(a',s',t')^T + b_{(a,s,t)},
$$

where

$$
A_{(a,s,t)} = \begin{pmatrix} a & 0 & 0 \\ 0 & |a|^{1-\frac{1}{d}}I_{d-1} & 0 \\ 0 & 0 & S_s A_a \end{pmatrix}, \quad b_{(a,s,t)} = \begin{pmatrix} 0 \\ s \\ t \end{pmatrix}.
$$

By Proposition 1.2.8 the left Haar-measure is now given by

$$
d(a,s,t) = |\det(A_{(a,s,t)})|^{-1}\, da\, ds\, dt = \frac{da}{|a|^{d+1}}\, ds\, dt. \tag{2.3.21}
$$

Moreover the modular function Δ is given by

$$
\Delta(a,s,t) = |a|^{-d}, \tag{2.3.22}
$$

see for example the proof of [20, Lemma 3.27].

The Continuous Shearlet Transform

In the next step we define a suitable representation associated to the shearlet group. For that consider the *translation*, *shearing* and *dilation* operators acting on a function f on $\mathbb{R}^d$ by

$$
T_t f(x) = f(x-t), \quad D_s f(x) = f(S_s^{-1}x) \quad \text{and} \quad D_a = |\det(A_a)|^{-\frac{1}{2}}f(A_a^{-1}x).
$$

Then the representation π of $\mathbb{S}$ on $\mathcal{H} = L_2(\mathbb{R}^d)$ is given by

$$
\pi(a,s,t)f(x) := T_t D_s D_a f(x) = |\det(A_a)|^{-\frac{1}{2}}f(A_a^{-1}S_s^{-1}(x-t)), \tag{2.3.23}
$$

$(a,s,t) \in \mathbb{S}$, $f \in L_2(\mathbb{R}^d)$. It is clear to see that the group structure of $\mathbb{S}$ is chosen exactly such that π is indeed a representation.

Proposition 2.3.6. *The mapping $\pi \colon \mathbb{S} \to \mathcal{U}(L_2(\mathbb{R}^d))$ defined in (2.3.23) is a unitary continuous and irreducible representation of $\mathbb{S}$.*

Proof. We first show that π is compatible with the group structure, i.e., a homomorphism. For this let $(a, s, t), (a', s', t') \in \mathbb{S}$ and $f \in L_2\mathbb{R}^d)$, then using (2.3.19)

$$
\begin{aligned}
\pi(a, s, t)&[\pi(a', s', t')f] \\
&= \pi(a, s, t)[|\det(A_a)|^{-\frac{1}{2}}f(A_a^{-1}S_s^{-1}(x - t))] \\
&= |\det(A_a)|^{-\frac{1}{2}}|\det(A_{a'})|^{-\frac{1}{2}}f(A_{a'}S_{s'}^{-1}A_a^{-1}S_s^{-1}(x - t - S_sA_at')) \\
&= |\det(A_{aa'})|^{-\frac{1}{2}}f(A_{aa'}^{-1}S_{s+|a|^{1-\frac{1}{d}}s'}^{-1}(x - t - S_sA_at')) \\
&= \pi(aa', s + |a|^{1-\frac{1}{d}}s', S_sA_at')f(x) \\
&= [\pi(a, s, t) \circ \pi(a', s', t')]f(x).
\end{aligned}
$$

To prove that π is unitary we note that $\det(S_s) = 1$ and calculate

$$
\|\pi(a, s, t)f\|_{L_2}^2 = \int_{\mathbb{R}^d} |\det(S_sA_a)|^{-1}|f(A_a^{-1}S_s^{-1}(x - t))|^2\,\mathrm{d}x = \int_{\mathbb{R}^d} |f(x)|^2\,\mathrm{d}x = \|f\|_{L_2}^2
$$

for all $(a, s, t) \in \mathbb{S}$, $f \in L_2(\mathbb{R}^d)$. For brevity we set $f_{(a,s,t)} := \pi(a, s, t)f$ and note that we have $\det(A_a) = a^{2-\frac{1}{d}}$. Then, using the Fourier transform $\mathcal{F}$ for functions in d variables, we have

$$
\mathcal{F}(f_{(a,s,t)})(\xi) = |\det(A_a)|^{\frac{1}{2}}e^{-2\pi i\langle t, \xi\rangle}\widehat{f}(A_a^T S_s^T \xi) = |a|^{1-\frac{1}{2d}}e^{-2\pi i\langle t, \xi\rangle}\widehat{f}(A_a S_s^T \xi) \tag{2.3.24}
$$

as well as

$$
\langle f, g_{(a,s,t)}\rangle_{L_2} = (f * g_{(a,s,0)}^*)(t) = |\det(A_a)|^{\frac{1}{2}}\mathcal{F}^{-1}(\widehat{f}(\cdot)\overline{\widehat{g}(A_a S_s^T \cdot)})(t) \tag{2.3.25}
$$

for all $f, g \in L_2(\mathbb{R}^d)$, $(a, s, t) \in \mathbb{S}$. Using this for non-zero functions $f, g \in L_2(\mathbb{R}^d)$ and applying Plancherel's formula and Fubini's theorem we have

$$
\begin{aligned}
\int_{\mathbb{S}} |\langle f, g_{(a,s,t)}\rangle_{L_2}|^2\,d(a, s, t) &= \int_{\mathbb{R}^*} \int_{\mathbb{R}^{d-1}} \int_{\mathbb{R}^d} |\det(A_a)| \cdot |\mathcal{F}^{-1}(\widehat{f}(\cdot)\overline{\widehat{g}(A_a S_s^T \cdot)})(t)|^2\,\mathrm{d}t\,\mathrm{d}s\,\frac{\mathrm{d}a}{|a|^{d+1}} \\
&= \int_{\mathbb{R}^*} \int_{\mathbb{R}^{d-1}} \int_{\mathbb{R}^d} |\det(A_a)| \cdot |\widehat{f}(t)|^2 \cdot |\widehat{g}(A_a S_s^T t)|^2\,\mathrm{d}t\,\mathrm{d}s\,\frac{\mathrm{d}a}{|a|^{d+1}} \\
&= \int_{\mathbb{R}^d} \left(\int_{\mathbb{R}^*} \int_{\mathbb{R}^{d-1}} |\widehat{g}(A_a S_s^T t)|^2\,\mathrm{d}s\,\frac{\mathrm{d}a}{|a|^{d-1+\frac{1}{d}}} \right) |\widehat{f}(t)|^2\,\mathrm{d}t. \tag{2.3.26}
\end{aligned}
$$

We focus on the expression in the parentheses for fixed $t \in \mathbb{R}^d$. We use the notation $\widetilde{\omega} = (\omega_2, \ldots, \omega_d)$ and substitute $\widetilde{\omega} = \mathrm{sgn}(a)|a|^{\frac{1}{d}}I_{d-1}(\widetilde{t} + t_1 s)$ with $\mathrm{d}\omega = |\det(\mathrm{sgn}(a)|a|^{\frac{1}{d}}t_1I_{d-1})|\mathrm{d}(a, s)$, then $A_a S_s^T t = (\omega_1 t_1, \widetilde{\omega})^T$, and

$$
\int_{\mathbb{R}^*} \int_{\mathbb{R}^{d-1}} |\widehat{g}(A_a S_s^T t)|^2\,\mathrm{d}s\,\frac{\mathrm{d}a}{|a|^{d-1+\frac{1}{d}}} = \int_{\mathbb{R}^d} \left|\widehat{g}\left(\genfrac{}{}{0pt}{}{t_1\omega_1}{\widetilde{\omega}}\right)\right|^2 \frac{\mathrm{d}\omega}{|t_1|^{d-1} \cdot |\omega_1|^d} = \int_{\mathbb{R}^d} \frac{|\widehat{g}(\xi)|^2}{|\xi_1|^d}\,\mathrm{d}\xi.
$$

Together with (2.3.26) we conclude that

$$
\int_{\mathbb{S}} |\langle f, g_{(a,s,t)}\rangle_{L_2}|^2\,d(a, s, t) = \int_{\mathbb{R}^d} |\widehat{f}(t)|^2\,\mathrm{d}t \cdot \int_{\mathbb{R}^d} \frac{|\widehat{g}(\xi)|^2}{|\xi_1|^d}\,\mathrm{d}\xi. \tag{2.3.27}
$$

Assuming that the map $(a, s, t) \mapsto \langle f, g_{(a,s,t)}\rangle_{L_2}$ is identical to zero implies $f = g = 0$, so by Proposition 1.4.5 the map π is indeed an irreducible representation of $\mathbb{S}$. $\qquad\square$

As an immediate consequence of (2.3.27) a function $\psi \in L_2(\mathbb{R}^d)$ is admissible if and only if it fulfills the *(shearlet) admissibility condition*

$$0 < C_\psi := \int_{\mathbb{R}^d} \frac{|\hat{\psi}(\omega)|^2}{|\omega_1|^d} \, d\omega < \infty. \tag{2.3.28}$$

We call a function ψ fulfilling (2.3.28) an *admissible shearlet*. It is known that both band-limited [28] and compactly supported admissible shearlets [29] exist, and therefore the representation π is square-integrable. For any admissible shearlet $\psi \in L_2(\mathbb{R}^d)$ the corresponding voice transform $\mathcal{SH}_\psi \colon L_2(\mathbb{R}^d) \to L_2(\mathbb{R}^d)$ is defined by

$$\mathcal{SH}_\psi f(a,s,t) := V_\psi f(a,s,t) = \langle f, \pi(a,s,t)\psi\rangle_{L_2}, \quad f \in L_2(\mathbb{R}^d), (a,s,t) \in \mathbb{S}. \tag{2.3.29}$$

We call $\mathcal{SH}_\psi f$ the *(continuous) shearlet transform* of f.

Shearlet Coorbit Spaces

We consider the weight function $w_\rho(a,s,t) = w_\rho(a) = |a|^\rho + |a|^{-\rho}$ for $\rho \in \mathbb{R}$ fixed, which fulfills the conditions (2.1.11) and (2.1.12). Furthermore we fix $m_\sigma(a,s,t) = m_\sigma(a) = |a|^{-\sigma}$ for some $\sigma \leqslant \rho$. It is easy to verify that m_σ is w_ρ-moderate. In order to construct the coorbit spaces related to the shearlet group we have to ensure that there exist admissible shearlets that are integrable, i.e., $K_\psi = \mathcal{SH}_\psi \psi \in L_{1,w_\rho}(\mathbb{S})$. For this we first consider a certain class of band-limited functions.

Theorem 2.3.7. *Let ψ be a Schwartz function such that* $\operatorname{supp} \hat{\psi} \subseteq ([-a_1,-a_0] \cup [a_0,a_1]) \times Q_b$ *for some $0 < a_0 < a_1$ and $Q_b = [-b_1,b_1] \times \ldots \times [-b_{d-1},b_{d-1}]$ for $b \in \mathbb{R}_+^{d-1}$. Then $\mathcal{SH}_\psi \psi \in L_{1,w_\rho}(\mathbb{S})$.*

Proof. Using (2.3.25) we have

$$\|\mathcal{SH}_\psi\psi\|_{L_{1,w_\rho}} = \int_{\mathbb{R}^*} \int_{\mathbb{R}^{d-1}} \int_{\mathbb{R}^d} |\mathcal{F}^{-1}(\hat{\psi}\overline{\hat{\psi}_{(a,s,0)}})(t)| \cdot w_\rho(a) \, dt \, ds \, \frac{da}{|a|^{d+1}}$$

$$= \int_{\mathbb{R}^*} \int_{\mathbb{R}^{d-1}} \|\mathcal{F}^{-1}(\hat{\psi}\overline{\hat{\psi}_{(a,s,0)}})\|_{L_1} w_\rho(a) \, ds \, \frac{da}{|a|^{d+1}}.$$

By [28, Lemma 3.1] if $\hat{\psi}\overline{\hat{\psi}_{(a,s,0)}} \not\equiv 0$, or equivalently $\mathcal{F}^{-1}(\hat{\psi}\overline{\hat{\psi}_{(a,s,0)}}) \not\equiv 0$, then

$$a \in [-a_1/a_0, -a_0/a_1] \cup [a_0/a_1, a_1/a_0] =: A$$

and $s \in Q_c = [-c_1,c_1] \times \ldots \times [-c_{d-1},c_{d-1}]$ where $c := b \cdot (1 + (a_1/a_0)^{1/d})/a_0 \in \mathbb{R}^{d-1}$. Obviously the set $A \times Q_c$ is of finite Lebesgue measure, therefore

$$\|\mathcal{SH}_\psi\psi\|_{L_{1,w_\rho}} = \int_A \int_{Q_c} \|\mathcal{F}^{-1}(\hat{\psi}\overline{\hat{\psi}_{(a,s,0)}})\|_{L_1} w_\rho(a) \, ds \, \frac{da}{|a|^{d+1}}$$

is finite. $\qquad\square$

In [22] it has been proven that we can also find compactly supported admissible shearlets ψ, such that $\mathcal{SH}_\psi\psi \in L_{1,w_\rho}(\mathbb{S})$, if the Fourier transform $\hat{\psi}$ satisfies certain decay conditions.

Finally we define the coorbit spaces related to the shearlet group for a given analyzing function ψ by

$$\mathcal{SC}_p^\sigma(\mathbb{R}^d) := \operatorname{Co}(L_{p,m_\sigma}) = \{T \in \mathcal{H}'_{1,w_\rho} : V_{e,\psi}T \in L_{p,m_\sigma}(\mathbb{S})\}, \quad 1 \leqslant p \leqslant \infty, \ \sigma \leqslant \rho.$$

We call the spaces $\mathcal{SC}_p^\sigma(\mathbb{R}^d)$ *(homogeneous) shearlet coorbit spaces*. The wording of homogeneity is lended from the definition of the homogeneous Besov spaces in the previous subsection. Since the shearlet group has a similar structure, this wording is reasonable.

Discretization

In the next step we intend to use the machinery from Section 2.2 to obtain atomic decompositions and Banach frames for the shearlet coorbit spaces $\mathcal{SC}_p^\sigma(\mathbb{R}^d)$. For this we construct suitable U-dense and relatively separated sets. Let $e_{\mathbb{S}} \in U \subset \mathbb{S}$ be a compact set with non-void interior. Let $\alpha > 1$, $\beta, \tau > 0$ be defined such that

$$U_0 := \left[\alpha^{\frac{1}{d}-1}, \alpha^{\frac{1}{d}}\right) \times \left[-\frac{\beta}{2}, \frac{\beta}{2}\right)^{d-1} \times \left[-\frac{\tau}{2}, \frac{\tau}{2}\right)^d \subseteq U, \tag{2.3.30}$$

then the set

$$\mathcal{X} := \left\{ x_{\varepsilon,j,k,m} := (\varepsilon \alpha^j, \beta \alpha^{j(1-\frac{1}{d})}k, S_{\beta \alpha^{j(1-\frac{1}{d})}k} A_{\varepsilon \alpha^j} \tau m) : \varepsilon \in \{-1,1\}, j \in \mathbb{Z}, k \in \mathbb{Z}^d, m \in \mathbb{Z}^d \right\} \tag{2.3.31}$$

is U-dense and relatively separated. For a detailed proof see [28], the calculations therein are elementary. By the definition of U_0, the set U can be chosen arbitrarily small for a suitable choice of α, β and τ. To apply Theorem 2.2.1 and Theorem 2.2.3 we need to ensure that we can choose U such that $\mathrm{osc}_U K_\psi$ suffices (2.2.3) and (2.2.5). It has been proven that this is indeed possible for band-limited shearlets [28, Theorem 3.7] and for compactly supported shearlets with decay conditions on the Fourier transform [22]. This paves the way for a suitable discretization for an admissible shearlet ψ with U small enough.

Then, by Theorem 2.2.1, every $f \in \mathcal{SC}_p^\sigma(\mathbb{R}^d)$ can be decomposed into

$$f = \sum_{\varepsilon \in \{-1,1\}} \sum_{j \in \mathbb{Z}} \sum_{k \in \mathbb{Z}^{d-1}} \sum_{m \in \mathbb{Z}^d} c(f)_{\varepsilon,j,k,m} \cdot \pi(x_{\varepsilon,j,k,m})\psi, \tag{2.3.32}$$

where $c(f) \in \ell_{p,m_\sigma}(\{-1,1\} \times \mathbb{Z} \times \mathbb{Z}^{d-1} \times \mathbb{Z}^d)$ is a sequence depending on f satisfying

$$\sum_{\varepsilon \in \{-1,1\}} \sum_{j \in \mathbb{Z}} \sum_{k \in \mathbb{Z}^{d-1}} \sum_{m \in \mathbb{Z}^d} |c(f)_{\varepsilon,j,k,m}|^p |\alpha|^{j\sigma p} \lesssim \|f\|_{\mathcal{SC}_p^\sigma}^p. \tag{2.3.33}$$

Conversely, if $d \in \ell_{p,m_\sigma}(\{-1,1\} \times \mathbb{Z} \times \mathbb{Z}^{d-1} \times \mathbb{Z}^d)$, then

$$g = \sum_{\varepsilon \in \{-1,1\}} \sum_{j \in \mathbb{Z}} \sum_{k \in \mathbb{Z}^{d-1}} \sum_{m \in \mathbb{Z}^d} d_{\varepsilon,j,k,m} \cdot \pi(x_{\varepsilon,j,k,m})\psi \in \mathcal{SC}_p^\sigma(\mathbb{R}^d) \tag{2.3.34}$$

and

$$\|g\|_{\mathcal{SC}_p^\sigma}^p \lesssim \sum_{\varepsilon \in \{-1,1\}} \sum_{j \in \mathbb{Z}} \sum_{k \in \mathbb{Z}^{d-1}} \sum_{m \in \mathbb{Z}^d} |d_{\varepsilon,j,k,m}|^p |\alpha|^{j\sigma p}. \tag{2.3.35}$$

Chapter 3

Coorbit Theory with Non-Integrable Kernel

In the previous chapter we have seen that coorbit spaces are well-defined for analyzing vectors $\psi \in \mathcal{H}$. This means we need an admissible function $\psi \in \mathcal{H}$ for which the kernel K_ψ is integrable, i.e., contained in $L_{1,w}(G)$ for a suitable weight w. Among others, this assumption is crucial for the definition of the test space $\mathcal{H}_{1,w}$, whose dual space serves as a reservoir for the coorbit spaces, as well as for the definition of the coorbit spaces themselves. In Section 2.2 we have also discussed the importance of the integrability of the kernel for discretization results for coorbit spaces. The fundamental idea of this chapter is the following question: What happens if the kernel is *not* contained in $L_{1,w}(G)$?

The answer to this question can be split into several parts. First, we extend the notion of analyzing vectors to functions that are contained in certain Fréchet spaces. Similarly, a substitute for the test space $\mathcal{H}_{1,w}$ is given in (3.1.4), whose dual space again serves as a reservoir for the coorbit spaces. Proposition 3.1.8 then shows that these coorbit spaces are well-defined and have the desired properties.

It remains to answer the question of discretization results for the new coorbit spaces. As we will show, this question is difficult to answer already for coorbit spaces $\mathrm{Co}(L_{r,m})$ associated with weighted Lebesgue spaces. We are able to give first discretization results in Theorem 3.2.17, but this does not give us proper atomic decompositions or Banach frames and is therefore rather unsatisfactory. In addition, we assume the space $\mathrm{span}\{\pi(x)\psi\}_{x \in G}$ to be dense in $\mathrm{Co}(L_{r,m})$, which is not necessarily fulfilled. To make matters worse, in Theorem 3.3.1 we show that discretization results similar to the ones in Chapter 2 always require the density condition above.

There is, however, a silver lining. If we require the existence of a second kernel W on G that is sufficiently smooth and acts as the identity by left or right convolution, respectively, on the reproducing kernel of the representation, then the existence of both Banach frames and atomic decompositions can be shown in Theorem 3.4.8 and Theorem 3.4.15, respectively. We illustrate all theoretical results by providing a reoccuring application giving both positive and negative examples, namely the Paley-Wiener spaces.

This chapter is organized as follows. We start in Section 3.1 by defining coorbit spaces for kernels that are contained in certain Fréchet spaces and, more specifically, for non-integrable kernels. Next, in Section 3.2, we present a first discretization result for these coorbit spaces under an additional assumption on the kernel. While this result is unsatisfactory, we show in Section 3.3 that we cannot hope to find stable decompositions of the coorbit spaces. Finally, in Section 3.4 we present satisfactory discretization results with the aid of an additional kernel W.

3.1   Coorbit Spaces

The goal of this section is to extend the idea of coorbit spaces to kernels that are not necessarily contained in $L_1(G)$, as in the previous chapter, but possibly in a Fréchet space. The general setting for kernels in Fréchet spaces is presented in Subsection 3.1.1 and then applied to the more specific setting of non-integrabel kernels, i.e., kernels that are not contained in $L_1(G)$, in Subsection 3.1.2. This section is based on [18] and the proofs can be found therein.

As in Chapter 2, let G be a locally compact topological group with Haar measure dx and modulation function Δ. We denote with $\lambda(x)$ and $\rho(x)$ the left- and right translation operators acting on a function f on G by $x \in G$. Furthermore, let π be a unitary irreducible representation of G on a Hilbert space $\mathcal{H}$ as in Definition 2.1.3 and denote with $\psi \in \mathcal{H}$ an admissible vector. The irreducibility of π is not needed in general, but we restrict ourselves to this setting. Let us emphasize that the results in Propositions 2.1.5, 2.1.6 and 2.1.7 hold true in this setting. We denote with V_ψ the corresponding *voice transform*

$$V_\psi \colon \mathcal{H} \to L_2(G), \quad V_\psi f(x) := \langle f, \pi(x)\psi \rangle_\mathcal{H} \tag{3.1.1}$$

and as usual with

$$K_\psi(x) := V_\psi \psi(x) = \langle \psi, \pi(x)\psi \rangle_\mathcal{H} \tag{3.1.2}$$

the *reproducing kernel function*.

3.1.1   Kernel in a Fréchet Space

Let us now choose a Fréchet space $\mathcal{T}$ with the following properties:

(i) there is a continuous embedding $j \colon \mathcal{T} \hookrightarrow L_0(G)$;

(ii) there is a continuous representation ℓ of G acting on $\mathcal{T}$ such that it intertwines with λ, that is $j\ell(x) = \lambda(x)j$ for all $x \in G$;

(iii) there is a continuous involution $f \mapsto \overline{f}$ such that $\overline{j(f)} = j(\overline{f})$.

From now on we will drop j for the sake of simplicity and implicitly use the embedding, at the cost of a slight abuse of notation. This implies $\ell = \lambda$ above. Typical examples of $\mathcal{T}$ are (weighted) L_p-spaces, or $\mathcal{C}^\infty$ when G is a Lie-group. The classic coorbit theory in Chapter 2 corresponds to the choice $\mathcal{T} = L_{1,w}(G)$.

The space $\mathcal{T}$ will serve as our target space, opposed to Definition 2.1.13. We denote with

$$\mathcal{M}^\mathcal{T} := \{f \in \mathcal{T} : f * K_\psi = f\} \tag{3.1.3}$$

the *reproducing kernel space* associated with $\mathcal{T}$. The following assumptions are crucial and trivially satisfied for $L_1(G)$. Recall that the Köthe-dual of a space of functions Y on G is given by

$$Y^\sharp = \{g \text{ measurable} : gf \in L_1(G) \text{ for all } f \in Y\}.$$

Assumption 3.1.1.

(i) The kernel K_ψ is in $\mathcal{T}$ and $f \cdot K_\psi \in L_1(G)$ for all $f \in \mathcal{T}$, i.e., $K_\psi \in \mathcal{T} \cap \mathcal{T}^\sharp$.

(ii) For all $f \in \mathcal{M}^\mathcal{T}$ and all $g \in \mathcal{H}$ we have $f \cdot V_\psi g \in L_1(G)$, i.e., $V_\psi \mathcal{H} \subseteq (\mathcal{M}^\mathcal{T})^\sharp$.

(iii) The space $\mathrm{span}\{\ell(x)K_\psi : x \in G\}$ is dense in $\mathcal{M}^\mathcal{T}$.

The fact that $\text{span}\{\ell(x)K_\psi : x \in G\}$ is a subspace of $\mathcal{M}^{\mathcal{T}}$ can be derived from (i), but (iii) is strengthening (i) by requiring density. Moreover, items (i) and (ii) should be compared with hypotheses (R2) and (R3) of [14], where a different generalization of coorbit theory is presented and where the test space is given a-priori.

We now define the substitute for $\mathcal{H}_{1,w}$ in (2.1.14), namely the *test space*

$$\mathcal{S} := \{f \in \mathcal{H} : V_\psi f \in \mathcal{T}\}. \tag{3.1.4}$$

We topologize $\mathcal{S}$ by means of a restriction of V_ψ to $\mathcal{S} \subset \mathcal{H}$, that is by equiping $\mathcal{S}$ with the initial topology induced by $V_\psi|_\mathcal{S} : \mathcal{S} \to \mathcal{T}$. Also, item (i) of Assumption 3.1.1 implies $\psi \in \mathcal{S}$, since $V_\psi \psi \in \mathcal{T}$.

The following theorem collects some facts about $\mathcal{S}$.

Theorem 3.1.2 (Theorem 3.1 of [18]). *The space $\mathcal{S}$ is a Fréchet space isomorphic to $\mathcal{M}^{\mathcal{T}}$ via V_ψ, and $\mathcal{M}^{\mathcal{T}} \subset L_2(G)$. Furthermore, $\mathcal{S}$ is continuously and densely embedded into $\mathcal{H}$ and $\mathcal{H}'$ is continuously and densely embedded into the (strong) dual $\mathcal{S}'$. The representation π leaves $\mathcal{S}$ invariant, its restriction to $\mathcal{S}$ is a continuous representation of G acting on $\mathcal{S}$ and ψ is a cyclic vector of the restricted representation.*

Similarly as in Definition 2.1.11 we can now define the *extended voice transform* on $\mathcal{S}'$ by setting

$$V_{e,\psi}T(x) := \langle T, \pi(x)\psi \rangle_{\mathcal{S}' \times \mathcal{S}} \quad \text{for } T \in \mathcal{S}' \text{ and } x \in G. \tag{3.1.5}$$

This definition makes sense, because of the Gelfand-triple setting established in Theorem 3.1.2 and since $\mathcal{S}$ is π-invariant.

Let us collect a few facts about the extended voice transform.

Theorem 3.1.3 (Theorem 3.2 of [18]).

(i) *The restriction of the voice transform $V_\psi : \mathcal{S} \to \mathcal{T}$ is a injective continuous linear map with closed image $\mathcal{M}^{\mathcal{T}}$ and it intertwines π and λ.*

(ii) *The extended voice transform $V_{e,\psi}$ is injective and continuous from $\mathcal{S}'$ to $\mathcal{C}(G)$, where both spaces are endowed with the topology of compact convergence.*

(iii) *For all $\Phi \in \mathcal{T}^\sharp \subseteq \mathcal{T}'$ we have*

$$V_{e,\psi} V_\psi^* \Phi = \Phi * K_\psi. \tag{3.1.6}$$

The next assumption assures the reproducing formula similarly to part (i) of Proposition 2.1.12.

Assumption 3.1.4. *For all $T \in \mathcal{S}'$ we have $K_\psi \cdot V_{e,\psi}T \in L_1(G)$ and $V_{e,\psi}T * K_\psi = V_{e,\psi}T$.*

The first part of the assumption, namely the requirement $K_\psi \cdot V_{e,\psi}T \in L_1(G)$ assures that $V_{e,\psi}T * K_\psi$ exists and it is always fulfilled if $V_{e,\psi}T \in \mathcal{T}^\sharp$ since $K_\psi \in \mathcal{T}$. Let us give a sufficient condition for the assumption above.

Proposition 3.1.5 (Proposition 3.4 of [18]). *Assume that $\mathcal{T}^\sharp = \mathcal{T}'$ and suppose that $|K_\psi| * |K_\psi|$ exists and belongs to $\mathcal{T}$. Then, $V_{e,\psi}T * K_\psi = V_{e,\psi}T$ for all $T \in \mathcal{S}'$.*

We now fix a Banach function space Y on G, which will serve as our *target space* analogously to Definition 2.1.13. We assume that Y is a two-sided translation invariant space continuously embedded into $L_0(G)$. This assumption can be weakend, but since we are interested in (weighted) L_p-spaces for fixed p, it will always be fulfilled and therefore suffices. As in (2.1.24) we write

$$\mathcal{M}^Y := \{f \in Y : f * K_\psi = f\} \tag{3.1.7}$$

for the *reproducing kernel space* associated with Y. We assume the following.

Assumption 3.1.6.

(i) *For all $f \in Y$ we have $f \cdot K_\psi \in L_1(G)$, i.e., $K_\psi \in Y^\sharp$.*

(ii) *For all $f \in Y$ and all $g \in \mathcal{S}$ we have $f \cdot V_\psi g \in L_1(G)$, i.e., $V_\psi \mathcal{S} \subset (\mathcal{M}^Y)^\sharp$.*

Item (i) implies that $\mathcal{M}^Y$ is a closed and λ-invariant subspace of Y. With this at hand we define the coorbit spaces.

Definition 3.1.7. The *coorbit space* $\mathrm{Co}(Y)$ associated with Y is defined as

$$\mathrm{Co}(Y) := \{T \in \mathcal{S}' : V_{e,\psi}T \in Y\}. \tag{3.1.8}$$

As in Section 2.1 we equip these spaces with the natural norm $\|T\|_{\mathrm{Co}(Y)} := \|V_{e,\psi}T\|_Y$. Then, as the next proposition shows, this definition of coorbit spaces yields meaningful Banach spaces.

Proposition 3.1.8 (Proposition 3.5 of [18]). *Assume that Assumptions 3.1.1, 3.1.4 and 3.1.6 are fulfilled. Then, we have the following.*

(i) *The space $\mathrm{Co}(Y)$ is a π-invariant Banach space with its natural norm.*

(ii) *The extended voice transform induces an isometric isomorphism $V_{e,\psi} \colon \mathrm{Co}(Y) \to \mathcal{M}^Y$, that is, the reproducing identity*

$$V_{e,\psi}T * K_\psi = V_{e,\psi}T, \quad T \in \mathrm{Co}(Y), \tag{3.1.9}$$

holds true and

$$\mathrm{Co}(Y) = \{\pi(f)\psi : f \in \mathcal{M}^Y\}.$$

(iii) *We have*

$$V_{e,\psi}\pi(f)\psi = f, \quad \text{for } f \in \mathcal{M}^Y,$$
$$\pi(V_{e,\psi}T)\psi = T, \quad \text{for } T \in \mathrm{Co}(Y).$$

Here and in the following the notation $\pi(f)u$ is motivated by the following fact.

Remark 3.1.9. In the framework of abstract harmonic analysis, any function $f \in L_1(G)$ defines a bounded operator $\pi(f)$ on $\mathcal{H}$, which is weakly given by

$$\langle \pi(f)v, v' \rangle_\mathcal{H} = \int_G f(x) \langle \pi(x)v, v' \rangle_\mathcal{H} \, dx, \quad v, v' \in \mathcal{H},$$

see for example Section 3.2 of [51]. However, if $f \notin L_1(G)$, then in general $\pi(f)v$ is well defined only if $v = u$, where u is an admissible vector for the representation π.

As in the classical case in Section 2.1, we can express the basic spaces in terms of coorbit spaces. More precisely, $L_2(G)$ satisfies Assumption 3.1.6 and we have $\mathrm{Co}(L_2(G)) = \mathcal{H}$. Furthermore, under Assumption 3.1.4 we have $\mathrm{Co}(\mathcal{T}) = \mathcal{S}$. If we assume $V_{e,\psi}\mathcal{S}' \subset \mathcal{T}^\sharp$, which is in general stronger than Assumption 3.1.4, we also have $\mathrm{Co}(\mathcal{T}') = \mathcal{S}'$.

3.1.2   Non-Integrable Kernel

In this subsection we apply the machinery developed in the preceding subsection to the case of target spaces that are the intersection of $L_{p,w}(G)$-spaces for $p > 1$. This is of particular interest, since we will specifically exclude $p = 1$ and therefore allow the kernel K_ψ to be non-integrable, i.e., not an element of $L_{1,w}(G)$.

Let $w : G \to \mathbb{R}_+$ denote a fixed *weight*, that is, a continuous function satisfying conditions (2.1.11) and (2.1.12):

$$w(xy) \leqslant w(x)w(y), \tag{3.1.10}$$

$$w(x) = w(x^{-1}) \tag{3.1.11}$$

for all $x, y \in G$. As usual, we denote with $L_{p,w}(G)$ the weighted Lebesgue-spaces on G. Before we check the assumptions of the previous subsection, we have the following characterization of the Köthe-dual of weighted Lebesgue spaces.

Lemma 3.1.10 (Lemma 4.1 of [18]). *Fix $1 \leqslant p < \infty$ and denote by $q = \frac{p}{p-1}$ the dual exponent. Then,*

$$L_{p,w}(G)^\sharp = L_{q,w^{-1}}(G).$$

Furthermore, the map $g \mapsto \langle g, \cdot \rangle_{L_2}$ is an isomorphism from $L_{q,w^{-1}}(G)$ onto $L_{p,w}(G)^\sharp$.

We now choose the space

$$\mathcal{T}_w := \bigcap_{1 < p < \infty} L_{p,w}(G) \tag{3.1.12}$$

as the *target space* for the coorbit space theory. We recall some basic properties of $\mathcal{T}_w$; the proofs can be found in Theorem 4.3 of [18], which is based on results in [33]. We endow $\mathcal{T}_w$ with the (unique) topology such that a sequence $(f_n)_{n \in \mathbb{N}} \subset \mathcal{T}_w$ converges to 0 if and only if $\lim_{n \to \infty} \|f_n\|_{L_{p,w}} = 0$ for all $1 < p < \infty$. With this topology, $\mathcal{T}$ becomes a reflexive Fréchet space. The (anti-)linear dual of $\mathcal{T}_w$ can be identified with

$$\mathcal{U}_w := \bigcup_{1 < q < \infty} L_{q,w^{-1}}(G) \tag{3.1.13}$$

under the pairing

$$\int_G \Phi(x)\overline{f(x)}\, dx = \langle \Phi, f \rangle_w, \quad \text{for } \Phi \in \mathcal{U}_w, f \in \mathcal{T}_w. \tag{3.1.14}$$

Remark 3.1.11. The space $\mathcal{U}_w$ is endowed with one of the following equivalent topologies, both compatible with the pairing (3.1.14).

(i) The finest topology making the inclusions $L_{q,w^{-1}}(G) \hookrightarrow \mathcal{U}_w$ continuous for all $1 < q < \infty$.

(ii) The topology induced by the family of semi-norms $(\|\cdot\|_{p,r})_{1 < p < r < \infty}$, where

$$\|\Phi\|_{p,r} = \sup\left\{ |\langle \Phi, f \rangle_w| \ : \ f \in \mathcal{T}_w \text{ and } \max\left\{ \|f\|_{L_{p,w}}, \|f\|_{L_{r,w}} \right\} \leqslant 1 \right\},$$

for $\Phi \in \mathcal{U}_w$.

The spaces $\mathcal{T}_w$ and $\mathcal{U}_w$ are both λ-invariant and λ acts continuously on $\mathcal{T}_w$. Take $g \in \mathcal{T}_w$ with $\check{g} \in \mathcal{T}_w$. For all $f \in \mathcal{T}_w$ the convolution $f * g$ is in $\mathcal{T}_w$ and the map

$$f \mapsto f * g$$

is continuous from $\mathcal{T}_w$ into $\mathcal{T}_w$. Furthermore, for all $\Phi \in \mathcal{U}_w$ the convolution $\Phi * g$ is in $\mathcal{U}_w$ and the map

$$\Phi \mapsto \Phi * g$$

is continuous from $\mathcal{U}_w$ into $\mathcal{U}_w$.

We now state the following basic assumption that motivates this chapter.

Assumption 3.1.12. *We assume* $K_\psi \in \mathcal{T}_w$, *i.e.,*

$$K_\psi \in L_{p,w}(G) \quad \text{for all } 1 < p < \infty. \tag{3.1.15}$$

We add some remarks.

Remark 3.1.13.   (i) Since $w(x) \geqslant 1$, Assumption (3.1.12) implies $K_\psi \in L_p(G)$ for all $p > 1$. If π is irreducible, this last fact gives that V_ψ is an isometry up to a constant, so that π is always a reproducing representation. If π is reducible, condition (3.1.15) is not sufficient to ensure that π is reproducing; however if $K_\psi * K_\psi = K_\psi$, then π is always reproducing.

  (ii) If w^{-1} belongs to $L_q(G)$ for some $1 < q < \infty$, then Hölder's inequality shows $K_\psi \in L_1(G)$, but in general $K \notin L_{1,w}(G)$. However in many interesting examples w is independent of one or more variables, so that $w^{-1} \notin L_q(G)$ for all $1 < q < \infty$.

  (iii) The assumption made in (3.1.15) is indeed weaker than the assumption $K_\psi \in L_{1,w}(G)$ made in Chapter 2, at least in the unweighted case. This can be seen as follows. By part (i) of Lemma 2.1.2 the kernel K_ψ is a bounded continuous function on G, that is $K_\psi \in L_\infty(G)$. If we additionally assume $K_\psi \in L_1(G)$, then obviously $K_\psi \in L_p(G)$ for all $1 \leqslant p \leqslant \infty$, which implies (3.1.15).

Parallel to (3.1.4) we now define the *test space* $\mathcal{S}_w$ as

$$\mathcal{S}_w = \{f \in \mathcal{H} : V_\psi f \in L_{p,w}(G) \text{ for all } 1 < p < \infty\}, \tag{3.1.16}$$

which becomes a locally convex topological vector space under the family of semi-norms

$$\|f\|_{p,\mathcal{S}_w} = \|V_\psi f\|_{L_{p,w}}. \tag{3.1.17}$$

We recall the main properties of $\mathcal{S}_w$.

Theorem 3.1.14 (Theorem 4.4 of [18]). *Under Assumption 3.1.12 the following hold:*

  (i) the space $\mathcal{S}_w$ is a reflexive Fréchet space, continuously and densely embedded in $\mathcal{H}$;

  (ii) the representation π leaves $\mathcal{S}_w$ invariant and its restriction to $\mathcal{S}_w$ is a continuous representation;

  (iii) the space $\mathcal{H}$ is continuously and densely embedded into the (anti-)linear dual $\mathcal{S}_w'$, where both spaces are endowed with the weak topology;

  (iv) the restriction of the voice transform $V_\psi : \mathcal{S}_w \to \mathcal{T}_w$ is a topological isomorphism from $\mathcal{S}_w$ onto the closed subspace $\mathcal{M}^{\mathcal{T}_w}$ of $\mathcal{T}_w$, given by

$$\mathcal{M}^{\mathcal{T}_w} = \{f \in \mathcal{T}_w \mid f * K_\psi = f\},$$

and it intertwines π and λ;

(v) for every $f \in \mathcal{T}_w$, there exists a unique element $\pi(f)\psi \in \mathcal{S}_w$ such that

$$\langle \pi(f)\psi, v \rangle_{\mathcal{H}} = \int_G f(x) \langle \pi(x)\psi, v \rangle_{\mathcal{H}} \, dx = \int_G f(x) \overline{V_\psi v(x)} \, dx, \quad v \in \mathcal{H}.$$

Furthermore, it holds that

$$V_\psi \pi(f)\psi = f * K_\psi,$$

and the map

$$\mathcal{T}_w \ni f \mapsto \pi(f)\psi \in \mathcal{S}_w$$

is continuous and its restriction to $\mathcal{M}^{\mathcal{T}_w}$ is the inverse of V_ψ.

Recalling that the (anti-)dual of $\mathcal{T}_w$ is $\mathcal{U}_w$ under the pairing (3.1.14), we denote by ${}^t V_\psi$ the contragradient map ${}^t V_\psi : \mathcal{U}_w \to \mathcal{S}'_w$ given by

$$\langle {}^t V_\psi \Phi, v \rangle_{\mathcal{S}'_w \times \mathcal{S}_w} = \langle \Phi, V_\psi v \rangle_w, \qquad \Phi \in \mathcal{U}_w, \, v \in \mathcal{S}_w.$$

As usual, we extend the voice transform from $\mathcal{H}$ to the (anti-)dual $\mathcal{S}'_w$ of $\mathcal{S}_w$, where $\mathcal{S}'_w$ plays the role of the space of distributions. For all $T \in \mathcal{S}'_w$ we set

$$V_{e,\psi} T(x) = \langle T, \pi(x)\psi \rangle_{\mathcal{S}'_w \times \mathcal{S}_w}, \qquad x \in G, \tag{3.1.18}$$

which is a continuous function on G by item (ii) of the previous theorem and $\langle \cdot, \cdot \rangle_{\mathcal{S}_w}$ denotes the pairing between $\mathcal{S}_w$ and $\mathcal{S}'_w$, whereas $\langle \cdot, \cdot \rangle_w$ is the pairing between $\mathcal{T}_w$ and $\mathcal{U}_w$.

We summarize the main properties of the extended voice transform in the following theorem.

Theorem 3.1.15 (Theorem 4.4 of [18]). *Under Assumption 3.1.12, the following hold:*

(i) for every $\Phi \in \mathcal{U}_w$ there exists a unique element $\pi(\Phi)\psi \in \mathcal{S}'_w$ such that

$$\langle \pi(\Phi)\psi, v \rangle_{\mathcal{S}'_w \times \mathcal{S}_w} = \int_G \Phi(x) \langle \pi(x)\psi, v \rangle_{\mathcal{H}} \, dx = \int_G \Phi(x) \overline{V_\psi v(x)} \, dx, \qquad v \in \mathcal{S}_w.$$

Furthermore, it holds that

$$V_{e,\psi} \pi(\Phi)\psi = \Phi * K_\psi;$$

(ii) for all $T \in \mathcal{S}'_w$ the voice transform $V_{e,\psi} T$ is in $\mathcal{U}_w$ and satisfies

$$V_{e,\psi} T = V_{e,\psi} T * K_\psi, \tag{3.1.19}$$
$$\langle T, v \rangle_{\mathcal{S}'_w \times \mathcal{S}_w} = \langle V_{e,\psi} T, V_\psi v \rangle_w, \qquad v \in \mathcal{S}_w; \tag{3.1.20}$$

(iii) the extended voice transform $V_{e,\psi}$ is injective and continuous from $\mathcal{S}'_w$ into $\mathcal{U}_w$ (when both spaces are endowed with the strong topology), its range is the closed subspace

$$\mathcal{M}^{\mathcal{U}_w} = \{\Phi \in \mathcal{U}_w \, : \, \Phi * K_\psi = \Phi\} = \operatorname{span} \bigcup_{p \in (1,\infty)} \mathcal{M}^{L_{p,w}(G)} \subset L_{\infty, w^{-1}}(G) \tag{3.1.21}$$

and it intertwines the contragradient representation of $\pi|_{\mathcal{S}_w}$ and $\lambda|_{\mathcal{U}_w}$;

(iv) the map

$$\mathcal{M}^{\mathcal{U}_w} \ni \Phi \mapsto \pi(\Phi)\psi \in \mathcal{S}'_w$$

is the left inverse of $V_{e,\psi}$ and coincides with the restriction of the map ${}^t V_\psi$ to $\mathcal{M}^{\mathcal{U}_w}$, namely

$$V_{e,\psi}({}^t V_\psi \Phi) = V_{e,\psi} \pi(\Phi)\psi = \Phi, \qquad \Phi \in \mathcal{M}^{\mathcal{U}_w}; \tag{3.1.22}$$

(v) regarding $\mathcal{S}_w \hookrightarrow \mathcal{H} \hookrightarrow \mathcal{S}'_w$, we have

$$\mathcal{S}_w = \left\{ T \in \mathcal{S}'_w \ : \ V_{e,\psi} T \in \mathcal{T}_w \right\} = \left\{ \pi(f)\psi \ : \ f \in \mathcal{M}^{\mathcal{T}_w} \right\}.$$

Item (ii) of the previous theorem states that the voice transform of any distribution $T \in \mathcal{S}'_w$ satisfies the reproducing formula (3.1.19) and uniquely defines the distribution T by means of the reconstruction formula (3.1.20), i.e.,

$$T = \int_G \langle T, \pi(x)u \rangle_{\mathcal{S}'_w \times \mathcal{S}_w} \pi(x)u \, dx,$$

where the integral is a Dunford-Pettis integral with respect to the duality between $\mathcal{S}_w$ and $\mathcal{S}'_w$, see, for example, Appendix 3 of [51].

With the previous theorems at hand we can show that Assumptions 3.1.1 and 3.1.4 are fulfilled.

Theorem 3.1.16. *If Assumption 3.1.12 is fulfilled, i.e., $K \in \mathcal{T}_w$, then Assumptions 3.1.1 and 3.1.4 are fulfilled. This means that the coorbit theory developed in the previous subsection is applicable for suitable target spaces Y.*

Proof. Since $K_\psi \in L_{p,w}(G)$ for some fixed $1 < p < \infty$ and since it holds that $K_\psi = \widetilde{K_\psi}$ we have $K_\psi \in L_{q,w^{-1}}(G) = L_{p,w}(G)^\sharp$, where q denotes the dual exponent of p. So, for every $f \in \mathcal{T}_w$ we have $f \cdot K_\psi \in L_1(G)$, which shows item (i) of Assumption 3.1.1. To show item (ii) we note that $\mathcal{T}_w \subset L_{2,w}(G) \subseteq L_2(G)$, so that $\mathcal{M}^{\mathcal{T}_w} \subset L_2(G)$. The reconstruction formula (3.1.20) clarifies that ψ is a cyclic vector for the representation τ, which is equivalent to $V_\psi \psi = K_\psi$ being a cyclic vector for the representation λ restricted to $\mathcal{M}^{\mathcal{T}_w}$. This in turn is equivalent to item (iii). Finally, Assumption 3.1.4 is a direct consequence of the reproducing formula (3.1.19). □

We are interested in suitable target spaces for the coorbit theory, which will be weighted Lebesgue spaces $L_{r,m}(G)$, as in Section 2.1. We fix an exponent $1 \leqslant r < \infty$ and a *w-moderate weight* m on G, i.e., a continuous function $m \colon G \to \mathbb{R}_+$ such that equations similar to (2.1.20) are fulfilled, namely

$$m(xy) \leqslant w(x)m(y) \quad \text{and} \quad m(xy) \leqslant m(x)w(y) \quad \text{for all } x, y \in G. \tag{3.1.23}$$

Up to a constant $w(e)$ this is equivalent to

$$m(xyz) \leqslant w(x)m(y)w(z), \quad \text{for all } x, y, z \in G. \tag{3.1.24}$$

The result of the following lemma is used multiple times in this chapter.

Lemma 3.1.17. *If m is a w-moderate weight on G, then so is m^{-1}.*

Proof. To prove the estimates in (3.1.23) for m^{-1} we fix $x, y \in G$, then by the w-moderateness of m it holds

$$m(y) = m(x^{-1}xy) \leqslant w(x^{-1}) \cdot m(xy) = w(x) \cdot m(xy),$$

which implies $m(xy)^{-1} \leqslant w(x) \cdot m(y)^{-1}$. Similarly we observe that

$$m(x) = m(xyy^{-1}) \leqslant m(xy) \cdot w(y^{-1}) = m(xy) \cdot w(y),$$

which in turn implies $m(xy)^{-1} \leqslant m(x)^{-1} \cdot w(y)$. □

Using a fixed w-moderate weight m, we choose as a *target space* for the coorbit theory developed in the previous Subsection the Banach space $Y = L_{r,m}(G)$ with $1 < r < \infty$. The corresponding coorbit space is defined as in Definition 3.1.7 via

$$\mathrm{Co}(L_{r,m}) = \{T \in \mathcal{S}'_w : V_{e,\psi}T \in L_{r,m}(G)\} \tag{3.1.25}$$

endowed with the natural norm

$$\|T\|_{\mathrm{Co}(L_{r,m})} = \|V_{e,\psi}T\|_{L_{r,m}}. \tag{3.1.26}$$

Before we show that this definition is indeed meaningful, we need to check Assumption 3.1.6.

Proposition 3.1.18. *If Assumption 3.1.12 is fulfilled, i.e., $K_\psi \in \mathcal{T}_w$, then Assumption 3.1.6 is fulfilled. This implies that Proposition 3.1.8 is applicable.*

Proof. We first note that $L_{r',m^{-1}}(G) = L_{r,m}(G)^\sharp$ by Lemma 3.1.10, where r' denotes the dual exponent of r. Furthermore, by (3.1.23) and (3.1.11), for any $x \in G$ we have

$$m(e) = m(xx^{-1}) \leqslant m(x)w(x^{-1}) = m(x)w(x),$$

and hence $m(x)^{-1}m(e) \leqslant w(x)$. Therefore $L_{r,w}(G) \subseteq L_{r,m^{-1}}(G)$ and we have $K_\psi \in L_{r,m}(G)^\sharp$, which shows part (i) of Assumption 3.1.6.

To show (ii) let $f \in L_{r,m}(G)$ and $g \in \mathcal{S}_w$, then by item (iv) of Theorem 3.1.14 we have $V_\psi g \in \mathcal{T}_w$. But we have $L_{r,m}(G) \subseteq L_{r,w^{-1}}(G) \subset \mathcal{U}_w$ by the calculations above and $\mathcal{U}_w = \mathcal{T}'_w$ under the pairing (3.1.14) and therefore $f \cdot V_\psi g \in L_1(G)$.   □

Now that the definition of the coorbit spaces in (3.1.25) is meaningful, we summarize a few properties.

Proposition 3.1.19. *The space $\mathrm{Co}(Y)$ is a Banach space invariant under the action of the contragradient representation of $\pi|_{\mathcal{S}_w}$. The extended voice transform is an isometry from $\mathrm{Co}(L_{r,m})$ onto the λ-invariant closed subspace*

$$\mathcal{M}^{L_{r,m}} = \{F \in L_{r,m}(G) : F * K_\psi = F\} \subset \mathcal{U}_w,$$

and we have

$$\mathrm{Co}(L_{r,m}) = \left\{\pi(F)\psi : F \in \mathcal{M}^{L_{r,m}}\right\}.$$

Furthermore

$$V_e\pi(F)\psi = F, \qquad F \in \mathcal{M}^Y, \tag{3.1.27}$$
$$\pi(V_eT)\psi = T, \qquad T \in \mathrm{Co}(L_{r,m}). \tag{3.1.28}$$

Proof. The proof is a direct application of Proposition 3.1.8, taking into account that Assumptions 3.1.1, 3.1.4 and 3.1.6 are fulfilled by Theorem 3.1.16 and Proposition 3.1.18.   □

3.2   Discretization

In this section we intend to establish certain atomic decompositions for the coorbit spaces described in Section 3.1. Before we start let us briefly examine the difficulty of this task compared to the discretization of classic coorbit spaces in Section 2.2. For the classic coorbit spaces a reoccuring condition for the kernel is given in (2.2.3) in the case of atomic decompositions and in (2.2.5) in the case of Banach frames, where we always need the integrability condition $\|K_\psi\|_{L_{1,w}} < \infty$ plus integrability conditions on the oscillator $\mathrm{osc}K_\psi$. However, in the approach of this chapter, we only assume $K_\psi \in L_{p,w}(G)$ for all $1 < p < \infty$, which is a weaker assumption. We therefore cannot hope to use the same techniques as in Section 2.2. We will, however, present a weaker form of atomic decompositions in Theorem 3.2.17 for coorbit spaces $\mathrm{Co}(L_{r,m})$ defined in Section 3.1, using integrability conditions on the kernel with respect to weighted L_p-spaces for $p < r$.

This section is organized as follows. After recalling some basic facts, in Subsection 3.2.1 we present a very basic assumption on the kernel that is needed for a first atomic decomposition for coorbit spaces with non-integrable kernel, which we will then present and prove in Subsection 3.2.2. Finally, in Subsection 3.2.3 we apply the theory of Subsection 3.1.2 to the example of Paley-Wiener spaces, i.e., the space of L_p-functions with compact support in the frequency domain, and use the atomic decomposition to discretize these spaces.

We recall from the previous section that

$$\mathcal{T}_w = \bigcap_{1<p<\infty} L_{p,w}(G), \qquad \mathcal{T}'_w = \mathcal{U}_w = \mathrm{span} \bigcup_{1<q<\infty} L_{q,w^{-1}}(G) \tag{3.2.1}$$

and for some fixed $1 < r < \infty$,

$$Y = L_{r,m}(G). \tag{3.2.2}$$

Proposition 3.1.19 shows that the correspondence principle holds, i.e., the extended voice transform $V_{e,\psi}$ is an isometry from the associated coorbit space

$$\mathrm{Co}(L_{r,m}) := \left\{ T \in \mathcal{S}'_w : V_{e,\psi}(T) \in L_{r,m}(G) \right\} \tag{3.2.3}$$

onto the corresponding reproducing kernel Banach space

$$\mathcal{M}_{r,m} := \mathcal{M}^{L_{r,m}} = \{ f \in L_{r,m}(G) : f * K_\psi = f \}. \tag{3.2.4}$$

Remark 3.2.1. Assumption 3.1.12 on the kernel K_ψ and the fact that m is w-moderate imply that for all $f \in L_{r,m}(G)$ the convolution $f * K_\psi$ is well-defined; see the extension of Young's inequality in Proposition 1.3.11.

In this setting we can characterize the anti-dual $\mathcal{M}'_{r,m}$ of the reproducing kernel space.

Lemma 3.2.2. *The anti-dual $\mathcal{M}'_{r,m}$ of $\mathcal{M}_{r,m}$ is canonically isomorphic to $L_{r',m^{-1}}(G)/\mathcal{M}^\perp_{r,m}$, where*

$$\mathcal{M}^\perp_{r,m} = \left\{ \widetilde{F} \in L_{r',m^{-1}}(G) : \langle \widetilde{F}, F \rangle_{L_2} = 0 \text{ for all } F \in \mathcal{M}_{r,m} \right\} \tag{3.2.5}$$

and r' denotes the dual exponent of r. Hence, for every $\Gamma \in \mathcal{M}'_{r,m}$ there is a $\widetilde{F} \in L_{r',m^{-1}}(G)$ such that $\Gamma(F) = \langle \widetilde{F}, F \rangle_{L_2}$ for all $F \in \mathcal{M}_{r,m}$.

Proof. Since $\mathcal{M}_{r,m}$ is a closed subspace of $L_{r,m}(G)$, [85, Proposition 1.4] yields that $\mathcal{M}'_{r,m}$ is canonically isomorphic to $L'_{r,m}(G)/\mathcal{M}^\perp_{r,m}$. The claim follows because $L'_{r,m}(G)$ is canonically isomorphic to $L_{r',m^{-1}}(G)$. $\square$

We need some more preparations. In the following, let $Q \subset G$ denote a compact neighborhood of e with $Q = \overline{\mathrm{int}\, Q}$. Then, as in (2.1.22) and (2.1.23), the *local maximal functions (with respect to the right regular representation)* $M^R_Q f$ of $f \in L_\infty(G)$ are defined by

$$M^R_Q f(x) := \| f \cdot \rho(x) \chi_Q \|_{L_\infty}, \quad \text{whence} \quad \widetilde{M}^R_Q f(x) := M^R_Q f(x^{-1}) = \| f \|_{L_\infty(Qx)}. \tag{3.2.6}$$

Then, for a function space Y on G, we define the *Wiener-type spaces* as

$$\mathcal{M}_Q^R(Y) := \mathcal{M}_Q^R(L_\infty, Y) = \left\{ f \in L_\infty(G) : \widetilde{M}_Q^R f \in Y \right\}. \tag{3.2.7}$$

We recall the definition of the *Q-oscillation* of a function f on G with respect to Q as

$$\operatorname{osc}_Q f(x) := \sup_{y \in Q} |f(yx) - f(x)| \quad \text{for } x \in G. \tag{3.2.8}$$

As in Section 2.2, the decay property of the Q-oscillation plays an important role in view of the discretization of coorbit spaces of any kind. In Lemma 2.2.2 we observed conditions under which the decay is rapid enough such that $\operatorname{osc}_Q f \in L_{1,w}(G)$. A similar result holds for $L_{p,w}(G)$, where $1 < p < \infty$, as we state in the following lemma. The proof is a simple generalization of the proof of [68, Lemma 4.6].

Lemma 3.2.3. *Let w be a weight on G, let $1 < p < \infty$, and assume that f is a continuous function on G and that $f \in \mathcal{M}_{Q_0}^R(L_{p,w})$ for some compact unit neighborhood Q_0 with $Q_0 = \overline{\operatorname{int} Q_0}$. Then the following hold:*

(i) $\|\operatorname{osc}_{Q_0} f\|_{L_{p,w}} < \infty$.

(ii) For arbitrary $\varepsilon > 0$, there is a unit neighborhood $Q_\varepsilon \subset Q_0$ such that for each unit neighborhood $Q \subset Q_\varepsilon$, we have $\|\operatorname{osc}_Q f\|_{L_{p,w}} < \varepsilon$. Put briefly,

$$\lim_{Q \to \{e\}} \|\operatorname{osc}_Q f\|_{L_{p,w}} = 0.$$

Proof. We start with an auxiliary observation: We claim that $\|g\|_{L_\infty(Qx)} = \sup_{y \in Qx} |g(y)|$ if $g : G \to \mathbb{C}$ is continuous and if $Q \subset G$ is a compact unit neighborhood with $Q = \overline{\operatorname{int} Q}$.

Indeed, the inequality "$\leqslant$" is trivial. Conversely, if we set $\alpha := \|g\|_{L_\infty(Qx)}$, then the set $M := \{y \in G : |g(y)| > \alpha\}$ is open, and $M \cap Qx$ is a null-set. Hence, $M \cap (\operatorname{int} Q)x = \varnothing$, since this is an open null-set. In other words, $|g(y)| \leqslant \alpha$ for all $y \in (\operatorname{int} Q)x$. By continuity of g and since $Q \subset \overline{\operatorname{int} Q}$, we see $|g(y)| \leqslant \alpha$ for all $y \in Qx$. In particular, this implies $\widetilde{M}_Q^R g(x) = \sup_{q \in Q} |g(qx)|$, and thus, because of $e \in Q$, $\widetilde{M}_Q^R g \geqslant |g|$.

To prove (i) we note that $\widetilde{M}_{Q_0}^R f \in L_{p,w}(G)$ which implies $f \in L_{p,w}(G)$, since we just saw that $\widetilde{M}_{Q_0}^R f \geqslant |f|$. We intend to show $\|\operatorname{osc}_{Q_0} f\|_{L_{p,w}} < \infty$. But we have

$$\operatorname{osc}_{Q_0} f(x) = \sup_{q \in Q_0} |f(qx) - f(x)| \leqslant \sup_{q \in Q_0} |f(qx)| + |f(x)| \leqslant |f(x)| + \widetilde{M}_{Q_0}^R f(x).$$

Therefore,

$$\|\operatorname{osc}_{Q_0} f\|_{L_{p,w}} \leqslant \|\widetilde{M}_{Q_0}^R f\|_{L_{p,w}} + \|f\|_{L_{p,w}} < \infty. \tag{3.2.9}$$

It remains to prove (ii). For this we first note that $\operatorname{osc}_Q f \leqslant \operatorname{osc}_{Q_0} f$ if $Q \subset Q_0$. Furthermore, by part (i) we have $\operatorname{osc}_{Q_0} f \in L_{p,w}(G)$. Hence, since G is σ-compact, for any $\varepsilon > 0$, there exists a compact set $K \subset G$ of positive measure such that

$$\int_{G \setminus K} |\operatorname{osc}_Q f(x) w(x)|^p \, dx \leqslant \int_{G \setminus K} |\operatorname{osc}_{Q_0} f(x) w(x)|^p \, dx < \frac{\varepsilon}{2} \tag{3.2.10}$$

for all unit neighborhoods $Q \subset Q_0$.

Next, we observe that since f is continuous, it is uniformly continuous on K in the following sense: For every $\delta > 0$ there is a unit neighborhood $U_\delta \subset G$ with $|f(x) - f(ux)| < \delta$ for all $x \in K$ and $u \in U_\delta$.

The uniform continuity described above simply means $\mathrm{osc}_{U_\delta} f(x) \leqslant \delta$ for all $x \in K$. Choosing $\delta := \varepsilon^{1/p}/([2 \cdot |K|]^{1/p} \sup_{y \in K} w(y))$, we see for every unit neighborhood $Q \subset Q_0 \cap U_\delta$ that

$$\int_K |\mathrm{osc}_Q f(x) w(x)|^p \, dx \overset{?}{\leqslant} \int_K \frac{\varepsilon}{2|K|} \cdot \frac{w(x)^p}{\sup_{y \in K} w(y)^p} \, dx \leqslant \int_K \frac{\varepsilon}{2|K|} \, dx = \frac{\varepsilon}{2}. \tag{3.2.11}$$

Equations (3.2.10) and (3.2.11) yield $\|\mathrm{osc}_Q f\|_{L_{p,w}}^p < \varepsilon$, which concludes the proof. □

3.2.1 An Assumption on the Kernel

As we will show throughout this chapter, it is not sufficient to assume $K_\psi \in L_{p,w}(G)$ for all $1 < p < \infty$ for obtaining an atomic decomposition of coorbit spaces. Instead, we need an additional assumption on the reproducing kernel space $\mathcal{M}_{r,m}$ which we state in the following assumption.

Assumption 3.2.4. *Assumption 3.1.12 is satisfied, i.e., $K_\psi \in L_{p,w}(G)$ for all $1 < p < \infty$, and* $\mathrm{span}\{\lambda(x) K_\psi\}_{x \in G}$ *is dense in* $\mathcal{M}_{r,m}$.

Remark 3.2.5. The necessity of the assumption above can be seen as follows. First, the density of $\mathrm{span}\{\lambda(x) K_\psi\}_{x \in G}$ in $\mathcal{M}_{r,m}$ is equivalent to the density of $\mathrm{span}\{\pi(x) \psi\}_{x \in G}$ in $\mathrm{Co}(L_{r,m})$ by the correspondence principle, see Proposition 3.1.19. When looking at the discretization techniques in Section 2.2, we always decompose and reconstruct using functions $\pi(x_i) \psi$ for some $x_i \in G$. Following the same ideas, if the set $\mathrm{span}\{\pi(x) \psi\}_{x \in G}$ was not dense in $\mathrm{Co}(L_{r,m})$, we would have no hope in finding a meaningful discretization as the one in Chapter 2 by means of functions of the type $\pi(x) \psi$.

Assumption 3.2.4 is similar to the density of $\mathrm{span}\{\pi(x) K - \psi\}_{x \in G}$ in $\mathcal{H}$—which is equivalent to K_ψ being a cyclic vector for the representation π on $\mathcal{H}$—and in $\mathcal{M}^{T_w}$, which is item (iii) of Assumption 3.1.1 and fulfilled in our setting, as can be seen by combining Theorems 3.1.15 and 3.1.16. Therefore Assumption 3.2.4 is an assumption on the kernel and the density can be shown via properties of K_ψ.

In the following, for the sake of breviety, we will denote with RC_K the right convolution operator $RC_K f := f * K_\psi$, where the space on which RC_K acts may vary depending on the context.

Before we provide a sufficient condition under which Assumption 3.2.4 is fulfilled in Lemma 3.2.9, we need a couple of auxiliary results.

Proposition 3.2.6. *Assume that for all $f \in L_{r,m}(G)$, f and K_ψ are convolvable in the sense that $f \cdot \lambda(x) \check{K}_\psi \in L_1(G)$ for almost all $x \in G$ and $f * K_\psi \in L_{r,m}(G)$, then the right convolution operator acting on $L_{r,m}(G)$*

$$RC_K : L_{r,m}(G) \to L_{r,m}(G), \qquad RC_K f = f * K_\psi, \tag{3.2.12}$$

is bounded.

Proof. For $r = 2$ the result is stated in [73, Proposition 3.10], whose proof holds true for any p. Indeed, by the closed graph theorem, it is enough to show that RC_K is a closed operator. Take a sequence $(f_n)_{n \in \mathbb{N}}$ converging to $f \in L_{r,m}(G)$ such that $(RC_K f_n)_{n \in \mathbb{N}}$ converges to $g \in L_{r,m}(G)$. By a sharp version of the Riesz-Fischer theorem, see [3, Theorem 13.6], there exists a positive function $g \in L_{r,m}(G)$ such that, possibly passing twice to a subsequence, there exist two null sets E, F such that for all $y \in G \backslash E$ and $x \in G \backslash F$ we have

$$|f_n(y)| \leqslant g(y),$$
$$\lim_{n \to \infty} f_n(y) = f(y),$$
$$\lim_{n \to \infty} RC_K f_n(x) = g(x).$$

Furthermore, by definition of convolution and after possibly re-defining the null set F, we get that for all $x \in G\backslash F$ and all $n \in \mathbb{N}$ the mappings

$$y \mapsto f_n(y)K_\psi(y^{-1}x), \qquad y \mapsto g(y)K_\psi(y^{-1}x)$$

are integrable. Then, given $x \in G\backslash F$, for all $y \in G\backslash E$

$$|f_n(y)K_\psi(y^{-1}x)| \leqslant |g(y)K_\psi(y^{-1}x)|, \quad \lim_{n\to\infty} f_n(y)K_\psi(y^{-1}x) = f(y)K_\psi(y^{-1}x).$$

For $x \in G\backslash F$, the function $y \mapsto g(y)K_\psi(y^{-1}x)$ is integrable, so that by dominated convergence we obtain

$$g(x) = \lim_{n\to\infty} \int_G f_n(y)K_\psi(y^{-1}x)\,dy = \int_G f(y)K_\psi(y^{-1}x)\,dy = f * K_\psi(x),$$

so RC_K is indeed closed. □

Here and in the following, the duality pairing for $f \in L_{r,m}(G)$ and $g \in L_{r,m}(G)' = L_{r',m^{-1}}(G)$ is the sesqui-linear form

$$\langle f, g \rangle_{L_2} = \int_G f(x)\overline{g(x)}\,dx,$$

see Lemma 3.1.10.

Proposition 3.2.7. *Denote by r' the dual exponent $1/r + 1/r' = 1$. Assume that the right convolution operator*

$$RC_K : L_{r,m}(G) \to L_{r,m}(G), \qquad RC_K f = f * K_\psi,$$

is bounded, then

(i) *the right convolution RC_K operator is bounded on $L_{r',m^{-1}}(G)$ and it coincides with the adjoint of RC_K;*

(ii) *the right convolution operator RC_K is a projection from $L_{r,m}(G)$ onto the reproducing kernel Banach space $\mathcal{M}_{r,m}$.*

Proof. Since RC_K is a bounded operator on $L_{r,m}(G)$, the adjoint is a bounded operator on $L_{r,m}(G)' = L_{r',m^{-1}}(G)$. Take $g \in L_{r',m^{-1}}(G)$ and $f \in C_c(G) \subset L_{r,m}(G)$, then

$$\langle RC_K^* g, f \rangle_{L_2} = \langle g, RC_K f \rangle_{L_2} = \int_G g(x) \left(\int_G \overline{f(y)K_\psi(y^{-1}x)}\,dy \right) dx$$

$$= \int_G \left(\int_G g(x)K_\psi(x^{-1}y)\,dx \right) \overline{f(y)}\,dy$$

$$= \langle g * K_\psi, f \rangle_{L_2},$$

where $\overline{K_\psi(y^{-1}x)} = K_\psi(x^{-1}y)$. Note that we can interchange the integral in the second line above by Fubini's theorem since

$$\int_G |g(x)| \left(\int_G |f(y)K_\psi(y^{-1}x)|\,dy \right) dx \leqslant \|g\|_{L_{r',m^{-1}}} \cdot \||f| * |K_\psi|\|_{L_{r,m}}$$

and $|f| * |K| \in L_{r,m}(G)$ by Young's inequality, see Proposition 1.3.9, with $q = 1$ and $p = r$, $f \in L_{1,m}(G)$ and $g = K \in L_{r,w}(G)$. Note further that Fubini's theorem shows that

$$\int |f(y)| \cdot (|g| * |K_\psi|)(y)\,dy < \infty.$$

Since this holds for any $f \in \mathcal{C}_c(G)$, we see $|g| * |K_\psi| < \infty$ almost everywhere, so that g and K_ψ are convolvable. By density of $\mathcal{C}_c(G)$ in $L_{r,m}(G)$ we get that $RC_K^* g = g * K_\psi$, so that $g * K_\psi \in L_{r',m^{-1}}(G)$. Hence the convolution operator acts continuously on $L_{r',m^{-1}}(G)$ and it coincides with RC_K^*.

To show the second claim, observe first that for any $f \in \mathcal{C}_c(G) \subset \mathcal{T}_w \subset L_{r',m^{-1}}(G)$, since $K_\psi \in \mathcal{T}_w$, both $|f| * |K_\psi|$ and $(|f| * |K_\psi|) * |K_\psi|$ exist, so that by (77d) of [18] the convolution is associative and

$$RC_K^2 f = (f * K_\psi) * K_\psi = f * (K_\psi * K_\psi) = f * K_\psi = RC_K f.$$

By density, and since RC_K is bounded on $L_{r,m}(G)$ by assumption, we get that $RC_K^2 = RC_K$ and hence Ran $RC_K \subseteq \mathcal{M}_{r,m}$. The other inclusion is trivial. $\quad\square$

As a consequence of the above result, we get the following corollary.

Corollary 3.2.8. *Denote by r' the dual exponent $1/r + 1/r' = 1$ and assume the right convolution operator RC_K is bounded on $L_{r,m}(G)$. The sesqui-linear pairing on $\mathrm{Co}(L_{r,m}) \times \mathrm{Co}(L_{r',m^{-1}})$ given by*

$$\langle T, T' \rangle_{\mathrm{Co}(L_{r,m})} = \langle V_{e,\psi} T, V_{e,\psi} T' \rangle_{L_2} \tag{3.2.13}$$

is such that the linear map

$$T' \mapsto \left(T \mapsto \overline{\langle T, T' \rangle_{\mathrm{Co}(L_{r,m})}} \right)$$

is an isomorphism of $\mathrm{Co}(L_{r',m^{-1}})$ onto the anti-linear dual of $\mathrm{Co}(L_{r,m})$.

Proof. We identify $\mathrm{Co}(L_{r,m})$ with $\mathcal{M}_{r,m}$ by the extended voice transform $V_{e,\psi}$, so that the pairing becomes

$$\langle f, g \rangle_{L_2} = \int_G f(x)\overline{g(x)}\,dx, \qquad f \in \mathcal{M}_{r,m},\ g \in \mathcal{M}_{r',m^{-1}}.$$

Since $g \in L_{r',m^{-1}}(G)$, clearly $f \mapsto \overline{\langle f, g \rangle_{L_2}}$ is a continuous anti-linear map, which we denote by Γ_g, on $\mathcal{M}_{r,m}$ whose norm is

$$
\begin{aligned}
\|\Gamma_g\| &= \sup\left\{ |\langle f, g \rangle_{L_2}| : f \in \mathcal{M}_{r,m},\ \|f\|_{L_{r,m}} \leqslant 1 \right\} \\
&\leqslant \sup\left\{ |\langle h, g \rangle_{L_2}| : h \in L_{r,m}(G),\ \|h\|_{L_{r,m}} \leqslant 1 \right\} \\
&= \|g\|_{L_{r',m^{-1}}}.
\end{aligned}
$$

Next, since $L_{r,m}(G)$ is the dual of $L_{r',m^{-1}}(G)$, there is $h \in L_{r,m}(G)$ with $\|h\|_{L_{r,m}} \leqslant 1$ such that $\|g\|_{L_{r',m^{-1}}} = \langle h, g \rangle_{L_2}$. Now, setting $c := \|RC_K\|_{L_{r,m} \to L_{r,m}}$ and $f = c^{-1} \cdot RC_K h$, we have $\|f\|_{L_{r,m}} \leqslant 1$ and

$$\langle f, g \rangle_{L_2} = c^{-1}\langle RC_K h, g \rangle_{L_2} = c^{-1}\langle h, RC_K g \rangle_{L_2} = c^{-1}\langle h, g \rangle_{L_2} = c^{-1}\|g\|_{L_{r',m^{-1}}}.$$

Hence, $c^{-1} \cdot \|g\|_{L_{r',m^{-1}}} \leqslant \|\Gamma_g\| \leqslant \|g\|_{L_{r',m^{-1}}}$, which implies that the map $g \to \Gamma_g$ is injective.

We finally prove that the map $g \mapsto \Gamma_g$ is surjective. Take Γ in the anti-linear dual of $\mathcal{M}_{r,m}$. Since $\mathcal{M}_{r,m}$ is a subspace of $L_{r,m}(G)$ there exists $g' \in L_{r',m^{-1}}(G)$ such that $\Gamma(f) = \overline{\langle f, g' \rangle_{L_2}}$ for all $f \in \mathcal{M}_{r,m}$. By setting $g = RC_K g' \in \mathcal{M}_{r',m^{-1}}$, as above

$$\Gamma(f) = \overline{\langle f, g' \rangle_{L_2}} = \overline{\langle RC_K f, g' \rangle_{L_2}} = \overline{\langle f, g \rangle_{L_2}} = \Gamma_g(f) \quad \text{for all } f \in \mathcal{M}_{r,m},$$

thus $\Gamma = \Gamma_g$. $\quad\square$

Now we can finally prove that in the following setting Assumption 3.2.4 is fulfilled.

Lemma 3.2.9. *Assume that Assumption 3.1.12 is fulfilled, i.e., $K_\psi \in L_{p,w}(G)$ for all $1 < p < \infty$, and that the right convolution operator RC_K is bounded on $L_{r,m}(G)$. Then the sets $\mathrm{span}\{\pi(x)\psi\}_{x\in G}$ and $\mathrm{span}\{\lambda(x)K_\psi\}_{x\in G}$ are dense in $\mathrm{Co}(L_{r,m})$ and $\mathcal{M}_{r,m}$, respectively. Thus, Assumption 3.2.4 is fulfilled.*

Proof. By the correspondence principle, it is enough to show the second claim. Let $\Gamma \in \mathcal{M}'_{r,m}$ be such that for all $x \in G$,

$$\Gamma(\lambda(x)K) = 0.$$

By the above corollary, there exists $g \in \mathcal{M}_{r',m^{-1}}$ such that $\Gamma(f) = \langle g, f\rangle_{L_2}$ for all $f \in \mathcal{M}_{r,m}$. In particular,

$$0 = \Gamma\big(\lambda(x)K_\psi\big) = \langle g, \lambda(x)K_\psi\rangle_{L_2} = g * K_\psi(x) = RC_K g(x)$$

for all $x \in G$, that is, $RC_K g = 0$. Since $g \in \mathcal{M}_{r',m^{-1}}$, this implies $g = 0$ and then $\Gamma = 0$. Since this holds for any $\Gamma \in \mathcal{M}'_{r,m}$ such that $\Gamma(\lambda(x)K_\psi) = 0$ for all $x \in G$, we see that $\mathrm{span}\{\lambda(x)K_\psi\}_{x\in G}$ is dense in $\mathcal{M}_{r,m}$. $\qquad\square$

By a weighted version of Young's inequality, see Proposition 1.3.9, we know that the $L_1(G)$-integrability of $K_\psi \cdot w$ implies that the (right) convolution operator RC_K is a bounded operator acting on $L_{p,m}(G)$ for all $1 < p < \infty$. However for general $K_\psi \in \mathcal{T}_w$ this question is unclear. As we will show in Section 3.2.3 there are kernels that act boundedly on all $L_p(G)$ — when G is the additive group $\mathbb{R}$ — without being integrable. This will be shown by exploiting specific properties of $\mathbb{R}$. But in Section 3.3 we also show that there exist kernels for a very similar setting that are contained in $\mathcal{T}_w$ but that do *not* give rise to bounded operators on $L_{p,m}(G)$.

3.2.2 Atomic Decompositions

In this subsection we are interested in finding atomic decompositions of coorbit spaces $\mathrm{Co}(L_{r,m})$, provided that Assumption 3.2.4 is fulfilled. The main result of this subsection will be stated in Theorem 3.2.17.

Let us start by introducing some notation. First, for each $n \in \mathbb{N}$, we choose a countable subset $Y_n = \{x_{j,n}\}_{j\in I_n} \subset G$ such that

$$Y_n \subset Y_{n+1}, \tag{3.2.14}$$

$$\overline{\bigcup_{n\in\mathbb{N}} Y_n} = G. \tag{3.2.15}$$

Moreover, for every $n \in \mathbb{N}$, we assume that there exists a compact neighborhood Q_n of the identity $e \in G$, such that Y_n is Q_n-*dense* in G, i.c.,

$$G = \bigcup_{j\in I_n} x_{j,n} Q_n. \tag{3.2.16}$$

Additionally we assume each Y_n to be uniformly relatively Q_n-separated, i.e., there exists an integer $\mathcal{I}$, independent of n, and subsets $Z_{n,i} \subset Y_n$, $1 \leqslant i \leqslant \mathcal{I}$, such that

$$Y_n = \bigcup_{i=1}^{\mathcal{I}} Z_{n,i} \tag{3.2.17}$$

and for all $x, y \in Z_{n,i}$, $1 \leqslant i \leqslant \mathcal{I}$, it holds that $xQ_n \cap yQ_n \neq \varnothing$ if and only if $x = y$.

By $\Phi_n = \{\varphi_{n,x}\}_{x \in Y_n}$ we denote a *partition of unity* subordinate to the Q_n-dense set Y_n, i.e.,

$$0 \leqslant \varphi_{n,x} \leqslant 1, \tag{3.2.18}$$

$$\sum_{x \in Y_n} \varphi_{n,x} \equiv 1, \tag{3.2.19}$$

$$\mathrm{supp}(\varphi_{n,x}) \subseteq x Q_n. \tag{3.2.20}$$

We also assume that the family $\Phi_n = \{\varphi_{n,x}\}_{x \in Y_n}$ is linearly independent as a.e. defined functions, i.e., for any finite subset $X \subset Y_n$ and $(\alpha_x)_{x \in X} \in \mathbb{C}^X$, the condition

$$\sum_{x \in X} \alpha_x \varphi_{n,x}(y) = 0$$

for almost all $y \in G$ implies that $\alpha_x = 0$ for all $x \in X$.

We now denote with X_n a *finite* subset of Y_n, such that

$$X_n \subset X_{n+1}, \tag{3.2.21}$$

$$\overline{\bigcup_{n \in \mathbb{N}} X_n} = G. \tag{3.2.22}$$

Therefore, for every $n \in \mathbb{N}$, the finite set of functions $\{\varphi_{n,x}\}_{x \in X_n}$ is similar to a partition of unity subordinate to the family $(xQ_n)_{x \in X_n}$.

For each $n \in \mathbb{N}$ and $1 < r < \infty$, set

$$T_n \colon L_{r,m}(G) \to \mathcal{M}_{r,m}, \qquad T_n F := \sum_{x \in X_n} \langle F, \varphi_{n,x} \rangle_{L_2} \lambda(x) K_\psi. \tag{3.2.23}$$

We first observe that this operator is well-defined. Since the sum is finite, we only have to verify that each term of the sum is a well-defined element of $\mathcal{M}_{r,m}$. It is easy to verify that the reproducing identity holds for $\lambda(x)K_\psi$, since it holds for K_ψ. Moreover we have $\lambda(x)K_\psi \in L_{r,m}(G)$ by Assumption 3.1.12 and by translation invariance of the spaces $L_{r,m}(G)$; thus, $\lambda(x)K_\psi \in \mathcal{M}_{r,m}$. Finally, the pairing

$$\langle F, \varphi_{n,x} \rangle_{L_2} = \int_G F(y) \varphi_{n,x}(y)\, dy$$

is well-defined for all $x \in X_n$, since $\varphi_{n,x}$ is bounded with compact support, so that $\varphi_{n,x}$ is an element of $L_{r',m^{-1}}(G)$.

Let us define the set $V_n = \mathrm{Ran}\, T_n$, which is a finite dimensional subspace of $\mathcal{M}_{r,m}$, as well as $\widetilde{V}_n = V_{e,\psi}^{-1}(V_n)$, which is a finite dimensional subspace of $\mathrm{Co}(L_{r,m})$ by the correspondence principle, see Proposition 3.1.19. We show some basic properties of V_n in the following lemma.

Lemma 3.2.10. *Assume that Assumption 3.2.4 is fulfilled. Then, the following holds for all $n \in \mathbb{N}$:*

$$V_n = \mathrm{span}\left\{\lambda(x)K_\psi\right\}_{x \in X_n}, \tag{3.2.24}$$

$$V_n \subset V_{n+1}, \tag{3.2.25}$$

$$\overline{\bigcup_{n \geqslant 1} V_n} = \mathcal{M}_{r,m}. \tag{3.2.26}$$

Proof. We start by showing (3.2.24). By the construction we made above, we have the inclusion $V_n \subseteq \mathrm{span}\left\{\lambda(x)K_\psi\right\}_{x \in X_n}$.

We first observe that the map

$$F \mapsto \left(\langle F, \varphi_{n,x} \rangle_{L_2(G)} \right)_{x \in X_n}$$

is surjective from $L_{r,m}(G)$ to $\mathbb{C}^{X_n}$. Indeed, if this was not true, there would be a nonzero family $(\alpha_x)_{x \in X_n} \in \mathbb{C}^{X_n}$ satisfying $\sum_{x \in X_n} \alpha_x \langle F, \varphi_{n,x} \rangle_{L_2} = 0$ for all $F \in L_{r,m}(G)$. This means $\sum_{x \in X_n} \alpha_x \varphi_{n,x} = 0$ in $L_{r',m^{-1}}(G)$ and, hence, almost everywhere as well. By the linear independence of Φ_n this implies $\alpha_x = 0$ for all $x \in X_n$ which is a contradiction. It follows that $\mathrm{span}\,\{\lambda(x)K_\psi\}_{x \in X_n} \subseteq V_n$ and (3.2.24) holds true.

Equation (3.2.25) is an easy consequence of (3.2.21) and (3.2.24).

It remains to show (3.2.26). Since the sequence $(V_n)_{n \in \mathbb{N}}$ is an increasing family of subspaces, and since $V_n \subset \mathcal{M}_{r,m}$ for all $n \in \mathbb{N}$, the set $\overline{\bigcup_{n \geq 1} V_n}$ is a subspace of the closed space $\mathcal{M}_{r,m}$. Hence, by the Hahn-Banach theorem, condition (3.2.26) is equivalent to the following condition: If $\Gamma \in \mathcal{M}'_{r,m}$ satisfies

$$\langle \Gamma, F \rangle_{\mathcal{M}'_{r,m} \times \mathcal{M}_{r,m}} = 0, \quad \text{for all } F \in V_n, \ n \in \mathbb{N},$$

then $\Gamma = 0$ in $\mathcal{M}'_{r,m}$. By Lemma 3.2.2 we can write $\langle \Gamma, F \rangle_{\mathcal{M}'_{r,m} \times \mathcal{M}_{r,m}} = \langle g, F \rangle_{L_2}$ for all $F \in \mathcal{M}_{r,m}$, for a suitable $g \in L_{r',m^{-1}}(G)$. Since $\lambda(x)K_\psi \in \mathcal{M}_{r,m}$, $x \in G$, for every $f \in L_{r',m^{-1}}(G)$ with $f - g \in \mathcal{M}^\perp_{r,m}$, it holds for all $x \in G$,

$$(g * K_\psi)(x) = \langle g, \lambda(x)K_\psi \rangle_{L_2} = \langle \Gamma, \lambda(x)K_\psi \rangle_{\mathcal{M}'_{r,m} \times \mathcal{M}_{r,m}}.$$

Now, with $F = T_n f$ for some $f \in L_{r,m}(G)$, we obtain

$$
\begin{aligned}
0 &= \langle \Gamma, T_n f \rangle_{\mathcal{M}'_{r,m} \times \mathcal{M}_{r,m}} \\
&= \sum_{x \in X_n} \langle \varphi_{n,x}, f \rangle_{L_2} \cdot \langle \Gamma, \lambda(x)K_\psi \rangle_{\mathcal{M}'_{r,m} \times \mathcal{M}_{r,m}} \\
&= \sum_{x \in X_n} \langle \varphi_{n,x}, f \rangle_{L_2} \cdot (g * K_\psi)(x) \\
&= \langle \sum_{x \in X_n} (g * K_\psi)(x) \varphi_{n,x}, f \rangle_{L_2}.
\end{aligned}
$$

Since this holds for any $f \in L_{r,m}(G)$, we get $\sum_{x \in X_n} (g * K_\psi)(x) \varphi_{n,x} = 0$ in $L_{r',m^{-1}}(G)$ for all $n \in \mathbb{N}$. Because the finite family $\{\varphi_{n,x}\}_{x \in X_n}$ is linearly independent as elements of $L_{r',m^{-1}}(G)$, we have $(g * K_\psi)(x) = 0$ for all $x \in X_n$ and $n \in \mathbb{N}$. Therefore, by (3.2.22), the function $g * K_\psi$ vanishes on a dense subset of G. But since we have $g * K_\psi(x) = \langle g, \lambda(x)K_\psi \rangle_{L_2}$ with $g \in L_{r',m^{-1}}(G)$, and since the map $G \to L_{r,m}(G), x \mapsto \lambda(x)K_\psi$ is continuous, we see that $g * K_\psi : G \to \mathbb{C}$ is a continuous functions, so that we get $g * K_\psi \equiv 0$, i.e., $\langle \Gamma, \lambda(x)K_\psi \rangle_{\mathcal{M}'_{r,m} \times \mathcal{M}_{r,m}} = 0$ for all $x \in G$.

By Assumption 3.2.4, i.e., $\mathrm{span}\{\lambda(x)K_\psi\}_{x \in G}$ is dense in $\mathcal{M}_{r,m}$, this implies $\Gamma = 0$ as an element of $\mathcal{M}'_{r,m}$, which proves (3.2.26). $\qquad \square$

Remark 3.2.11. By the correspondence principle, analogous results to (3.2.24), (3.2.25) and (3.2.26) hold true for $\widetilde{V}_n = V_{e,\psi}^{-1}(V_n)$. This can be seen as follows: Since it holds that $V_{e,\psi}\pi(x)\psi = \lambda(x)K_\psi$ for all $x \in X_n$, by (3.2.24) we obtain

$$\widetilde{V}_n = \mathrm{span}\,\{\pi(x)\psi\}_{x \in X_n}. \tag{3.2.27}$$

Hence, the nesting property $\widetilde{V}_n \subset \widetilde{V}_{n+1}$ analogous to (3.2.25) is straightforward. By the correspondence principle, it follows from (3.2.25) that

$$\overline{\bigcup_{n \in \mathbb{N}} \widetilde{V}_n} = \mathrm{Co}(L_{r,m}). \tag{3.2.28}$$

The spaces V_n will serve as substitutes for $\mathcal{M}_{r,m}$, in the sense that we will discretize functions in V_n instead of the whole space $\mathcal{M}_{r,m}$. By virtue of Lemma 3.2.10 this is reasonable.

To do this, we are interested in projections from $\mathcal{M}_{r,m}$ onto V_n and their properties. To this end, let $\pi_n \colon \mathcal{M}_{r,m} \to V_n$ be the metric projection defined by

$$\pi_n(F) = \operatorname{argmin}_{g \in V_n} \|F - g\|_{\mathcal{M}_{r,m}}. \tag{3.2.29}$$

Since $\mathcal{M}_{r,m}$ is a closed subspace of $L_{r,m}(G)$ with $1 < r < \infty$, the space $\mathcal{M}_{r,m}$ is a uniformly convex Banach space and every V_n is convex and closed; therefore $\pi_n(F)$ is a well-defined and unique function, see [64, Proposition 3.1]. Similarly, we define the projection $\widetilde{\pi}_n \colon \operatorname{Co}(L_{r,m}) \to \widetilde{V}_n$ by setting $\widetilde{\pi}_n T = V_{e,\psi}^{-1} \pi_n V_{e,\psi} T$, $T \in \operatorname{Co}(L_{r,m})$.

The following lemma yields a first norm estimate for the metric projection π_n.

Lemma 3.2.12. *Assume that Assumption 3.2.4 is fulfilled. Given $\varepsilon > 0$ and $F \in \mathcal{M}_{r,m}$, there exists $n^* = n^*_{F,\varepsilon} \in \mathbb{N}$ such that for all $n \geqslant n^*$ we have*

$$\|F - \pi_n(F)\|_{\mathcal{M}_{r,m}} \leqslant \varepsilon, \tag{3.2.30}$$

$$\|\pi_n(F)\|_{\mathcal{M}_{r,m}} \leqslant (1 + \varepsilon)\|F\|_{\mathcal{M}_{r,m}}. \tag{3.2.31}$$

Proof. If $F = 0$ the claim is clear since $0 \in V_n$ so that $\pi_n(F) = 0$. Hence, we can assume that $F \neq 0$. Let $\delta := \min\left\{1, \|F\|_{\mathcal{M}_{r,m}}\right\} \cdot \varepsilon > 0$. By (3.2.26) there exists $n^* \geqslant 1$ and $g \in V_{n^*}$ such that $\|F - g\|_{\mathcal{M}_{r,m}} \leqslant \delta$. For all $n \geqslant n^*$, by (3.2.25) we have $g \in V_n$ and, by definition of the metric projection,

$$\|F - \pi_n(F)\|_{\mathcal{M}_{r,m}} \leqslant \|F - g\|_{\mathcal{M}_{r,m}} \leqslant \delta \leqslant \varepsilon.$$

The triangle inequality gives

$$\|\pi_n(F)\|_{\mathcal{M}_{r,m}} \leqslant \|F - \pi_n(F)\|_{\mathcal{M}_{r,m}} + \|F\|_{\mathcal{M}_{r,m}} \leqslant \delta + \|F\|_{\mathcal{M}_{r,m}} \leqslant (1 + \varepsilon)\|F\|_{\mathcal{M}_{r,m}},$$

which concludes the proof.   □

As a direct consequence of the lemma above, for every $\varepsilon > 0$ and $F \in \mathcal{M}_{r,m}$ there exists n^* such that for all $n \geqslant n^*$ there is a $f \in V_n$ with $\|F - f\|_{L_{r,m}} \leqslant \varepsilon$ and $\|f\|_{L_{r,m}} \leqslant (1 + \varepsilon)\|F\|_{L_{r,m}}$, i.e., arbitrarily close to F.

Next, in the following auxiliary result, we establish an upper bound for certain coefficients related to functions $F \in \mathcal{M}_{r,m}$. These coefficients will be used for the atomic decomposition in this subsection.

Proposition 3.2.13. *For any $F \in L_{r,m}(G)$ and $n \in \mathbb{N}$, let the coefficients $c_{n,x} \in \mathbb{C}$, $x \in X_n$, be defined via*

$$c_{n,x} := \langle F, \varphi_{n,x} \rangle_{L_2} = \int_G F(y)\varphi_{n,x}(y)\,dy.$$

Then the inequality

$$\left(\sum_{x \in X_n} |c_{n,x}|^r m(x)^r \right)^{1/r} \leqslant |Q_n|^{1/r'} \cdot \sup_{q \in Q_n} w(q) \cdot \|F\|_{L_{r,m}} \tag{3.2.32}$$

holds true, where $|Q_n|$ denotes the Haar measure of the set Q_n and r' denotes the dual exponent of r.

Proof. We first note that, since $\varphi_{n,x}$ is compactly supported and bounded, the coefficient $c_{n,x}$ is well-defined for every $x \in X_n$ and $n \in \mathbb{N}$.

Next, we observe that if $\varphi_{n,x}(y) \neq 0$ for fixed $x \in X_n$ and $n \in \mathbb{N}$, then $y = xq_n$ for some $q_n \in Q_n$, and hence $m(x) = m(xq_nq_n^{-1}) \leqslant m(xq_n) \cdot w(q_n^{-1}) \leqslant m(y) \cdot \sup_{q \in Q_n} w(q)$. This shows

$$m(x) \cdot |c_{n,x}| \leqslant m(x) \cdot \int_G |F(y)| \cdot \varphi_{n,x}(y) \, dy$$

$$\leqslant \sup_{q \in Q_n} w(q) \cdot \int_G |(mF)(y)| \cdot \varphi_{n,x}(y) \, dy. \tag{3.2.33}$$

We will now further estimate the integral on the right-hand side of the equation above, setting $F_0 := m \cdot F$ for brevity.

To this end, for $x \in X_n$, we define the measure $d\mu_x$ on G by setting

$$d\mu_x(y) = \frac{\varphi_{n,x}(y)}{\|\varphi_{n,x}\|_{L_1}} \, dy$$

and readily observe that $\int_G 1 \, d\mu_x = 1$. Thus, by Jensen's inequality, see [36, Theorem 10.2.6], we obtain

$$\left(\int_G |F_0(y)| \frac{\varphi_{n,x}(y)}{\|\varphi_{n,x}\|_{L_1}} \, dy \right)^r = \left(\int_G |F_0(y)| \, d\mu_x(y) \right)^r$$

$$\leqslant \int_G |F_0(y)|^r \, d\mu_x(y)$$

$$= \int_G |F_0(y)|^r \frac{\varphi_{n,x}(y)}{\|\varphi_{n,x}\|_{L_1}} \, dy.$$

By the properties of Φ_n, see (3.2.18), (3.2.19) and (3.2.20), we obtain

$$\|\varphi_{n,x}\|_{L_1} = \int_G \varphi_{n,x}(y) \, dy \leqslant \int_{xQ_n} 1 \, dy = \int_{Q_n} 1 \, dy = |Q_n|.$$

Recalling (3.2.33), we thus see

$$\sum_{x \in X_n} (m(x) \cdot |c_{n,x}|)^r \leqslant \sup_{q \in Q_n} w(q)^r \cdot \sum_{x \in X_n} \|\varphi_{n,x}\|_{L_1}^r \left(\int_G |F_0(y)| \frac{\varphi_{n,x}(y)}{\|\varphi_{n,x}\|_{L_1}} \, dy \right)^r$$

$$\leqslant \sup_{q \in Q_n} w(q)^r \cdot \sum_{x \in X_n} \|\varphi_{n,x}\|_{L_1}^r \int_G |F_0(y)|^r \frac{\varphi_{n,x}(y)}{\|\varphi_{n,x}\|_{L_1}} \, dy$$

$$\leqslant \sup_{q \in Q_n} w(q)^r \cdot \sup_{x \in X_n} \|\varphi_{n,x}\|_{L_1}^{r-1} \cdot \sum_{x \in X_n} \int_G |F_0(y)|^r \varphi_{n,x}(y) \, dy$$

$$\leqslant \sup_{q \in Q_n} w(q)^r \cdot |Q_n|^{r-1} \cdot \|F\|_{L_{r,m}}^r,$$

which concludes the proof. $\square$

Using this estimate, we are able to present a first atomic decomposition of functions $F \in V_n$, $n \in \mathbb{N}$, as well as an estimate for the norm of the coefficients involved.

Lemma 3.2.14. *Given $n \in \mathbb{N}$, for all $F \in V_n$ the following atomic decomposition holds true:*

$$F = \sum_{x \in X_n} c(F)_{n,x} \lambda(x) K_\psi, \tag{3.2.34}$$

where the coefficients $c(F)_{n,x}$ are of the form

$$c(F)_{n,x} = \langle S_n F, \varphi_{n,x} \rangle_{L_2}, \tag{3.2.35}$$

where S_n denotes any linear right inverse of $T_n : L_{r,m}(G) \to V_n$. In particular, the coefficients depend linearly on F and they satisfy

$$\left(\sum_{x \in X_n} |c(F)_{n,x}|^r m(x)^r \right)^{1/r} \leqslant C_n \cdot \|F\|_{\mathcal{M}_{r,m}}, \tag{3.2.36}$$

with $C_n = \|S_n\| \cdot |Q_n|^{1/r'} \cdot \sup_{q \in Q_n} w(q)$.

Proof. We first observe that the operator T_n admits a bounded right inverse $S_n : V_n \to L_{r,m}(G)$. Indeed, by [8, Theorem 2.12] the existence of a bounded right inverse is equivalent to the existence of a topological supplement of the kernel of T_n. However, since the spaces V_n are finite dimensional, such a topological supplement exists, see [8, Example 2.4.2].

In the remainder of the proof, we denote by S_n an arbitrary linear right inverse of T_n. Thus, for all $F \in V_n$ we have the decomposition

$$F = T_n S_n F = \sum_{x \in X_n} \langle S_n F, \varphi_{n,x} \rangle_{L_2} \lambda(x) K_\psi,$$

so that (3.2.34) holds true if we define the coefficients $c(F)_{n,x}$ as in (3.2.35). With this notation the coefficients depend linearly on F. By applying (3.2.32) we obtain the estimate

$$\left(\sum_{x \in X_n} |c(F)_{n,x}|^r m(x)^r \right)^{1/r} \leqslant |Q_n|^{1/r'} \cdot \sup_{q \in Q_n} w(q) \cdot \|S_n F\|_{\mathcal{M}_{r,m}}$$

$$\leqslant |Q_n|^{1/r'} \cdot \sup_{q \in Q_n} w(q) \cdot \|S_n\| \cdot \|F\|_{\mathcal{M}_{r,m}},$$

where $\|S_n\|$ is the operator norm of S_n as an operator from V_n into $L_{r,m}(G)$. This proves (3.2.36). $\square$

Remark 3.2.15. Note that if the sequence $(|Q_n|^{1/r'} \cdot \sup_{q \in Q_n} w(q) \cdot \|S_n\|)_{n \in \mathbb{N}}$ is bounded, then the constant C_n in (3.2.36) can be estimated from above independently of n. Naturally the question arises under which conditions this really is the case. To answer this question, it is necessary to determine the asymptotic behaviour of the operator norm of S_n. As we will show in Section 3.2.3 this task is already non-trivial for a very simple setting. Still, in Lemma 3.2.21 at the end of this subsection we give a partial answer, as we present a technique to characterize the operator-norm of S_n in a different manner.

Before we can prove the main result of this subsection, we need the following auxiliary lemma. We recall that the integer $\mathcal{I}$ is defined through assumption (3.2.17).

Lemma 3.2.16. *Let $1 \leqslant p \leqslant \infty$ and $(d_x)_{x \in Y_n} \in \ell_{p,m}(Y_n)$ for some $n \in \mathbb{N}$, then*

$$\left\| \sum_{x \in Y_n} |d_x| \chi_{x Q_n} \right\|_{L_{p,m}} \leqslant \mathcal{I}^{1 - \frac{1}{p}} \cdot \sup_{q \in Q_n} w(q) \cdot |Q_n|^{\frac{1}{p}} \cdot \|(d_x)_{x \in Y_n}\|_{\ell_{p,m}} \tag{3.2.37}$$

with the convention $\frac{1}{\infty} := 0$.

Proof. Let $1 \leqslant p < \infty$ and $(d_x)_{x \in Y_n} \in \ell_{p,m}(Y_n)$, then we first note that for all $x \in G$ we have

$$\int_{xQ_n} m(y)^p \, dy = \int_{Q_n} m(xy)^p \, dy \leqslant m(x)^p \int_{Q_n} w(y)^p \, dy$$
$$\leqslant \sup_{q \in Q_n} w(q)^p \cdot |Q_n| \cdot m(x)^p.$$

With this at hand and since Y_n is relatively Q_n-separated, as stated in (3.2.17), we derive

$$\left\| \sum_{x \in Y_n} |d_x| \chi_{xQ_n} \right\|_{L_{p,m}} \leqslant \sum_{i=1}^{\mathcal{I}} \left\| \sum_{x \in Z_{n,i}} |d_x| \chi_{xQ_n} \right\|_{L_{p,m}}$$

$$= \sum_{i=1}^{\mathcal{I}} \left(\sum_{x \in Z_{n,i}} |d_x|^p \int_{xQ_n} m(y)^p \, dy \right)^{1/p}$$

$$\leqslant \sum_{i=1}^{\mathcal{I}} \left(\sum_{x \in Z_{n,i}} |d_x|^p m(x)^p \right)^{1/p} \cdot \sup_{q \in Q_n} w(q) \cdot |Q_n|^{\frac{1}{p}}$$

$$\leqslant \mathcal{I}^{1-\frac{1}{p}} \cdot \sup_{q \in Q_n} w(q) \cdot |Q_n|^{\frac{1}{p}} \cdot \|(d_x)_{x \in Y_n}\|_{\ell_{p,m}}.$$

It remains to prove the case $p = \infty$. Similarly as above, we see that

$$\left\| \sum_{x \in Y_n} |d_x| \chi_{xQ_n} \right\|_{L_{\infty,m}} \leqslant \sum_{i=1}^{\mathcal{I}} \left\| \sum_{x \in Z_{n,i}} |d_x| \chi_{xQ_n} \right\|_{L_{\infty,m}}$$

$$= \sum_{i=1}^{\mathcal{I}} \sup_{x \in Z_{n,i}} \left(|d_x| \cdot \sup_{y \in xQ_n} m(y) \right)$$

$$\leqslant \mathcal{I} \cdot \left(\sup_{x \in Y_n} |d_x| \cdot m(x) \right) \cdot \sup_{y \in Q_n} w(y)$$

$$= \mathcal{I} \cdot \sup_{q \in Q_n} w(q) \cdot \|(d_x)_{x \in Y_n}\|_{\ell_{\infty,m}},$$

which concludes the proof. $\qquad\square$

Now we are finally ready to state and prove our main result.

Theorem 3.2.17. *We assume that K_ψ satisfies Assumption 3.2.4, i.e., $K_\psi \in L_{t,m}(G)$ for all $1 < t < \infty$ and span$\{\lambda(x)K_\psi\}$ is dense in $\mathcal{M}_{r,m}$. Assume further that there exists $p < r$ such that*

$$K_\psi \in L_{p,w\Delta^{-1/p}}(G), \tag{3.2.38}$$

$$\operatorname{osc}_{Q_n} K_\psi \in L_{p,w}(G) \cap L_{p,w\Delta^{-1/p}}(G), \tag{3.2.39}$$

for all $n \in \mathbb{N}$.

(i) Fix $\varepsilon > 0$; then for any $T \in \operatorname{Co}(L_{r,m})$ there exists $n^* = n^*_{T,\varepsilon} \in \mathbb{N}$ such that for all $n \geqslant n^*$

$$\left\| T - \sum_{x \in X_n} c(T)_{n,x} \pi(x)\psi \right\|_{\operatorname{Co}(L_{r,m})} \leqslant \varepsilon, \tag{3.2.40}$$

where the family $(c(T)_{n,x})_{x \in X_n}$ *satisfies*

$$\|(c(T)_{n,x})_{x \in X_n}\|_{\ell_{r,m}} \leqslant C_n(1+\varepsilon)\|T\|_{\mathrm{Co}(L_{r,m})}, \tag{3.2.41}$$

with $C_n = |Q_n|^{1/r'} \cdot \sup_{q \in Q_n} w(q) \cdot \|S_n\|$, *where* S_n *denotes any linear right inverse to the operator* $T_n \colon L_{r,m}(G) \to V_n$ *defined in* (3.2.23).

(ii) Let $n \in \mathbb{N}$ *and* $d = (d_x)_{x \in Y_n} \in \ell_{q,m}(Y_n)$. *Then* $T = \sum_{x \in Y_n} d_x \pi(x)\psi$ *is in* $\mathrm{Co}(L_{r,m})$. *Furthermore the estimate*

$$\|T\|_{\mathrm{Co}(L_{r,m})} \leqslant D_n \|(d_x)_{x \in Y_n}\|_{\ell_{q,m}} \tag{3.2.42}$$

holds, where $1/q + 1/p = 1 + 1/r$, *and*

$$D_n := |Q_n|^{\frac{1}{q}-1} \cdot \mathcal{I}^{1-\frac{1}{q}} \cdot \sup_{q \in Q_n} w(q) \cdot \theta_n \tag{3.2.43}$$

with $\theta_n := \max \left\{ \|\mathrm{osc}_{Q_n} K_\psi + |K_\psi| \, \|_{L_{p,w}}, \, \|\mathrm{osc}_{Q_n} K_\psi + |K_\psi| \, \|_{L_{p,w\Delta^{-1/p}}} \right\}$.

Proof. To prove (i), choose $n^* = n^*_{F,\varepsilon}$ as in Lemma 3.2.12 with $F = V_{e,\psi}T \in \mathcal{M}_{r,m}$. By applying (3.2.34) and (3.2.35) to $\pi_n(F) \in V_n$ we obtain the atomic decomposition

$$\widetilde{\pi}_n(T) = V_{e,\psi}^{-1}\pi_n(F)$$

$$= V_{e,\psi}^{-1}\left(\sum_{x \in X_n} \langle S_n \pi_n(F), \varphi_{n,x}\rangle_{L_2} \cdot \lambda(x)K_\psi \right)$$

$$= \sum_{x \in X_n} \langle S_n \pi_n V_{e,\psi}(T), \varphi_{n,x}\rangle_{L_2} \cdot V_{e,\psi}^{-1}\lambda(x)K_\psi$$

$$= \sum_{x \in X_n} c(T)_{n,x}\pi(x)\psi,$$

where $c(T)_{n,x} = \langle S_n \pi_n V_{e,\psi}(T), \varphi_{n,x}\rangle_{L_2}$. Using (3.2.30) and the correspondence principle we derive

$$\left\| T - \sum_{x \in X_n} c(T)_{n,x}\pi(x)\psi \right\|_{\mathrm{Co}(L_{r,m})} = \left\| T - \widetilde{\pi}_n(T) \right\|_{\mathrm{Co}(L_{r,m})}$$

$$= \left\| V_{e,\psi}T - V_{e,\psi}\widetilde{\pi}_n(T) \right\|_{\mathcal{M}_{r,m}}$$

$$= \left\| F - \pi_n(F) \right\|_{\mathcal{M}_{r,m}} \leqslant \varepsilon.$$

Now (3.2.36) and (3.2.31) yield the estimate

$$\left(\sum_{x \in X_n} |c(T)_{n,x}|^r m(x)^r \right)^{\frac{1}{r}} \leqslant C_n \|\pi_n(V_{e,\psi}T)\|_{\mathcal{M}_{r,m}} \leqslant C_n(1+\varepsilon)\|V_{e,\psi}T\|_{\mathcal{M}_{r,m}}$$

$$= C_n(1+\varepsilon)\|T\|_{\mathrm{Co}(L_{r,m})}$$

for any $n \geqslant n^*$.

It remains to prove (ii). In [20, Chap. 3, p. 100] the following pointwise estimate for $y \in G$ has been established:

$$\left| \sum_{x \in Y_n} d_x \lambda(x)K_\psi(y) \right| \leqslant \left(\sum_{x \in Y_n} |d_x|\frac{\chi_{xQ_n}}{|Q_n|} \right) * (\mathrm{osc}_{Q_n} K_\psi + |K_\psi|)(y). \tag{3.2.44}$$

Let now $q > 1$ such that $1/q + 1/p = 1 + 1/r$. By using Young's inequality, see Proposition 1.3.11, and Lemma 3.2.16, we obtain

$$\left\| \sum_{x \in Y_n} d_x \pi(x) \psi \right\|_{\mathrm{Co}(L_{r,m})}$$

$$= \left\| \sum_{x \in Y_n} d_x \lambda(x) K_\psi \right\|_{L_{r,m}}$$

$$\leqslant \left\| \left(\sum_{x \in Y_n} |d_x| \chi_{xQ_n} \right) * (\mathrm{osc}_{Q_n} K_\psi + |K_\psi|) \right\|_{L_{r,m}} \cdot |Q_n|^{-1}$$

$$\leqslant \left\| \sum_{x \in Y_n} |d_x| \chi_{xQ_n} \right\|_{L_{q,m}} \cdot \max \left\{ \|\mathrm{osc}_{Q_n} K_\psi + |K_\psi|\|_{L_{p,w}}, \|\mathrm{osc}_{Q_n} K_\psi + |K_\psi|\|_{L_{p,w\Delta^{-1/p}}} \right\} \cdot |Q_n|^{-1}$$

$$\leqslant |Q_n|^{\frac{1}{q}-1} \cdot \mathcal{I}^{1-\frac{1}{q}} \cdot \sup_{q \in Q_n} w(q) \cdot \|(d_x)\|_{\ell_{q,m}}$$

$$\cdot \max \left\{ \|\mathrm{osc}_{Q_n} K_\psi + |K_\psi|\|_{L_{p,w}}, \|\mathrm{osc}_{Q_n} K_\psi + |K_\psi|\|_{L_{p,w\Delta^{-1/p}}} \right\}.$$

By the assumption (3.2.39) the expression on the right-hand side is finite. $\qquad\square$

Remark 3.2.18. (i) The coefficients $c(T)_{n,x}$, $x \in X_n$, in Theorem 3.2.17 (i) depend linearly on T if and only if the projection π_n from (3.2.29) is linear.

(ii) The proof of Theorem 3.2.17 (ii) shows, that the second part of Assumption 3.2.4, i.e. the density of span$\{\lambda(x)K_\psi\}_{x \in X}$ in $\mathcal{M}_{r,m}$, is not used. The reconstruction of a function in $\mathrm{Co}(L_{r,m})$ therefore does not require the assumptions developed in Subsection 3.2.1.

The following proposition presents a slight variation of Theorem 3.2.17.

Proposition 3.2.19. *Under the same assumptions as in Theorem 3.2.17 the following holds true: Fix $\varepsilon > 0$ and $T \in \mathrm{Co}(L_{r,m})$; then there exists $n^* = n^*_{T,\varepsilon} \in \mathbb{N}$ such that for all $n \geqslant n^*$*

$$\frac{1}{\tau_n(1+\varepsilon)} \cdot \|(c(T)_{n,x})_{x \in X_n}\|_{\ell_{q,m}} \leqslant \|T\|_{\mathrm{Co}(L_{r,m})} \tag{3.2.45}$$

and

$$\|T\|_{\mathrm{Co}(L_{r,m})} \leqslant \varepsilon + D_n \cdot \|(c(T)_{n,x})_{x \in X_n}\|_{\ell_{q,m}}, \tag{3.2.46}$$

with D_n as in (3.2.43) and $\tau_n := C_n \cdot |X_n|^{\frac{1}{q}-\frac{1}{r}}$, where $|X_n|$ denotes the cardinality of X_n and $1/q + 1/p = 1 + 1/r$.

Proof. Throughout this proof we use the same notations as in Theorem 3.2.17. We first note that for any finite sequence $(d_x)_{x \in X_n}$, $n \in \mathbb{N}$, by Hölder's inequality, it holds that

$$\|(d_x)_{x \in X_n}\|_{\ell_{q,m}} \leqslant \|(d_x)_{x \in X_n}\|_{\ell_{r,m}} \cdot \|1_{X_n}\|_{\ell_{\frac{rq}{r-q}}},$$

where 1_{X_n} is a sequence of ones only. Furthermore we have

$$\|1_{X_n}\|_{\ell_{\frac{rq}{r-q}}} = |X_n|^{\frac{1}{q}-\frac{1}{r}}.$$

With $\tau_n := C_n \cdot |X_n|^{\frac{1}{q}-\frac{1}{r}}$ we then obtain from Theorem 3.2.17 (i) the estimate

$$\frac{1}{\tau_n(1+\varepsilon)} \|(c(T)_{n,x})_{x \in X_n}\|_{\ell_{q,m}} \leqslant \frac{1}{C_n(1+\varepsilon)} \|(c(T)_{n,x})_{x \in X_n}\|_{\ell_{r,m}} \leqslant \|T\|_{\mathrm{Co}(L_{r,m})},$$

which proves (3.2.45).

It remains to show the second inequality (3.2.46). For this we note that the sequence $(c(T)_{n,x})_{x \in X_n}$ can be understood as a sequence over the index set Y_n with only finitely many non-zero entries. Therefore, by (3.2.30) and Theorem 3.2.17 (ii), this yields

$$\|T\|_{\mathrm{Co}(L_{r,m})} \leqslant \varepsilon + \left\| \sum_{x \in X_n} c(T)_{n,x} \pi(x)\psi \right\|_{\mathrm{Co}(L_{r,m})} \leqslant \varepsilon + |Q_n|^{\frac{1}{q}-1} \cdot \theta_n \cdot \|(c(T)_{n,x})_{x \in X_n}\|_{\ell_{q,m}},$$

which concludes the proof. $\qquad\qquad\qquad\qquad\qquad\qquad\qquad\qquad\qquad\qquad\qquad\qquad\qquad$ $\square$

Before we close this subsection, let us discuss the results of Theorem 3.2.17. The conditions posed in (3.2.38) and (3.2.39) are identical to the conditions (2.2.3) on the kernel in the classic coorbit theory, when substituting $r = 1$. In part (i) of Theorem 3.2.17 we are, however, only able to obtain a bounded coefficient operator for functions in finite dimensional subspaces $\widetilde{V}_n$, $n \in \mathbb{N}$, not the entire coorbit space $\mathrm{Co}(L_{r,m})$. While this still works up to an arbitrarily small $\varepsilon > 0$, the cost is the constant $C_n = |Q_n|^{1/r'} \cdot \sup_{q \in Q_n} w(q) \cdot \|S_n\|$.

Let us examine this constant in detail. First, $\sup_{q \in Q_n} w(q)$ will be a number close to 1 for large n and therefore be negligible. Hence, for the sequence $(C_n)_{n \in \mathbb{N}}$ to be bounded it is necessary that $\|S_n\|^{-1} \sim |Q_n|^{1/r'}$ holds. This needs to be checked individually, but we cannot expect it to be true.

Part (ii) of Theorem 3.2.17 fortunately works for unbounded subsets Y_n, but poses a different problem. Due to the fact that K_ψ is not integrable with respect to $L_{1,w}(G)$, the integrability coefficient r does not match the corresponding sequence coefficient q; in fact we have $q < r$. This means (3.2.42) is weaker than the reconstruction for classic coorbit spaces in Theorem 2.2.1. Moreover, we have an additional constant $D_n = |Q_n|^{1/q-1} \cdot \mathcal{I}^{1-1/q} \cdot \sup_{q \in Q_n} w(q) \cdot \theta_n$. In this case, however, even though the constant can blow up — $|Q_n|^{1/q-1}$ cannot be bounded from above for large n — all we need is n to be large enough such that condition (3.2.39) is fulfilled. So a reconstrution of functions is possible, at the cost of a different parameter q.

The following proposition presents a way to obtain reconstruction operators with $q = r$, if we assume in addition that the oscillator of K_ψ and K_ψ itself define bounded operators on $L_{r,m}(G)$.

Proposition 3.2.20. *We assume that Assumption 3.1.12 is fulfilled, i.e., $K_\psi \in L_{p,w}(G)$ for all $1 < p < \infty$, and that for some fixed $n \in \mathbb{N}$ the right convolution with $\mathrm{osc}_{Q_n} K_\psi$ and $|K_\psi|$ define bounded operators on $L_{r,m}(G)$, that is*

$$\|F * \mathrm{osc}_{Q_n} K_\psi\|_{L_{r,m}} \leqslant C \|F\|_{L_{r,m}}, \qquad\qquad\qquad (3.2.47)$$

$$\|F * |K_\psi|\|_{L_{r,m}} \leqslant C \|F\|_{L_{r,m}} \qquad\qquad\qquad (3.2.48)$$

for all $F \in L_{r,m}(G)$. Let $d = (d_x)_{x \in Y_n} \in \ell_{r,m}(Y_n)$; then $T = \sum_{x \in Y_n} d_x \pi(x)\psi$ is in $\mathrm{Co}(L_{r,m})$ and

$$\|T\|_{\mathrm{Co}(L_{r,m})} \leqslant 2C \cdot \mathcal{I}^{1-1/r} \cdot |Q_n|^{1/r-1} \cdot \sup_{q \in Q_n} w(q) \cdot \|(d_x)_{x \in Y_n}\|_{\ell_{r,m}} \qquad (3.2.49)$$

Proof. The proof is similar to the proof of Theorem 3.2.17. Again we use (3.2.44), Young's inequality, see Proposition 1.3.11, and Lemma 3.2.16, to obtain

$$
\begin{aligned}
&\left\| \sum_{x \in Y_n} d_x \pi(x) \psi \right\|_{\mathrm{Co}(L_{r,m})} \\
&= \left\| \sum_{x \in Y_n} d_x \lambda(x) K_\psi \right\|_{L_{r,m}} \\
&\leqslant \left\| \Big(\sum_{x \in Y_n} |d_x| \chi_{xQ_n} \Big) * (\mathrm{osc}_{Q_n} K_\psi + |K_\psi|) \right\|_{L_{r,m}} \cdot |Q_n|^{-1} \\
&\leqslant \left\| \Big(\sum_{x \in Y_n} |d_x| \chi_{xQ_n} \Big) * \mathrm{osc}_{Q_n} K_\psi \right\|_{L_{r,m}} \cdot |Q_n|^{-1} + \left\| \Big(\sum_{x \in Y_n} |d_x| \chi_{xQ_n} \Big) * |K_\psi| \right\|_{L_{r,m}} \cdot |Q_n|^{-1} \\
&\leqslant 2C \cdot |Q_n|^{-1} \cdot \left\| \sum_{x \in Y_n} |d_x| \chi_{xQ_n} \right\|_{L_{r,m}} \\
&\leqslant 2C \cdot \mathcal{I}^{1 - \frac{1}{r}} \cdot |Q_n|^{\frac{1}{r} - 1} \cdot \sup_{q \in Q_n} w(q) \cdot \|(d_x)_{x \in Y_n}\|_{\ell_{p,m}}.
\end{aligned}
$$

By the assumptions the expression on the right-hand side is finite. $\qquad\square$

By Lemma 3.2.9, the assumption (3.2.48) implies Assumption 3.2.4, it is therefore similar to the assumptions in Theorem 3.2.17. But the additional condition (3.2.47) is stronger than (3.2.39) in Theorem 3.2.17. In many applications it may be hard to verify the latter condition, but Proposition 3.2.20 shows that a proper reconstruction of signals is possible, even without an integrable kernel.

In conclusion, the results in this subsection are weaker than an atomic decomposition of $\mathrm{Co}(L_{r,m})$ with associated sequence space $\ell_{r,m}(Y_n)$, cf. Definition 1.9.2. The synthesis operator acts on a smaller sequence space than desired and we are only able to define a bounded coefficient operator on a finite dimensional subspace of the coorbit space. While this is unsatisfactory, in light of the results in the upcoming Section 3.3 we cannot expect results comparable to the ones for classic coorbit spaces presented in Section 2.2.

We close the subsection with a lemma that provides a possibility to compute the norm of the linear right inverse S_n in Theorem 3.2.17. The technique used here is similar to Babuška-Brezzi conditions in the theory of partial differential equations [5, 9].

Lemma 3.2.21. *Let A be a bounded and surjective linear operator that maps a Banach space W onto a Banach space V. Suppose that the kernel of A admits a complement L in W. Set*

$$
\varepsilon := \inf \Big\{ \sup \{ |\langle Ax, y \rangle_{V \times V^*}| \ : \ y \in V^*, \|y\|_{V^*} = 1 \} \ : \ x \in L, \|x\|_W = 1 \Big\}.
$$

Then the map $S := (A|_L)^{-1} \colon V \to L \subset W$ is a linear right inverse of A with

$$
\|S\| = \varepsilon^{-1}.
$$

Proof. It is straightforward that $A|_L : L \to V$ is a bijection. Therefore S is indeed a linear right inverse of A, and we have

$$\inf\left\{\sup\left\{|\langle Ax, y\rangle_{V \times V^*}| \ : \ y \in V^*, \|y\|_{V^*} = 1\right\} \ : \ x \in L, \|x\|_W = 1\right\}$$

$$= \inf\left\{\|Ax\|_V \ : \ x \in L, \|x\|_W = 1\right\}$$

$$= \inf\left\{\frac{\|Ax\|_V}{\|x\|_W} \ : \ x \in L\backslash\{0\}\right\}$$

$$= \inf\left\{\frac{\|ASv\|_V}{\|Sv\|_W} \ : \ v \in V\backslash\{0\}\right\}$$

$$= \left(\sup\left\{\frac{\|Sv\|_W}{\|ASv\|_V} \ : \ v \in V\backslash\{0\}\right\}\right)^{-1}$$

$$= \left(\sup\left\{\frac{\|Sv\|_W}{\|v\|_V} \ : \ v \in V\backslash\{0\}\right\}\right)^{-1}$$

$$= \|S\|^{-1},$$

which proves the claim. $\qquad\qquad\qquad\qquad\qquad\qquad\qquad\qquad\qquad\qquad\qquad\qquad\qquad\qquad\square$

3.2.3 An Example: Paley-Wiener Spaces

In this subsection we consider an example for our theory developed in this chapter. The case we are dealing with is the space of band-limited functions on the real line. As we will see in the following, the kernel that arises is the sinc function, which is not L_1-integrable. Therefore, although being a very elementary and natural example, this kernel cannot be handled using the classic coorbit theory from Chapter 2. The theory from this chapter, however, does handle it and it is therefore a suitable example for our setting.

In this subsection, G denotes the additive group $\mathbb{R}$ whose Haar measure is the Lebesgue measure dx. Since the group is abelian, $\mathbb{R}$ is unimodular. As the Hilbert space $\mathcal{H}$ we consider the Paley-Wiener space of functions with band in the fixed set $\Omega \subset \mathbb{R}$, namely

$$\mathcal{H} = B_\Omega^2 = \{v \in L_2(\mathbb{R}) \ : \ \mathrm{supp}(\hat{v}) \subseteq \Omega\}$$

equipped with the $L_2(\mathbb{R})$ scalar product. Strictly speaking, the elements of B_Ω^2 are not functions, but equivalence classes of functions. Considering, however, the Paley-Wiener-Schwartz theorem [77], each equivalence class in B_Ω^2 has a unique representative which is smooth. It is therefore allowed to identify each equivalence class with its smooth representative.

Let us now define the representation π for $b \in \mathbb{R}$ as

$$\pi(b)v(x) = v(x - b), \quad v \in B_\Omega^2, x \in \mathbb{R}.$$

Then, π is a unitary and not irreducible representation of the group $\mathbb{R}$ acting on B_Ω^2. On the frequency side, $\hat{\pi} = \mathcal{F}\pi\mathcal{F}^{-1}$ acts on $\mathcal{F}B_\Omega^2 = L_2(\Omega)$ by modulations:

$$\hat{\pi}(b)\hat{v}(\xi) = e^{2\pi i b \xi}\hat{v}(\xi), \quad v \in B_\Omega^2, \xi \in \Omega.$$

Using the Plancherel identity, the voice transform is explicitly given by

$$V_u v(b) = \langle v, \pi(b)u\rangle_{L_2} = \langle \hat{v}, \hat{\pi}(b)\hat{u}\rangle_{L_2} = \int_\mathbb{R} \hat{v}(\xi)\overline{\hat{u}(\xi)}e^{2\pi i b \xi}\,d\xi = \mathcal{F}^{-1}(\hat{v}\overline{\hat{u}})(b), \qquad (3.2.50)$$

where $u \in B_\Omega^2$ is an admissible function.

From now on we restrict ourselves to the case where Ω is a symmetrical interval, that is $\Omega = [-\omega, \omega]$. Proposition 4.6 in [18] then shows that $u \in B_\Omega^2$ is an admissible function according to Definition 2.1.3 if and only if $|\hat{u}| = 1$ almost everywhere on Ω. Hence $u = \mathcal{F}^{-1}\chi_\Omega \in B_\Omega^2$ is an admissible function and the resulting kernel K as defined in (3.1.2) using (3.2.50) is given by

$$K(b) = \mathcal{F}^{-1}(\chi_\Omega^2)(b) = \mathcal{F}^{-1}\chi_\Omega(b) = 2\omega \operatorname{sinc}(2\omega\pi b) = \frac{\sin(2\omega\pi b)}{\pi b}, \qquad (3.2.51)$$

where $\operatorname{sinc} x = \sin x / x$. Clearly, K is not in $L_1(\mathbb{R})$, but it belongs to $L_p(\mathbb{R})$ for every $p > 1$. Let us choose the weight $w = 1$ and take $\mathcal{T}$ as in (3.1.12),

$$\mathcal{T} = \bigcap_{1 < p < \infty} L_p(\mathbb{R}),$$

as a target space to construct coorbits. As in (3.1.13), the (anti)-dual of $\mathcal{T}$ can be identified with

$$\mathcal{T}' = \mathcal{U} = \operatorname{span} \bigcup_{1 < q < \infty} L_q(\mathbb{R}).$$

For $1 \leqslant p < \infty$, we define the Paley-Wiener p-spaces

$$B_\Omega^p := \{ f \in L_p(\mathbb{R}) : \operatorname{supp}(\mathcal{F}f) \subseteq \Omega \} .$$

Recall that the Fourier transform maps $L_p(\mathbb{R})$ to $L_{p'}(\mathbb{R})$ for $p \leqslant 2$, which follows from the Hausdorff-Young inequality, see for example [65, Proposition 2.2.16]. In contrast, for $p > 2$ the space $\mathcal{F}L_p(\mathbb{R})$ contains distributions that in general are not functions, see [75, Theorem 7.6.6].

The spaces B_Ω^p are sometimes defined in the literature as the spaces of the entire functions of fixed exponential type whose restriction to the real line is in $L_p(\mathbb{R})$, see [87]. This definition is equivalent to ours since a Paley-Wiener theorem holds for all $1 \leqslant p < \infty$, see for example [108, Chapter VI.4]. In particular, all these functions are infinitely differentiable on $\mathbb{R}$. Moreover, if $f \in B_\Omega^p$ with $p < \infty$, then $f(x) \to 0$ as $x \to \pm\infty$, and hence

$$B_\Omega^p \subset \mathcal{C}_0^\infty(\mathbb{R}) = \{ f \in \mathcal{C}^\infty(\mathbb{R}) : f(x) \to 0 \text{ as } x \to \pm\infty \}, \quad 1 \leqslant p < \infty.$$

Furthermore, the Paley-Wiener spaces are nested and increase with p:

$$B_\Omega^p \subseteq B_\Omega^q, \quad 1 \leqslant p \leqslant q < \infty.$$

The next Proposition shows that the Paley-Wiener p-spaces can be identified as coorbit spaces.

Proposition 3.2.22 (Proposition 4.8 of [18]). *Let $\Omega = [-\omega, \omega]$ and define $u := K := \mathcal{F}^{-1}\chi_\Omega$. The test space as defined in (3.1.16) is*

$$\mathcal{S} = \bigcap_{1 < p < \infty} B_\Omega^p$$

and its dual space is

$$\mathcal{S}' = \bigcup_{1 < q < \infty} B_\Omega^q.$$

The extended voice transform is the inclusion

$$V_{e,u} : \mathcal{S}' \hookrightarrow \mathcal{U}$$

and the following identification holds:

$$\operatorname{Co}(L_p(\mathbb{R})) = \mathcal{M}^p = B_\Omega^p.$$

This proposition shows that the Paley-Wiener p-spaces are a very elementary example for an application of our theory. Now that we have shown that we can consider these spaces to be coorbit spaces, we can apply the discretization strategy developed in Subsection 3.2.2. To do this, we first nee to show that Assumption 3.2.4 is fulfilled, that is $u \in L_p(\mathbb{R})$ for all $1 < p < \infty$ and $\mathrm{span}\{u(\cdot - b)\}_{b \in \mathbb{R}}$ is dense in B_Ω^p. While the first part is certainly true, for the second part, by Lemma 3.2.9, it suffices to show that the right convolution operator associated to K is a bounded operator on $L_p(\mathbb{R})$.

Corollary 3.2.23. *Let $1 < p < \infty$, then RC_K is a bounded operator on $L_p(\mathbb{R})$.*

Proof. Since $K = \mathcal{F}^{-1}\chi_\Omega$, the convolution with K is a bounded operator on $L_p(\mathbb{R})$ if and only if χ_Ω is a Fourier multiplier on $L_p(\mathbb{R})$. By [65, Example 2.5.15] this is true if and only if $\chi_{[0,1]}$ is a Fourier multiplier on $L_p(\mathbb{R})$. However, it is well-known that this is true because the Hilbert transform is bounded as an operator acting on $L_p(\mathbb{R})$, see [65, Theorem 5.1.7]. $\qquad\square$

In general, the question whether a convolution operator RC_K is bounded on $L_p(G)$ is much harder to answer. In this case, where the group is $\mathbb{R}$, we have the rich Fourier theory at hand, but for general groups the question is much harder, see also Chapter 6.

We will now apply the analysis outlined in Subsection 3.2.2 to obtain a discretization for the Paley-Wiener p-spaces. To this end, for $n \in \mathbb{N}$, let

$$Y_n := \{2^{-n}k : k \in \mathbb{Z}\} \subset \mathbb{R}. \tag{3.2.52}$$

Furthermore we fix the set

$$Q_n = \left[-2^{-n-1}, 2^{-n-1}\right], \tag{3.2.53}$$

which is a compact neighborhood of zero, and we set

$$\varphi_{n,k} := \chi_{[-2^{-n-1}, 2^{-n-1})}(\cdot - 2^{-n}k) \tag{3.2.54}$$

for $n \in \mathbb{N}$ and $k \in \mathbb{Z}$, where χ denotes the characteristic function. The system $\Phi_n = \{\varphi_{n,k}\}_{k \in \mathbb{Z}}$ forms a partition of unity subordinate to Y_n, i.e., (3.2.18), (3.2.19) and (3.2.20) are fulfilled. For later use we note that $|Q_n| = 2^{-n}$. Furthermore the system $\Phi_n = \{\varphi_{n,k}\}_{k \in \mathbb{Z}}$, $n \in \mathbb{N}$ fixed, is $L_2(\mathbb{R})$-orthogonal with $\|\varphi_{n,k}\|_{L_2(\mathbb{R})}^2 = |Q_n|$. As a finite subset of Y_n, $n \in \mathbb{N}$, we set

$$X_n := \{2^{-n}k : -N(n) \leqslant k \leqslant N(n)\}, \tag{3.2.55}$$

where $N(n) \in \mathbb{N}$ is chosen such that (3.2.21) and (3.2.22) are fulfilled. A possible choice is $N = N(n) = n \cdot 2^n$.

According to (3.2.23) the operator $T_n \colon B_\Omega^p \to V_n \subset B_\Omega^p$ is defined via

$$T_n f(x) = \sum_{k=-N(n)}^{N(n)} \langle f, \varphi_{n,k}\rangle_{L_2} K(x - 2^{-n}k),$$

for $f \in B_\Omega^p$, where

$$\langle f, \varphi_{n,k}\rangle_{L_2} = \int_{2^{-n}(k-1/2)}^{2^{-n}(k+1/2)} f(y)\, dy.$$

In light of (3.2.24) this means

$$V_n = \mathrm{span}\left\{\mathrm{sinc}(2\pi\omega(\cdot - 2^{-n}k)) : -N(n) \leqslant k \leqslant N(n)\right\}.$$

Before we apply Theorem 3.2.17, we need to check assumption (3.2.39). The second assumption (3.2.38) is already fulfilled.

Lemma 3.2.24. *Let $1 < r < \infty$. It holds that $K \in \mathcal{M}_{Q_n}^R(L_r)$, and therefore $\mathrm{osc}_{Q_n} K \in L_r(\mathbb{R})$ for all $n \in \mathbb{N}$.*

Proof. We have $|K(y)| \leqslant 2\omega$ for all $y \in \mathbb{R}$, and $|K(y)| \leqslant 1/(\pi|y|)$ for all $y \neq 0$. This implies

$$|K(y)| \leqslant \frac{1 + 4\omega}{1 + |y|}.$$

Indeed, for $|y| \leqslant 1$ we have $|K(y)| \leqslant 2\omega \leqslant \frac{4\omega}{1+|y|}$, while for $|y| \geqslant 1$, we have $1/|y| \leqslant 2/(1+|y|)$, and thus $|K(y)| \leqslant \frac{2}{\pi}\frac{1}{|y|} \leqslant \frac{1+4\omega}{1+|y|}$.

Now, for $y \in x + Q_n \subset x + [-1, 1]$ we have $1 + |x| \leqslant 2 + |y| \leqslant 2(1 + |y|)$, so that $|K(y)| \leqslant \frac{1+4\omega}{1+|y|} \leqslant \frac{2+8\omega}{1+|x|}$. Hence,

$$\sup_{y \in x + Q_n} |K(y)|^r \leqslant \left(\frac{2 + 8\omega}{1 + |x|}\right)^r,$$

and thus

$$\int_{\mathbb{R}} \sup_{y \in x + Q_n} |K(y)|^r \, dx < \infty,$$

concluding the proof. □

With this at hand all assumptions of Theorem 3.2.17 are proven and we can apply the theorem to discretize the Paley-Wiener p-spaces.

Proposition 3.2.25. *Let $1 < p < \infty$.*

(i) Fix $\varepsilon > 0$; then for any $f \in B_\Omega^p$ there exists an integer $n^* = n_{f,\varepsilon}^* \in \mathbb{N}$, such that for all $n \geqslant n^*$:

$$\left\| f - \sum_{k=-N(n)}^{N(n)} c(f)_{n,k} K(\cdot - 2^{-n}k) \right\|_{L_p} \leqslant \varepsilon,$$

where the family of coefficients $(c(f)_{n,k})_{-N(n) \leqslant k \leqslant N(n)}$ satisfies

$$\|(c(f)_{n,k})_{-N(n) \leqslant k \leqslant N(n)}\|_{\ell_p} \leqslant 2^{-n/p'}(1 + \varepsilon)\|S_n\| \cdot \|f\|_{L_p}.$$

Here, as before, S_n is a linear right inverse for the operator T_n defined in (3.2.23) and p' is the dual of p.

(ii) For any sequence $(d_x)_{x \in Y_n} \in \ell_q(Y_n)$, $n \in \mathbb{N}$, the function f defined by

$$f = \sum_{k \in \mathbb{Z}} d_{2^{-n}k} K(\cdot - 2^{-n}k)$$

is in B_Ω^p with

$$\|f\|_{L_p(\mathbb{R})} \leqslant C \cdot 2^{n(1-1/q)}\|(d_x)_{x \in Y_n}\|_{\ell_q},$$

where $C = C(p, q) > 0$ is a constant and $q < p$.

Proof. Item (i) is an application of Theorem 3.2.17 (i), with $|Q_n| = 2^{-n}$.

It remains to prove (ii). Again, we can apply Theorem 3.2.17 (ii) and note that, by Lemma 3.2.24, the assumption (3.2.39) is fulfilled. Moreover, Lemma 3.2.24 shows that the norm $\|\mathrm{osc}_{Q_n}(K) + |K|\|_{L_r}$ can be estimated from above by a constant $C > 0$ independent of $n \in \mathbb{N}$. □

As stated in Remark 3.2.15 the asymptotic behaviour of the operator norm of S_n is crucial. In the following we apply Lemma 3.2.21 to obtain a useful characterization of $\|S_n\|$.

For this we restrict ourselves to the case $p = 2$ and obtain with the notation of Lemma 3.2.21

$$
\begin{aligned}
\|S_n\|^{-1} = \varepsilon &= \inf\left\{ \frac{\|T_n f\|_{L_2}}{\|f\|_{L_2}} \; : \; f \in (\operatorname{Ker} T_n)^\perp \right\} \\
&= \inf\left\{ \frac{\langle T_n^* T_n f, f\rangle_{L_2}}{\langle f, f\rangle_{L_2}} \; : \; f \in (\operatorname{Ker} T_n)^\perp \right\}^{1/2} \\
&= \lambda_{\min}(U_n)^{1/2},
\end{aligned}
$$

where $\lambda_{\min}(U_n)$ denotes the smallest eigenvalue of the operator

$$
U_n := T_n^* T_n \colon (\operatorname{Ker} T_n)^\perp \to (\operatorname{Ker} T_n)^\perp.
$$

Here, we used the well-known inclusion $\operatorname{Ran} A^* \subset (\operatorname{Ker} A)^\perp$ which guarantees that U_n is well-defined.

We have thus shown that the asymptotic behaviour of the smallest eigenvalue of U_n is equivalent to the asymptotic behaviour of $\|S_n\|$.

By using

$$
T_n f = \sum_{j=-N(n)}^{N(n)} \langle f, \varphi_{n,j}\rangle_{L_2} K(\cdot - 2^{-n}j), \quad T_n^* g = \sum_{k=-N(n)}^{N(n)} \langle g, K(\cdot - 2^{-n}k)\rangle_{L_2} \varphi_{n,k},
$$

we can rewrite U_n as

$$
\begin{aligned}
U_n f &= \sum_{j,k=-N(n)}^{N(n)} \langle f, \varphi_{n,j}\rangle_{L_2(\mathbb{R})} \langle K(\cdot - 2^{-n}j), K(\cdot - 2^{-n}k)\rangle_{L_2(\mathbb{R})} \varphi_{n,k} \\
&= \sum_{j,k=-N(n)}^{N(n)} \langle f, \varphi_{n,j}\rangle_{L_2(\mathbb{R})} K(2^{-n}(k-j)) \varphi_{n,k}
\end{aligned}
$$

for $f \in (\operatorname{Ker} T_n)^\perp$.

We set $W_n := \operatorname{span}\{\varphi_{n,k} \; : \; -N(n) \leqslant k \leqslant N(n)\}$ and obtain the relation $W_n^\perp \subset \operatorname{Ker} T_n$; thus $(\operatorname{Ker} T_n)^\perp \subset W_n$. Next, we note that the family $\{\lambda(x)K\}_{x\in\mathbb{R}}$ is linearly independent; indeed, we have $\mathcal{F}(\lambda(x)K) = e^{-2\pi i x\cdot}\mathcal{F}K = e^{-2\pi i x\cdot}\chi_{[-\omega,\omega]}$, and by analyticity these functions are linearly independent if and only if the functions $(\mathbb{R} \to \mathbb{C}, \xi \mapsto e^{-2\pi i x\xi})$ are. But each of these functions is an eigenvector of the differential operator $d/d\xi$ with pairwise distinct eigenvalues $2\pi i x, x \in \mathbb{R}$, which yields the linear independence. By this and by Lemma 3.2.10 it holds that $\operatorname{Ran} T_n = V_n = \operatorname{span}\{\lambda(x)K\}_{x\in X_n}$ satisfies $\dim \operatorname{Ran} T_n = |X_n| = 1 + 2N(n)$. But since $T_n \colon (\operatorname{Ker} T_n)^\perp \to V_n$ is an isomorphism, we have $\dim(\operatorname{Ker} T_n)^\perp = 1 + 2N(n)$ as well, so that we finally see $W_n = (\operatorname{Ker} T_n)^\perp$ by comparing dimensions. Hence, $U_n \colon W_n \to W_n$.

Moreover, by the orthogonality of the family $\{\varphi_{n,k}\}$, we see that

$$
U_n \varphi_{n,k} = \|\varphi_{n,k}\|_{L_2}^2 \sum_{\ell=-N(n)}^{N(n)} K(2^{-n}(\ell - k)) \varphi_{n,\ell} \tag{3.2.56}
$$

for any $-N(n) \leqslant k \leqslant N(n)$. Since $\dim W_n = 2N(n) + 1 < \infty$, we may define an isomorphism

$$
P_n \colon W_n \to \mathbb{R}^{2N(n)+1}, \quad P_n(\varphi_{n,k}) = \|\varphi_{n,k}\|_{L_2} e_k, \tag{3.2.57}
$$

where e_k denotes the k-th canonical unit vector of $\mathbb{R}^{2N(n)+1}$. Note that P_n maps the orthonormal basis $(\varphi_{n,k}/\|\varphi_{n,k}\|_{L_2(\mathbb{R})})$ to the orthonormal basis $(e_k)_{k \in \mathbb{N}}$, so that P_n is unitary.

The linear map $P_n U_n P_n^{-1} \colon \mathbb{R}^{2N(n)+1} \to \mathbb{R}^{2N(n)+1}$ is represented by a matrix M_n, whose entries are given via

$$
\begin{aligned}
(M_n)_{j,k} &= \langle P_n U_n P_n^{-1} e_k, e_j \rangle_{\mathbb{R}^{2N(n)+1}} \\
&= \langle U_n \frac{\varphi_{n,k}}{\|\varphi_{n,k}\|_{L_2}}, \frac{\varphi_{n,j}}{\|\varphi_{n,j}\|_{L_2(\mathbb{R})}} \rangle_{L_2} \\
&= \frac{\|\varphi_{n,k}\|_{L_2}}{\|\varphi_{n,j}\|_{L_2}} \sum_{\ell=-N(n)}^{N(n)} K(2^{-n}(\ell-k)) \langle \varphi_{n,\ell}, \varphi_{n,j} \rangle_{L_2} \\
&= \|\varphi_{n,k}\|_{L_2} \|\varphi_{n,j}\|_{L_2} K(2^{-n}(j-k)) \\
&= 2^{-n} K(2^{-n}(j-k)),
\end{aligned}
$$

for $1 \leqslant j, k \leqslant 2N(n)+1$. Since K is real, the matrix M_n is a *symmetric Toeplitz matrix*, which means that the entries of M_n only depend on the quantity $|k-j|$, thus yielding a band-structure. Since the eigenvalues of M_n coincide with those of the map U_n, finding the smallest eigenvalue of U_n is equivalent to finding the smallest eigenvalue of the Toeplitz matrix M_n.

Unfortunately, this task is very difficult. To the best knowledge of the authors, it is not possible to properly characterize the asymptotic behaviour of the smallest eigenvalue of such a Toeplitz matrix. We further refer to [7], where the author was told that leading experts on the field of Toeplitz matrices are unaware of these asymptotics.

Since there are already big obstacles in understanding the asymptotic behaviour of $\|S_n\|$ in this rather simple setting, one cannot hope that easy answers are available when turning to more complex groups and their associated coorbit spaces.

3.3 Obstructions to Discretization

The aim of this section is to show that it is not possible to obtain discretization results for coorbit spaces with non-integrable kernel similar to the ones presented in Section 2.2. As we have seen in Section 3.2 above, specifically in Theorem 3.2.17, a discretization that is weaker than an atomic decomposition was obtained. Obviously, the major difference between classic coorbit spaces and the coorbit spaces in this chapter is the integrability of the kernel K_ψ. Let us examine how precisely this affects the discretization techniques.

As presented in Chapter 2, in the classic coorbit theory the kernel $K_\psi(x) = \langle \psi, \pi(x)\psi \rangle_{\mathcal{H}}$ is assumed to be integrable, i.e., it satisfies the condition $K_\psi \in L_{1,w}(G)$ for some suitable weight $w \geqslant 1$ on G. This assumptions implies two important properties. First, it ensures that the test space $\mathcal{H}_{1,w}$ introduced in (2.1.14) is well-defined, whose dual space is the reservoir for the coorbit spaces. Second, by Definition 2.1.13, it implies that the right convolution operator associated to the kernel $f \mapsto f * K_\psi$ is a bounded operator on the target space Y. This is especially true for the important example of $Y = L_{p,m}(G)$, where m is a w-moderate weight on G, by Young's inequality, cf. Proposition 1.3.11.

If we replace the condition $K_\psi \in L_{1,w}(G)$ with the condition $K_\psi \in L_{p,w}(G)$ for all $1 < p < \infty$ as we have done in Subsection 3.1.2 the first property, namely the existence of a suitable test space $\mathcal{S}_w$ in (3.1.16), can still be recovered. The space $\mathcal{S}_w$ is not a Banach space but a Fréchet space as shown in Theorem 3.1.14, but its dual space $\mathcal{S}'_w$ still serves as the reservoir for the coorbit spaces. The second property, however, cannot be transferred to the setting of non-integrable kernel. As we will show in Proposition 3.3.5 the assumption $K_\psi \in L_{p,w}(G)$ for all $1 < p < \infty$ does not imply that the right convolution operator associated to K_ψ is a bounded operator on $L_{r,m}(G)$. In fact, in light of Subsection 3.2.3 a given kernel K_ψ satisfying the weak integrability condition might or might not be associated to bounded right convolution operator on $L_{r,m}(G)$.

In the previous section we were able to obtain atomic decompositions of coorbit spaces in Theorem 3.2.17 by additionally assuming span$\{\lambda(x)K_\psi\}_{x\in G}$ to be dense in $\mathcal{M}_{r,m}$, which, by Lemma 3.2.9 is fulfilled if the right convolution operator RC_K is bounded on $L_{r,m}(G)$. Therefore the boundedness of the right convolution operator is sufficient for a first atomic decomposition of coorbit spaces. In the following subsection we show that this condition is also necessary for atomic decompositions and Banach frames for coorbit spaces.

3.3.1   Necessary Assumptions on the Kernel

Before we show necessary assumption on the kernel K_ψ for proper discretization results of coorbit spaces, let us briefly recall the results for classic coorbit spaces. For the general definition of atomic decompositions and Banach frames we refer to Section 1.9. If we assume that the kernel is well-behaved and that $X = (x_i)_{i\in I}$ is a U-dense and relatively separated set for some $U \subset G$, Theorem 2.2.1 shows that the synthesis operator

$$\text{Synth}: \ell_{p,m}(I) \to \text{Co}(L_{p,m}), \quad (c_i)_{i\in I} \mapsto \sum_{i\in I} c_i \cdot \pi(x_i)\psi$$

is well-defined and bounded for each $1 < p < \infty$ and w-moderate weights m. Moreover, there is a bounded linear right inverse of Synth, which we call coefficient operator. Finally, the family $(\pi(x_i)\psi)_{i\in I}$ is called a family of atoms and the construction above is an atomic decomposition of coorbit spaces.

Under very similar assumptions we have shown in Theorem 2.2.3 that there also exist Banach frames. This means that the analysis or sampling operator

$$\text{Samp}: \text{Co}(L_{p,m}) \to \ell_{p,m}(I), \quad T \mapsto (\langle T, \pi(x_i)\psi\rangle_{\mathcal{H}'_{1,w}\times\mathcal{H}_{1,w}})_{i\in I}$$

is well-defined and bounded and also admits a bounded linear left inverse, which we call the reconstruction operator. This implies that the family $(\pi(x_i)\psi)_{i\in I}$ is a Banach frame for $\text{Co}(L_{p,m})$ with coefficient space $\ell_{p,m}(I)$.

Since the preceding statements hold for all w-moderate weights m, they also hold for m^{-1}, which is w-moderate by Lemma 3.1.17. Therefore the properties above not only hold for $L_{r,m}(G)$, $1 < r < \infty$, but also for $L_{r',m^{-1}}(G)$. In the following we assume that weaker forms of atomic decompositions and Banach frames hold for $L_{r,m}(G)$ and $L_{r',m^{-1}}(G)$, respectively, in the setting of this chapter. The next theorem then shows that this implies that the right convolution operator associated to K_ψ must be bounded on $L_{r,m}(G)$. The boundedness of the convolution operator is therefore a necessary condition for atomic decompositions and Banach frames.

Theorem 3.3.1. *Let $1 < r < \infty$ be arbitrary. Assume that Assumption 3.1.12 is satisfied, i.e., $K_\psi \in L_{p,w}(G)$ for all $1 < p < \infty$, and let $m\colon G \to (0,\infty)$ be a w-moderate weight. Furthermore, assume that for some family $(x_i)_{i\in I}$ in G and for some weight $\theta = (\theta_i)_{i\in I}$ on the index set I, the following hold:*

(i) "Weak Banach frame condition for $\text{Co}(L_{r,m})$": The analysis map

$$A: \text{Co}(L_{r,m}) \to \ell_{r,\theta}(I), \quad \varphi \mapsto \big(\langle\varphi, \pi(x_i)\psi\rangle_{\mathcal{S}_w}\big)_{i\in I}$$

is well-defined and bounded, with

$$\|A\varphi\|_{\ell_{r,\theta}} \asymp \|\varphi\|_{\text{Co}(L_{r,m})} \qquad \text{for all } \varphi \in \text{Co}(L_{r,m}). \tag{3.3.1}$$

(ii) "Weak atomic decomposition condition for $\mathrm{Co}(L_{r',m^{-1}})$": The synthesis map

$$S: \ell_{r',\theta^{-1}}(I) \to \mathrm{Co}(L_{r',m^{-1}}), \quad (c_i)_{i\in I} \mapsto \sum_{i\in I} [c_i \cdot \pi(x_i)\,\psi]$$

is well-defined and bounded.

Then the right convolution operator $RC_K: f \mapsto f * K_\psi$ defines a bounded linear operator on $L_{r,m}(G)$.

The proof of this theorem can be split into multiple parts that we will prove individually in several technical lemmata. The main idea is to prove the theorem via a contradiction. To construct this contradiction we are interested in a voice transform on the (anti-)dual space $[\mathrm{Co}(L_{r,m})]'$. In Section 3.1 we have shown that, as usual, the voice transform V_ψ can be extended to the reservoir, and dual space, $\mathcal{S}'_w$ and therefore to the coorbit spaces $\mathrm{Co}(L_{r,m})$. The following lemma extends the voice transform with very similar techniques to the dual space $[\mathrm{Co}(L_{r,m})]'$.

Lemma 3.3.2. *If Assumption 3.1.12 is satisfied for $1 < r < \infty$, and if $m: G \to \mathbb{R}_+$ is w-moderate, then there is a constant $C = C(m, r, w, K) > 0$ such that*

$$\text{for all } x \in G: \quad \pi(x)\,\psi \in \mathrm{Co}(L_{r,m}) \quad \text{and} \quad \|\pi(x)\,\psi\|_{\mathrm{Co}(L_{r,m})} \leqslant C \cdot w(x).$$

Therefore, for any (antilinear) continuous functional $\varphi \in [\mathrm{Co}(L_{r,m})]'$, the special voice transform

$$V_{\mathrm{sp}}\,\varphi: G \to \mathbb{C}, \quad x \mapsto \varphi(\pi(x)\,\psi) = \langle \varphi\,, \pi(x)\psi\rangle_{[\mathrm{Co}(L_{r,m})]'\times\mathrm{Co}(L_{r,m})} \tag{3.3.2}$$

is a well-defined function.

Proof. First, let us set $C_1 := m(e)$, where e is the unit element of G. Since m is w-moderate, cf. (3.1.23), we have

$$m(x) = m(x \cdot e) \leqslant w(x) \cdot m(e) \leqslant C_1 \cdot w(x) \qquad \text{for all } x \in G.$$

Furthermore,

$$w(y) = w(xx^{-1}y) \leqslant w(x) \cdot w(x^{-1}y) = w(x) \cdot (\lambda(x)w)(y).$$

Now, recall from Section 3.1 the embedding $\mathcal{H} \hookrightarrow \mathcal{S}'_w$, and that the extended voice transform $V_{e,\psi}$ coincides with the usual voice transform on $\mathcal{H}$. Therefore, since $\pi(x)\,\psi \in \mathcal{H}$, and since $K_\psi = V_\psi\psi$, we get

$$\begin{aligned}
\|V_{e,\psi}\,[\pi(x)\psi]\|_{L_{r,m}} &= \|V_\psi\,[\pi(x)\psi]\|_{L_{r,m}} \leqslant C_1 \cdot \|V_\psi\,[\pi(x)\psi]\|_{L_{r,w}} \\
&= C_1 \cdot \|w \cdot \lambda(x)\,[V_\psi\psi]\|_{L_r} \leqslant C_1 \cdot w(x) \cdot \|\lambda(x)\,[w \cdot V_\psi\psi]\|_{L_r} \\
&= C_1 \cdot w(x) \cdot \|w \cdot V_\psi\psi\|_{L_r} = C_1 \cdot w(x) \cdot \|K_\psi\|_{L_{r,w}} = C \cdot w(x),
\end{aligned}$$

where $C := C_1 \cdot \|K\|_{L_{r,w}}$ is finite thanks to Assumption 3.1.12. This proves the first part of the lemma, which then trivially implies that $V_{\mathrm{sp}}\,\varphi$ is a well-defined function, for any $\varphi \in [\mathrm{Co}(L_{r,m})]'$. $\square$

Our next lemma shows that if the right convolution with K_ψ does *not* act boundedly on $L_{r',m^{-1}}(G)$, then there exist certain pathological functionals on $\mathrm{Co}(L_{r,m})$. This is first step towards the desired contradiction.

Lemma 3.3.3. *Assume that Assumption 3.1.12 is satisfied, and let $1 < r < \infty$. If the right convolution operator $RC_K \colon f \mapsto f * K_\psi$ does not yield a well-defined bounded linear operator on $L_{r',m^{-1}}(G)$, then there is an (antilinear) continuous functional $\varphi \in [\mathrm{Co}(L_{r,m})]'$ satisfying $V_{\mathrm{sp}}\,\varphi \notin L_{r',m^{-1}}(G)$.*

Proof. We first claim that there is some $\Phi \in L_{r',m^{-1}}(G)$ with $\Phi * K_\psi \notin L_{r',m^{-1}}(G)$; that is, we claim that $RC_K \colon L_{r',m^{-1}}(G) \to L_{r',m^{-1}}(G)$ is not well-defined.

To see this, recall from Assumption 3.1.12 that $K_\psi \in \bigcap_{1<p<\infty} L_{p,w}(G)$. Thus, since m^{-1} is w-moderate, see Lemma 3.1.17, Young's inequality, see Proposition 1.3.11, shows that the right convolution operator RC_K is bounded as a map $RC_K \colon L_{r',m^{-1}}(G) \to L_{q,m^{-1}}(G)$ for any $r' < q < \infty$. Therefore, if $RC_K \colon L_{r',m^{-1}}(G) \to L_{r',m^{-1}}(G)$ was well-defined, then the closed graph theorem would imply that $RC_K \colon L_{r',m^{-1}}(G) \to L_{r',m^{-1}}(G)$ is bounded, contradicting our assumptions. Hence, there is a function Φ as desired.

Now, define the antilinear functional

$$\varphi \colon \mathrm{Co}(L_{r,m}) \to \mathbb{C}, \quad f \mapsto \int_G \Phi(y) \cdot \overline{V_{e,\psi} f(y)}\, dy.$$

It is easy to see that φ is well-defined and bounded; in fact,

$$|\varphi(f)| \leqslant \|\Phi\|_{L_{r',m^{-1}}} \cdot \|V_{e,\psi} f\|_{L_{r,m}} = \|\Phi\|_{L_{r',m^{-1}}} \cdot \|f\|_{\mathrm{Co}(L_{r,m})}.$$

Finally, note for all $x \in G$ that

$$\begin{aligned}
V_{\mathrm{sp}}\,\varphi(x) &= \langle \varphi,\, \pi(x)\,\psi \rangle_{[\mathrm{Co}(L_{r,m})]' \times \mathrm{Co}(L_{r,m})} \\
&= \int_G \Phi(y) \cdot \overline{V_{e,\psi}\,[\pi(x)\,\psi]\,(y)}\, dy \\
&= \int_G \Phi(y) \cdot \overline{\langle \pi(x)\psi,\, \pi(y)\psi \rangle_{\mathcal{H}}}\, dy \\
&= \int_G \Phi(y) \cdot \langle \psi,\, \pi(y^{-1}x)\,\psi \rangle_{\mathcal{H}}\, dy \\
&= \int_G \Phi(y) \cdot K_\psi(y^{-1}x)\, dy = (\Phi * K_\psi)(x)
\end{aligned}$$

with $\Phi * K_\psi \in L_{q,m^{-1}}(G)$ for all $r' < q < \infty$. But by our choice of Φ, we have $V_{\mathrm{sp}}\,\varphi$, which is not in $\Phi * K_\psi \notin L_{r',m^{-1}}(G)$, as desired. $\qquad\square$

Our next lemma shows that the assumptions of Theorem 3.3.1 exclude the existence of pathological functionals as in the preceding lemma.

Lemma 3.3.4. *Under the assumptions of Theorem 3.3.1 and with notation as in Lemma 3.3.2, every antilinear continuous functional $\varphi \in [\mathrm{Co}(L_{r,m})]'$ satisfies $V_{\mathrm{sp}}\,\varphi \in L_{r',m^{-1}}(G)$.*

Proof. Let $\varphi \in [\mathrm{Co}(L_{r,m})]'$ be arbitrary, and let the analysis operator A be as in the assumptions of Theorem 3.3.1. Using this operator, we define the (antilinear) functional

$$\Lambda_0 \colon A\,(\mathrm{Co}(L_{r,m})) \to \mathbb{C}, Af \mapsto \varphi(f).$$

Note that this is well-defined, since (3.3.1) ensures that A is injective. Furthermore, with $A\,(\mathrm{Co}(L_{r,m}))$ considered as a subspace of $\ell_{r,\theta}(I)$, the functional Λ_0 is bounded, since (3.3.1) yields a constant $C > 0$ such that each $c = Af \in A\,(\mathrm{Co}(L_{r,m}))$ satisfies

$$\begin{aligned}
|\Lambda_0(c)| = |\varphi(f)| &\leqslant \|\varphi\|_{[\mathrm{Co}(L_{r,m})]'} \cdot \|f\|_{\mathrm{Co}(L_{r,m})} \\
&\leqslant C\|\varphi\|_{[\mathrm{Co}(L_{r,m})]'} \cdot \|Af\|_{\ell_{r,\theta}} = C\|\varphi\|_{[\mathrm{Co}(L_{r,m})]'} \cdot \|c\|_{\ell_{r,\theta}}.
\end{aligned}$$

With Λ_0 being bounded, an antilinear version of the Hahn-Banach theorem yields a bounded (antilinear) extension $\Lambda\colon \ell_{r,\theta}(I) \to \mathbb{C}$ of Λ_0. Therefore, an antilinear version of the Riesz representation theorem for the dual of $\ell_{r,\theta}(I)$ ensures the existence of $\varrho = (\varrho_i)_{i\in I} \in \ell_{r',\theta^{-1}}(I)$ satisfying $\Lambda(c) = \langle \varrho, c\rangle_{\ell_{r',\theta^{-1}} \times \ell_{r,\theta}}$ for all $c \in \ell_{r,\theta}(I)$. Here, the pairing between $\ell_{r',\theta^{-1}}(I)$ and $\ell_{r,\theta}(I)$ is given by $\langle (c_i)_{i\in I}, (e_i)_{i\in I}\rangle_{\ell_{r',\theta^{-1}} \times \ell_{r,\theta}} = \sum_{i\in I} c_i \cdot \overline{e_i}$.

Having constructed the sequence $\varrho \in \ell_{r',\theta^{-1}}(I)$, we can now apply the second assumption of Theorem 3.3.1 — the boundedness of the synthesis operator S — to define $g := S\varrho \in \mathrm{Co}(L_{r',m^{-1}})$. Furthermore, for arbitrary $x \in G$, we recall from Lemma 3.3.2 that $\pi(x)\, u \in \mathrm{Co}(L_{r,m})$, so that

$$
c^{(x)} = \left(c_i^{(x)}\right)_{i\in I} := A\left(\pi(x)\psi\right) = \left(\langle \pi(x)\,\psi\,,\,\pi(x_i)\,\psi\rangle_{\mathcal{H}}\right)_{i\in I} \in \ell_{r,\theta}(I)
$$

is well-defined. Combining our preceding observations, we see

$$
\begin{aligned}
V_{\mathrm{sp}}\,\varphi(x) &= \varphi(\pi(x)\,\psi) = \Lambda_0\big(A(\pi(x)\,\psi)\big) = \Lambda(c^{(x)}) = \langle \varrho, c^{(x)}\rangle_{\ell_{r',\theta^{-1}}, \times \ell_{r,\theta}} \\
&= \sum_{i\in I}\Big[\varrho_i \cdot \overline{\langle \pi(x)\,\psi\,,\,\pi(x_i)\,\psi\rangle_{\mathcal{H}}}\Big] = \sum_{i\in I}\Big[\varrho_i \cdot \langle \pi(x_i)\,\psi\,,\,\pi(x)\,\psi\rangle_{\mathcal{S}_w}\Big] \\
&\overset{\dagger}{=} \Big\langle \sum_{i\in I}\big(\varrho_i \cdot \pi(x_i)\,\psi\big)\,,\,\pi(x)\,\psi\Big\rangle_{\mathcal{S}_w} \\
&= \langle S\varrho\,,\,\pi(x)\,\psi\rangle_{\mathcal{S}_w} = [V_{e,\psi}\,g]\,(x).
\end{aligned}
\tag{3.3.3}
$$

This identity — which will be fully justified below — completes the proof, since then we have $g = S\varrho \in \mathrm{Co}(L_{r',m^{-1}})$, that is $V_{e,\psi}\,g \in L_{r',m^{-1}}(G)$. Therefore, equation (3.3.3) above implies $V_{\mathrm{sp}}\,\varphi = V_{e,\psi}\,g \in L_{r',m^{-1}}(G)$, as claimed.

It remains to justify the step marked with $\dagger$ in (3.3.3). At that step, we used on the one hand that $S\varrho = \sum_{i\in I}[\varrho_i \cdot \pi(x_i)\,u]$ with unconditional convergence in $\mathrm{Co}(L_{r',m^{-1}})$. To see that this indeed holds, recall that $r' < \infty$, so that $\varrho = \sum_{i\in I}\varrho_i\,\delta_i$, with unconditional convergence in $\ell_{r',\theta^{-1}}(I)$; by the boundedness of S, this implies the claimed identity. On the other hand, we also used at $(*)$ that $\mathrm{Co}(L_{r',m^{-1}}) \to \mathbb{C}, f \mapsto \langle f\,,\,\pi(x)\,\psi\rangle_{\mathcal{S}_w}$ is a bounded linear functional. Indeed, (3.1.20) and Lemma 3.3.2 imply

$$
\begin{aligned}
\left|\langle f\,,\,\pi(x)\,\psi\rangle_{\mathcal{S}_w}\right| &= \left|\langle V_{e,\psi}f\,,\,V\,[\pi(x)\,\psi]\rangle_{L_2}\right| \\
&\leqslant \|V_{e,\psi}f\|_{L_{r',m^{-1}}} \cdot \|V\,[\pi(x)\,\psi]\|_{L_{r,m}} \\
&= \|f\|_{\mathrm{Co}(L_{r',m^{-1}})} \cdot \|V\,[\pi(x)\,u]\|_{L_{r,m}} \\
&\leqslant C \cdot \|f\|_{\mathrm{Co}(L_{r',m^{-1}})} \cdot w(x),
\end{aligned}
$$

with $C = C(m, w, u, r)$. $\qquad\square$

We can now finally prove Theorem 3.3.1.

Proof of Theorem 3.3.1. Assume towards a contradiction that the right convolution operator $RC_K\colon L_{r,m}(G) \to L_{r,m}(G)$ is not bounded. By Proposition 3.2.7, and since the prerequisites of Theorem 3.3.1 include Assumption 3.1.12, this implies that $RC_K\colon L_{r',m^{-1}}(G) \to L_{r',m^{-1}}(G)$ is also not bounded. Therefore, Lemma 3.3.3 yields an antilinear continuous functional $\varphi \in [\mathrm{Co}(L_{r,m})]'$ with $V_{\mathrm{sp}}\varphi \notin L_{r',m^{-1}}(G)$. In view of Lemma 3.3.4, this yields the desired contradiction. $\qquad\square$

3.3.2 Revisiting Paley-Wiener Spaces I

In this subsection we revisit the Paley-Wiener spaces discussed in Subsection 3.2.3. We showed in Corollary 3.2.23 that for symmertical intervals $\Omega = [-\omega, \omega]$ the corresponding right convolution operator RC_K acts as a bounded operator on $L_p(\mathbb{R})$ for all $1 < p < \infty$. This, however, is not true for general sets Ω and $p \neq 2$. In fact, in Proposition 3.3.5 we show that there exists a compact set $\Omega \subset \mathbb{R}$, such that the right convolution operator associated to $\mathcal{F}^{-1}\chi_\Omega$ is not bounded for any $p \in (1, \infty)\backslash\{2\}$.

Let us briefly recall the setting from Subsection 3.2.3. We consider the Paley-Wiener space

$$\mathcal{H} = B_\Omega^2 = \{f \in L_2(\mathbb{R}) : \hat{f} \equiv 0 \text{ almost everywhere on } \mathbb{R}\backslash\Omega\} \tag{3.3.4}$$

for a fixed measurable subset $\Omega \subset \mathbb{R}$ of finite measure. As seen in Subsection 3.2.3, the group $G = \mathbb{R}$ acts on this space by translations; that is, if we set $\pi(x)f = \lambda(x)f$ for $f \in B_\Omega^2$, then π is a unitary representation of $\mathbb{R}$. Setting $u := \mathcal{F}^{-1}\chi_\Omega \in B_\Omega^2$, using Plancherel's theorem, and noting $\hat{f} = \hat{f} \cdot \chi_\Omega = \hat{f} \cdot \overline{\hat{u}}$ for $f \in B_\Omega^2$, we see by using (3.2.50) that the associated voice transform is given by

$$V_u v(b) = \int_\Omega \hat{v}(\xi) e^{2\pi i b \xi} \, d\xi = \mathcal{F}^{-1}\hat{v}(b) = v(b)$$

for all $v \in B_\Omega^2$. Thus, $V_u : B_\Omega^2 \to L_2(\mathbb{R})$ is an isometry and the reproducing kernel K_u is simply given by $K_u(x) = V_u u(x) = u(x)$ for $x \in \mathbb{R}$. As we have seen in Corollary 3.2.23, if $\Omega = [-\omega, \omega]$, then the convolution with $K_u = u$ is a bounded operator on $L_p(\mathbb{R})$ for all $1 < p < \infty$. But the following proposition shows that this is not always the case. More specifically, there is a reproducing kernel that satisfies Assumption 3.1.12, that is $K \in L_p(\mathbb{R})$ for all $1 < p < \infty$, but for which the associated right convolution operator does *not* act boundedly on $L_p(\mathbb{R})$ for any $p \neq 2$.

Proposition 3.3.5. *There is a compact set $C \subset [0, 1]$ with the following properties:*

(i) $\mathcal{F}^{-1}\chi_C \in \bigcap_{1 < p \leqslant \infty} L_p(\mathbb{R})$.

(ii) *For any $p \in (1, \infty)\backslash\{2\}$, the convolution operator $f \mapsto f * \mathcal{F}^{-1}\chi_C$ is not bounded, and by Proposition 3.2.6 not well-defined, as an operator on $L_p(\mathbb{R})$.*

Proof. We will construct $C \subset [0, 1]$ as a certain "fat Cantor set". In particular, we will show below that C has positive measure and fulfills the following two additional properties:

$$|C \cap B| < |B| \qquad \text{for all open intervals } \varnothing \neq B \subset \mathbb{R}, \tag{3.3.5}$$

and

$$C^c = \bigcup_{n=0}^\infty \bigcup_{j=0}^{2^n-1} B_j^n \quad \text{with} \quad B_j^n := \frac{a_j^{(n)} + b_j^{(n)}}{2} + \left(-\frac{\mu_{n+1}}{2}, \frac{\mu_{n+1}}{2}\right), \tag{3.3.6}$$

where the complement C^c is taken relative to $[0, 1]$, and where $a_j^{(n)}, b_j^{(n)} \in \mathbb{R}$ are suitable, while $\mu_n := \min\{4^{-n}, n^{-n}\}$ for $n \in \mathbb{N}$.

Before we provide the precise construction of such a set C, let us see how the properties (3.3.5) and (3.3.6) imply the properties of C that are stated in the proposition.

First, [83, Theorem 1] shows that if the operator $f \mapsto f * \mathcal{F}^{-1}\chi_C$ is bounded on $L_p(\mathbb{R})$ for some $p \in (1, \infty)\backslash\{2\}$, that is, if χ_C is an $L_p(\mathbb{R})$-Fourier multiplier, then C would be equivalent to an open set. In other words, there would be an open set $U \subset \mathbb{R}$ with $\chi_C = \chi_U$ Lebesgue almost everywhere. But since C has positive measure, this is only possible if U is a *nonempty* open set.

Therefore, U contains a nonempty open interval $B \subset U$. Since $\chi_C = \chi_U$ almost everywhere, this implies $|B \cap C| = |B \cap U| = |B|$, in contradiction to (3.3.5). In summary, we have thus shown that the convolution operator $f \mapsto f * \mathcal{F}^{-1}\chi_C$ is *not* bounded on any $L_p(\mathbb{R})$ space for $p \in (1, \infty)\backslash\{2\}$. But this even implies that $L_p(\mathbb{R}) \to L_p(\mathbb{R}), f \mapsto f * \mathcal{F}^{-1}\chi_C$ is not well-defined, by Proposition 3.2.6.

Second, we will see that equation (3.3.6) ensures $\mathcal{F}^{-1}\chi_{C^c} \in \bigcap_{1<p\leqslant\infty} L_p(\mathbb{R})$, which then implies $\mathcal{F}^{-1}\chi_C = \mathcal{F}^{-1}\chi_{(0,1)} - \mathcal{F}^{-1}\chi_{C^c} \in \bigcap_{1<p\leqslant\infty} L_p(\mathbb{R})$. Here, we used that $F := \mathcal{F}^{-1}\chi_{(0,1)} \in \bigcap_{1<p\leqslant\infty} L_p(\mathbb{R})$, since a direct computation shows $F(x) = \frac{e^{2\pi ix}-1}{2\pi ix}$ for $x \neq 0$, which implies $|F(x)| \lesssim (1 + |x|)^{-1}$. It remains to show $\mathcal{F}^{-1}\chi_{C^c} \in \bigcap_{1<p\leqslant\infty} L_p(\mathbb{R})$. To this end, we set $\xi_j^{(n)} := \frac{a_j^{(n)}+b_j^{(n)}}{2} - \frac{\mu_{n+1}}{2}$, recall the definition of the intervals $B_j^n = \xi_j^{(n)} + \mu_{n+1} \cdot (0,1)$ from (3.3.6), and use standard properties of the Fourier transform to compute

$$\mathcal{F}^{-1}\chi_{B_j^n} = \mu_{n+1} \cdot M_{\xi_j^{(n)}}\left[(\mathcal{F}^{-1}\chi_{(0,1)})(\mu_{n+1}\cdot)\right] = \mu_{n+1} \cdot M_{\xi_j^{(n)}}\left(F(\mu_{n+1}\cdot)\right),$$

where $(M_\xi f)(x) = e^{2\pi ix\xi}f(x)$ denotes the *modulation* with frequency ξ of a function f. Next, (3.3.6) shows

$$\mathcal{F}^{-1}\chi_{C^c} = \sum_{n=0}^{\infty}\sum_{j=0}^{2^n-1}\mathcal{F}^{-1}\chi_{B_j^n}.$$

Combining this with the triangle inequality for L_p and with the elementary identities $\|M_\xi f\|_{L_p} = \|f\|_{L_p}$ and $\|f(a\cdot)\|_{L_p(\mathbb{R})} = a^{-1/p}\|f\|_{L_p(\mathbb{R})}$ for $a > 0$ and $f \in L_p(\mathbb{R})$, we see because of $\mu_n \leqslant n^{-n}$ and $1 - p^{-1} > 0$ for each fixed $p \in (1, \infty]$ that

$$
\begin{aligned}
\|\mathcal{F}^{-1}\chi_{C^c}\|_{L_p} &\leqslant \sum_{n=0}^{\infty}\sum_{j=0}^{2^n-1}\mu_{n+1}\cdot\|M_{\xi_j^{(n)}}\left(F(\mu_{n+1}\cdot)\right)\|_{L_p} \\
&\leqslant \|F\|_{L_p}\cdot\sum_{n=0}^{\infty}\sum_{j=0}^{2^n-1}\mu_{n+1}^{1-p^{-1}} \leqslant \|F\|_{L_p}\cdot\sum_{\ell=1}^{\infty}2^{\ell-1}\cdot\ell^{-\ell(1-p^{-1})}.
\end{aligned}
\tag{3.3.7}
$$

But for $\ell \geqslant \ell_0 = \ell_0(p)$, we have $(1 - p^{-1})\cdot\log_2(\ell) \geqslant 2$, and thus

$$2^{\ell-1}\cdot\ell^{-\ell(1-p^{-1})} = \frac{1}{2}\cdot 2^\ell\cdot 2^{-\ell(1-p^{-1})\cdot\log_2(\ell)} \leqslant 2^{\ell\left(1-(1-p^{-1})\cdot\log_2(\ell)\right)} \leqslant 2^{-\ell},$$

so that the series on the right-hand side of (3.3.7) converges. Hence, $\mathcal{F}^{-1}\chi_{C^c} \in L_p(\mathbb{R})$ for every $p \in (1, \infty]$.

Finally, we note because of $\mu_n \leqslant 4^{-n}$ that property (3.3.6) also implies

$$|C^c| = \sum_{n=0}^{\infty}\sum_{j=0}^{2^n-1}|B_j^n| = \sum_{n=0}^{\infty}2^n\mu_{n+1} \leqslant \sum_{n=0}^{\infty}2^n\cdot 4^{-(n+1)} \leqslant \frac{1}{4}\cdot\sum_{n=0}^{\infty}2^{-n} = \frac{1}{2} < 1,$$

so that $C \subset [0,1]$ necessarily has positive measure if it satisfies properties (3.3.5) and (3.3.6). It remains to show that one can indeed construct a compact set $C \subset [0,1]$ that satisfies properties (3.3.5) and (3.3.6).

To this end, as for the construction of the classical Cantor set, we will set $C := \bigcap_{n=0}^{\infty}C^n$ where the sets $C^n := \bigcup_{j=0}^{2^n-1}C_j^n$ will be defined inductively.

For the start of the induction set $C_0^0 := [a_1^{(0)}, b_1^{(0)}] := [0,1]$.

For the induction step, assume for some $n \in \mathbb{N}_0$ that we have constructed closed intervals $C_\ell^n = [a_\ell^{(n)}, b_\ell^{(n)}] \subset [0,1]$, for $\ell = 0, \dots, 2^n - 1$, with

$$4^{-n} \leqslant b_\ell^{(n)} - a_\ell^{(n)} \leqslant 2^{-n} \quad \text{for all} \quad 0 \leqslant \ell < 2^n \tag{3.3.8}$$

and

$$b_\ell^{(n)} < a_{\ell+1}^{(n)} \quad \text{for} \quad 0 \leqslant \ell < 2^n - 1. \tag{3.3.9}$$

Now, for $0 \leqslant j < 2^{n+1}$ we can write $j = 2\ell + k$ with uniquely determined $k \in \{0,1\}$ and $0 \leqslant \ell < 2^n$. We then recall from after (3.3.6) that $\mu_{n+1} = \min\{4^{-(n+1)}, (n+1)^{-(n+1)}\}$, and define

$$
\begin{aligned}
C_j^{n+1} &:= [a_j^{(n+1)}, b_j^{(n+1)}] \\
&:= \begin{cases}
\left[a_\ell^{(n)}, \dfrac{a_\ell^{(n)} + b_\ell^{(n)}}{2} - \dfrac{\mu_{n+1}}{2}\right] \subset [a_\ell^{(n)}, b_\ell^{(n)}] = C_\ell^n & \text{if } k = 0, \\[3mm]
\left[\dfrac{a_\ell^{(n)} + b_\ell^{(n)}}{2} + \dfrac{\mu_{n+1}}{2}, b_\ell^{(n)}\right] \subset [a_\ell^{(n)}, b_\ell^{(n)}] = C_\ell^n & \text{if } k = 1.
\end{cases}
\end{aligned}
\tag{3.3.10}
$$

With this choice, we see from (3.3.8) and because of $\mu_{n+1} \leqslant 4^{-(n+1)}$ that

$$b_j^{(n+1)} - a_j^{(n+1)} = \frac{b_\ell^{(n)} - a_\ell^{(n)}}{2} - \frac{\mu_{n+1}}{2} \geqslant \frac{1}{2} \cdot \left(4^{-n} - 4^{-(n+1)}\right) = \frac{3}{8} \cdot 4^{-n} \geqslant 4^{-(n+1)}$$

and

$$b_j^{(n+1)} - a_j^{(n+1)} = \frac{b_\ell^{(n)} - a_\ell^{(n)}}{2} - \frac{\mu_{n+1}}{2} \leqslant \frac{1}{2}(b_\ell^{(n)} - a_\ell^{(n)}) \leqslant 2^{-(n+1)},$$

thereby proving (3.3.8) for $n+1$ instead of n.

For the proof of (3.3.9) for $0 \leqslant j < 2^{n+1} - 1$ with $j = 2\ell + k$ and $k \in \{0,1\}$, we distinguish two cases:

Case 1: $k = 0$. In this case, $j + 1 = 2\ell + 1$, and hence

$$b_j^{(n+1)} = \frac{a_\ell^{(n)} + b_\ell^{(n)}}{2} - \frac{\mu_{n+1}}{2} < \frac{a_\ell^{(n)} + b_\ell^{(n)}}{2} + \frac{\mu_{n+1}}{2} = a_{j+1}^{(n+1)}.$$

Case 2: $k = 1$. In this case, $2(\ell + 1) + 0 = j + 1 < 2^{n+1}$, so that $1 \leqslant \ell + 1 < 2^n$. Therefore, (3.3.9) shows $b_j^{(n+1)} = b_\ell^{(n)} < a_{\ell+1}^{(n)} = a_{j+1}^{(n+1)}$.

We have thus verified (3.3.9) for $n+1$ instead of n.

As indicated above, we define $C^n := \bigcup_{j=0}^{2^n - 1} C_j^n$ and observe as a consequence of (3.3.10) that each C^n is closed with $C^{n+1} \subset C^n$ for all $n \in \mathbb{N}_0$. Hence, $C := \bigcap_{n=0}^{\infty} C^n \subset C^0 = [0,1]$ is compact.

Having defined the set C, our first goal is to prove property (3.3.5). Let $B \subset \mathbb{R}$ be a nonempty open interval. If $C \cap B$ is a finite set, the inequality in (3.3.5) is trivially satisfied. Hence, we can assume that $C \cap B$ is infinite, so that there are $x, y \in C \cap B$ with $x < y$. Choose $n \in \mathbb{N}_0$ with $2^{-n} < y - x$ and note because of $x, y \in C \subset C^n = \bigcup_{j=0}^{2^n - 1} C_j^n$ that there are $j_x, j_y \in \{0, \dots, 2^n - 1\}$ with $x \in C_{j_x}^n$ and $y \in C_{j_y}^n$. In case of $j_y \leqslant j_x$, we would get because of $a_\ell^{(n)} \leqslant b_\ell^{(n)} \leqslant a_{\ell+1}^{(n)}$ for $0 \leqslant \ell < 2^n - 1$ and because of $b_\ell^{(n)} - a_\ell^{(n)} \leqslant 2^{-n}$ for $0 \leqslant \ell < 2^n$, see (3.3.8) and (3.3.9), that

$$2^{-n} < y - x \leqslant b_{j_y}^{(n)} - a_{j_x}^{(n)} \leqslant b_{j_y}^{(n)} - a_{j_y}^{(n)} \leqslant 2^{-n},$$

a contradiction. Hence, $j_y > j_x$, so that (3.3.8) and (3.3.9) show

$$B \ni x \leqslant b_{j_x}^{(n)} \leqslant b_{j_y - 1}^{(n)} < a_{j_y}^{(n)} \leqslant y \in B,$$

and thus $(b_{j_y-1}^{(n)}, a_{j_y}^{(n)}) \subset B \backslash C^n \subset B \backslash C$. But since this interval has positive measure, we see $|B| = |B \backslash C| + |B \cap C| > |B \cap C|$, thereby proving (3.3.5).

Finally, we prove the formula (3.3.6) for the complement C^c of C, with the complement taken relative to $[0,1]$. To see this, note $C^c = \bigcup_{n=0}^{\infty} (C^n)^c$. By disjointization, and since $(C^0)^c = \emptyset$ and $(C^n)^c \subset (C^{n+1})^c$, this yields

$$C^c = \bigcup_{n=1}^{\infty} (C^n)^c \backslash (C^{n-1})^c = \bigcup_{n=1}^{\infty} C^{n-1} \backslash C^n = \bigcup_{n=0}^{\infty} C^n \backslash C^{n+1}.$$

Next, recall $C^n = \bigcup_{j=0}^{2^n-1} C_j^n$ and also recall from (3.3.10) that $C_{2\ell+k}^{n+1} \subset C_\ell^n$ for $0 \leqslant \ell < 2^n$ and $k \in \{0,1\}$. Therefore, by (3.3.10) and the definition of B_j^n in (3.3.6) it holds

$$C_j^n \backslash C^{n+1} = \bigcap_{\ell=0}^{2^n-1} \bigcap_{k=0}^{1} C_j^n \backslash C_{2\ell+k}^{n+1} = \bigcap_{k=0}^{1} C_j^n \backslash C_{2j+k}^{n+1} = B_j^n.$$

Putting everything together, we see that (3.3.6) holds.   □

3.4 Improved Discretization Results Under Additional Assumptions

The preceding section has shown that a discretization of coorbit spaces $\mathrm{Co}(L_{p,m})$ involving atomic decompositions and Banach frames is only possible if the right convolution operator associated to the kernel K_ψ acts boundedly on $L_{p,m}(G)$. If this condition is not fulfilled, this approach fails.

But even if the right convolution operator associated to the kernel K_ψ does act boundedly on $L_{p,m}(G)$, the results in Section 3.2 show that we only obtain discretizations that are weaker than the results in classic coorbit theory in Chapter 2. It turns out that the missing integrability of the kernel limits our mathematical toolbox to an extend that atomic decompositions and Banach frames are out of reach.

There is, however, a silver lining, as we will show in this present section. We will show that discretization results similar to the ones in the classical setting are possible, under an additional assumption on the kernel. We assume there exists a second kernel $W : G \to \mathbb{C}$ which is integrable, i.e., $W \in L_{1,w}(G)$, and reproduces K_ψ in the following sense: for the construction of Banach frames we assume $K_\psi * W = K_\psi$ and for atomic decompositions we assume $W * K_\psi = K_\psi$.

Let us motivate this approach with the following example related to Paley-Wiener spaces and consider the real line $G = \mathbb{R}$. Let $K \neq 0$ be a function on $\mathbb{R}$ satisfying $K * K = K$, then by applying the Fourier transfrom this means $\widehat{K} \cdot \widehat{K} = \widehat{K}$, therefore $\widehat{K} = \chi_\Omega$ must be the indicator function of a measurable set $\Omega \subset \mathbb{R}$, see also [18]. Since $\widehat{K}$ is therefore non-continuous, this implies $K \notin L_1(\mathbb{R})$, which is the setting we are interested in. We can, however, find a Schwartz function $\varphi \in \mathcal{S}(\mathbb{R})$ that smoothly extends $\widehat{K}$ such that $\varphi \equiv 1$ on Ω. Then, $W := \mathcal{F}^{-1}\varphi \in \mathcal{S}(\mathbb{R})$ satisfies $\widehat{W * K} = \varphi \cdot \chi_\Omega = \chi_\Omega = \widehat{K}$ and therefore $W * K = K * W = K$. Obviously, this does not imply $W * W = W$ and we do not assume this identity, which gives us a larger freedom in the choice of W. It should be mentioned that related approaches have been established in [58] and [47, 48].

This section is structured as follows. We start by recalling some basic notations from the first chapter. In Subsection 3.4.1 we show the existence of Banach frames and specify what requirements are necessary. Similarly, the existence of atomic decompositions under certain assumptions on the kernel function W is discussed in Subsection 3.4.2. Finally, in the last

Subsection 3.4.3 we revisit the example of Paley-Wiener spaces one more time to show an application of the theoretical results of this section.

We should mention that most of the proofs in this section are heavily inspired by the original coorbit papers [44–46, 68]. The main novel ingredient here is the observation that instead of the idempotent reproducing formula $K_\psi * K_\psi = K_\psi$, it suffices to have $K_\psi * W = K_\psi$ or $W * K_\psi = K_\psi$ for potentially different kernels K_ψ, W.

Remark 3.4.1. Most of the results in this section can also be obtained for coorbit spaces $\mathrm{Co}(Y)$ where Y is a solid Banach space continuously embedded into $L_0(G)$. For simplicity, as usual we restrict our attention to the case $Y = L_{r,m}(G)$.

Before we turn to the discretization results we revisit some notion from Section 1.5, especially Definition 1.5.1 and Definition 1.5.3. In the following, let $U \subset G$ be a unit neighborhood, then we call $\mathcal{X} = (x_i)_{i \in I}$ a U-well-spread family if it is relatively separated, i.e., for every compact unit neighborhood $Q \subset G$ there is a constant $N > 0$, such that

$$\sum_{i \in I} \chi_{x_i Q}(x) \leqslant N \quad \text{for all } x \in G,$$

and it is U-dense, that is $G = \bigcup_{i \in I} x_i U$.

Given a U-well-spread family $\mathcal{X}$, we denote with $\Phi = (\varphi_i)_{i \in I}$ a bounded uniform partitio of unity subordinate to U. This family of functions of G satisfies $\varphi_i(x) \in [0, 1]$ for all $x \in G$ with $\varphi(x) = 0$ for all $x \notin U$ and $\sum_{i \in I} \varphi \equiv 1$ on G.

The following lemma points out an important property of relatively separated families that we will use time and time again:

Lemma 3.4.2. *Let $\mathcal{X} = (x_i)_{i \in I}$ be a relatively separated family and let $1 \leqslant r < \infty$. Then for every compact unit neighborhood $U \subset G$, the synthesis operator*

$$\mathrm{Synth}_{\mathcal{X}, U} \colon \ell_{r,m}(I) \to L_{r,m}(G), \quad (c_i)_{i \in I} \mapsto \sum_{i \in I} c_i \chi_{x_i U} \tag{3.4.1}$$

is well-defined and bounded, with pointwise absolute convergence of the defining series.

Furthermore, if $\Phi = (\varphi_i)_{i \in I}$ is a U-BUPU with localizing family $\mathcal{X}$, then the synthesis operator

$$\mathrm{Synth}_{\mathcal{X}, \Phi} \colon \ell_{r,m}(I) \to L_{r,m}(G), \quad (c_i)_{i \in I} \mapsto \sum_{i \in I} c_i \varphi_i \tag{3.4.2}$$

is well-defined and bounded, with pointwise absolute convergence of the defining series.

Proof. The second part of the lemma is a consequence of the first one: Since $0 \leqslant \varphi_i \leqslant 1$, and since φ_i vanishes outside of $x_i U$, we have

$$|(\mathrm{Synth}_{\mathcal{X}, \Phi} c)(x)| \leqslant \sum_{i \in I} |c_i| \, \varphi_i(x) \leqslant \sum_{i \in I} |c_i| \, \chi_{x_i U}(x) = (\mathrm{Synth}_{\mathcal{X}, U} |c|)(x) < \infty$$

for all $x \in G$ and all $c = (c_i)_{i \in I} \in \ell_{r,m}(I)$, where $|c| = (|c_i|)_{i \in I} \in \ell_{r,m}(I)$ with $\| \, |c| \, \|_{\ell_{r,m}} = \|c\|_{\ell_{r,m}}$, so that

$$\|\mathrm{Synth}_{\mathcal{X}, \Phi} c\|_{L_{r,m}} \leqslant \|\mathrm{Synth}_{\mathcal{X}, U} |c| \, \|_{L_{r,m}} \lesssim \| \, |c| \, \|_{\ell_{r,m}} = \|c\|_{\ell_{r,m}}.$$

Thus, it remains to prove the first part of the lemma.

By definition of a relatively separated family, there is $N = N(X, U) > 0$ with $\sum_{i \in I} \chi_{x_i U} \leqslant N$. On the one hand, this shows that for each $x \in G$ only finitely many terms of the series defining

$(\mathrm{Synth}_{\mathcal{X},U}c)(x)$ do no vanish; in particular, the defining series is pointwise absolutely convergent. On the other hand, we see

$$
\begin{aligned}
\left| (\mathrm{Synth}_{\mathcal{X},U}c)(x) \right|^r
&\leqslant \left(\sum_{i \in I} |c_i|\, \chi_{x_i U}(x)\, \chi_{x_i U}(x) \right)^r \\
&\leqslant \left(\sup_{j \in I} |c_j|\, \chi_{x_j U}(x) \cdot \sum_{i \in I} \chi_{x_i U}(x) \right)^r \\
&\leqslant N^r \cdot \sup_{j \in I} |c_j|^r\, \chi_{x_j U}(x) \\
&\leqslant N^r \cdot \sum_{i \in I} |c_i|^r\, \chi_{x_i U}(x)\,.
\end{aligned}
$$

Thus,

$$
\begin{aligned}
\| \mathrm{Synth}_{\mathcal{X},U}c \|_{L_{r,m}}^r
&\leqslant N^r \cdot \int_G m(x)^r \cdot \sum_{i \in I} |c_i|^r\, \chi_{x_i U}(x)\, dx \\
&\leqslant N^r \cdot \sum_{i \in I} \left(|c_i|^r \cdot \int_{x_i U} m(x)^r\, dx \right).
\end{aligned}
$$

But for $x = x_i u \in x_i U$, we have $m(x) = m(x_i u) \leqslant m(x_i) \cdot w(u) \leqslant C \cdot m(x_i)$ for $C := \sup_{u \in U} w(u)$, which is finite since U is compact and w is continuous. Overall, since $|x_i U| = |U|$ for all $i \in I$, where $|U|$ is the Haar-measure of U, we thus see

$$
\| \mathrm{Synth}_{\mathcal{X},U}c \|_{L_{r,m}}^r \leqslant N^r \cdot C^r \cdot |U| \cdot \sum_{i \in I} \left(m(x_i) \cdot |c_i| \right)^r,
$$

which easily yields the boundedness of $\mathrm{Synth}_{\mathcal{X},U}$. $\hfill\square$

3.4.1 Banach Frames

In this subsection we will prove the existence of Banach frames under certain assumptions on the kernel K_ψ. The main result will be stated in Theorem 3.4.8. From now on we assume, as usual, that $1 < r < \infty$ is fixed and that $m \colon G \to \mathbb{R}_+$ is a w-moderate weight.

As stated in the beginning of this section, we assume similar properties on K_ψ as in Subsection 3.2.1, and in addition we assume there is a function W on G which is integrable and reproduces K_ψ. This is made precise in the following assumption.

Assumption 3.4.3. *Assume that the kernel K_ψ from (3.1.2) satisfies Assumption 3.1.15, i.e., $K_\psi \in L_{p,w}(G)$ for all $1 < p < \infty$, and the right convolution operator $RC_K \colon L_{r,m}(G) \to L_{r,m}(G)$, $f \mapsto f * K_\psi$ is well-defined.*

Furthermore assume there is a kernel $W \colon G \to \mathbb{C}$ with the following properties:

(i) W *is continuous.*

(ii) $K_\psi * W = K_\psi$.

(iii) $M_{U_0}^L W \in L_{1,w}(G) \cap L_{1,w\Delta^{-1}}(G)$ *for some compact unit neighborhood $U_0 \subset G$, where*

$$
M_{U_0}^L W(x) = \| W \|_{L_\infty(x U_0)}, \quad x \in G \tag{3.4.3}
$$

is the left local maximum function defined in (2.1.23).

The motivation behind this assumption is made clear in the following lemma.

Lemma 3.4.4. *Let Assumption 3.4.3 be satisfied. Then the following holds true.*

(i) *Assumption 3.2.4 is satisfied, i.e.,* $\mathrm{span}\{\lambda(x)K_\psi\}_{x\in G}$ *is dense in the reproducing kernel space* $\mathcal{M}_{r,m}$ *defined in (3.2.4).*

(ii) *There is some unit neighborhood* $U_0 \subset G$ *such that for every unit neighborhood* $U \subset U_0$ *there is a constant* $C_U > 0$ *such that for all* $f \in \mathcal{M}_{r,m}$ *it holds that*

$$\|\mathrm{osc}_U^R f\|_{L_{r,m}} \leqslant C_U \cdot \|f\|_{L_{r,m}}. \tag{3.4.4}$$

Here,

$$\mathrm{osc}_U^R f(x) = \sup_{u\in U} |f(xu) - f(x)| \tag{3.4.5}$$

is the right oscillation of f.

(iii) *The constants* C_U *from the preceding point satisfy* $C_U \to 0$ *as* $U \to \{e\}$. *More precise, for every sufficiently small* $\varepsilon > 0$ *there is a unit neighborhood* $U_\varepsilon \subset U_0$ *with* $C_U \leqslant \varepsilon$ *for all unit neighborhoods* $U \subset U_\varepsilon$.

Proof. We first note that our assumptions imply $M_U^L W \in L_{1,w}(G) \cap L_{1,w\Delta^{-1}}(G)$ for *every* compact unit neighborhood $U \subset G$. Indeed, by compactness, and since $U \subset \bigcup_{x\in G} x\,\mathrm{int}(U_0)$, there is a finite family $(x_i)_{i=1,\dots,n}$ with $U \subset \bigcup_{i=1}^n x_i U_0$. Therefore, $xU \subset \bigcup_{i=1}^n xx_i U_0$, whence

$$M_U^L W(x) = \|W\|_{L_\infty(xU)} \leqslant \sum_{i=1}^n \|W\|_{L_\infty(xx_i U_0)}$$

$$= \sum_{i=1}^n M_{U_0}^L W(xx_i) = \sum_{i=1}^n \big[\rho(x_i)(M_{U_0}W)\big](x).$$

But since w and $w\Delta^{-1}$ are submultiplicative, the spaces $L_{1,w}(G)$ and $L_{1,w\Delta^{-1}}(G)$ are invariant under right translations, and hence $M_U^L W \in L_{1,w}(G) \cap L_{1,w\Delta^{-1}}(G)$.

Next, if V is an *open* precompact unit neighborhood and $U := \overline{V}$, then by continuity of W, we have

$$|W(x)| \leqslant \sup_{v\in V} |W(xv)| = \|W\|_{L_\infty(xV)} \leqslant M_U^L W(x)$$

for all $x \in G$. Therefore, $W \in L_{1,w}(G) \cap L_{1,w\Delta^{-1}}(G)$.

Since by assumption the right convolution operator RC_K acts boundedly on $L_{r,m}(G)$, Lemma 3.2.9 shows that the set $X_0 := \mathrm{span}\{\lambda(x)K\}_{x\in G}$ is dense in the reproducing kernel space $\mathcal{M}_{r,m}$, which proves part (i).

Furthermore, the assumption $K * W = K$ yields $(\lambda(x)K) * W = \lambda(x)(K * W) = \lambda(x)K$ for all $x \in G$, and thus $f * W = f$ for all $f \in X_0$. By density of X_0 in $\mathcal{M}_{r,m}$, and since the right convolution operator $f \mapsto f * W$ is continuous on $L_{r,m}(G)$ thanks to $W \in L_{1,w}(G) \cap L_{1,w\Delta^{-1}}(G)$ and Young's inequality, cf. Proposition 1.3.9, we see

$$f * W = f \quad \text{for all} \quad f \in \mathcal{M}_{r,m}. \tag{3.4.6}$$

We now use equation (3.4.6) to prove equation (3.4.4). To this end, let $U \subset G$ be an arbitrary compact unit neighborhood. Let $f \in \mathcal{M}_{r,m}$, $x \in G$ and $u \in U$ be arbitrary. Then

$$|f(xu) - f(x)| = |(f * W)(xu) - (f * W)(x)|$$

$$\leqslant \int_G |f(y)| \cdot |W(y^{-1}xu) - W(y^{-1}x)|\, dy$$

$$\leqslant \int_G |f(y)| \cdot (\mathrm{osc}_U^R W)(y^{-1}x)\, dy = \big(|f| * (\mathrm{osc}_U^R W)\big)(x).$$

Since this holds for every $u \in U$, we have $\mathrm{osc}_U^R f(x) \leqslant |f| * (\mathrm{osc}_U^R W)(x)$ for all $x \in G$. By solidity of $L_{r,m}(G)$ and in view of Young's inequality, cf. Proposition 1.3.9, this implies

$$\| \mathrm{osc}_U^R f \|_{L_{r,m}} \leqslant \|f\|_{L_{r,m}} \cdot \max\{ \| \mathrm{osc}_U^R W \|_{L_{1,w}}, \| \mathrm{osc}_U^R W \|_{L_{1,w\Delta^{-1}}} \}.$$

An easy generalization of Lemma 3.2.3, where we simply replace $\mathrm{osc}_U W$ by $\mathrm{osc}_U^R W$, shows that $\| \mathrm{osc}_U^R W \|_{L_{1,v}} \to 0$ as $U \to \{e\}$, for $v = w$ as well as for $v = w\Delta^{-1}$. Let us fix U_0 as in part (iii) of Assumption 3.4.3, then for every unit neighborhood $U \subset U_0$, then (3.4.4) holds true with $C_U = \max\{ \| \mathrm{osc}_U^R W \|_{L_{1,w}}, \| \mathrm{osc}_U^R W \|_{L_{1,w\Delta^{-1}}} \}$. Moreover, the constants above satisfy $C_U \to 0$ as $U \to \{e\}$ by the arguments above, which concludes the proof. $\qquad\square$

The next step is to show that Assumption 3.4.3 ensures a sufficiently fine discrete sampling of the continuous frame $(\pi(x)u)_{x \in G}$ provides a Banach frame for the coorbit space $\mathrm{Co}(L_{r,m})$. For this, as usual, we first turn to the reproducing kernel space $\mathcal{M}_{r,m}$, which is isomorphic to the coorbit space. We show that a Banach frame for $\mathcal{M}_{r,m}$ is given by $(f(x_i))_{i \in I}$ for a relatively separated family $(x_i)_{i \in I} \subset G$. By means of the correspondance principle this result is then transfered from the reproducing kernel space to the coorbit space.

The next proposition shows that certain sampling operators are bounded on the reproducing kernel space.

Lemma 3.4.5. *Let Assumption 3.4.3 be satisfied, and let $\mathcal{X} = (x_i)_{i \in I}$ be a relatively separated family in G. Then, the sampling operator*

$$\mathrm{Samp}_{\mathcal{X}} \colon \mathcal{M}_{r,m} \to \ell_{r,m}(I), \quad f \mapsto (f(x_i))_{i \in I} = \big(\langle f, \lambda(x_i) K_\psi \rangle_{L_2} \big)_{i \in I} \tag{3.4.7}$$

is well-defined and bounded.

Proof. We first recall that each $f \in \mathcal{M}_{r,m}$ satisfies $f = f * K_\psi$, and hence

$$f(x) = f * K_\psi(x) = \int_G f(y) \cdot K_\psi(y^{-1}x) \, dy = \int_G f(y) \cdot \overline{K_\psi(x^{-1}y)} \, dy = \langle f, \lambda(x) K_\psi \rangle_{L_2},$$

for all $x \in G$. But $K_\psi \in L_{r',w}(G)$, and thus also $\lambda(x) K_\psi \in L_{r',w}(G)$, since w is submultiplicative. Furthermore, since m is w-moderate, we have $m(e) = m(xx^{-1}) \leqslant m(x)w(x^{-1})$, and thus $[m(x)]^{-1} \leqslant w(x^{-1})/m(e) = w(x)/m(e)$, whence $L_{r',w}(G) \hookrightarrow L_{r',m^{-1}}(G)$. Thus, the dual pairing $\langle f, \lambda(x) K_\psi \rangle_{L_2} \in \mathbb{C}$ is well-defined for every $x \in \mathbb{R}$. Therefore, each entry $f(x_i)$ of the defining sequence of $\mathrm{Samp}_{\mathcal{X}} f = (f(x_i))_{i \in I}$ in (3.4.7) makes sense.

Now, let U be a compact unit neighborhood with $\| \mathrm{osc}_U^\rho f \|_{L_{r,m}} \leqslant C \cdot \|f\|_{L_{r,m}}$ for all $f \in \mathcal{M}_{r,m}$. Such a neighborhood exists by virtue of Lemma 3.4.4 (ii). Note that U^{-1} is also a compact unit neighborhood, so that by definition of a relatively separated family there is a constant $N > 0$ with $\sum_{i \in I} \chi_{x_i U^{-1}}(x) \leqslant N$ for all $x \in G$.

Next, fix any $i \in I$ and note that $\chi_{x_i U^{-1}}(x) \neq 0$ can only hold if $x = x_i u^{-1}$ and thus $x_i = xu$ for some $u \in U$. But in this case, we see by definition of the oscillation $\mathrm{osc}_U^R f$ that

$$|f(x_i)| \leqslant |f(x)| + |f(x_i) - f(x)| \leqslant |f(x)| + (\mathrm{osc}_U^R f)(x) =: F(x).$$

We have thus shown $|f(x_i)| \cdot \chi_{x_i U^{-1}}(x) \leqslant F(x) \cdot \chi_{x_i U^{-1}}(x)$ for all $x \in G$. Summing over $i \in I$, we see

$$\Theta_f(x) := \sum_{i \in I} |f(x_i)| \cdot \chi_{x_i U^{-1}}(x) \leqslant \left(\sum_{i \in I} \chi_{x_i U^{-1}}(x) \right) \cdot F(x) \leqslant N \cdot F(x)$$

for all $x \in G$.

Because of $r \geqslant 1$, we have $\ell_1(I) \hookrightarrow \ell_r(I)$, which implies $\sum_{i \in I} c_i^r \leqslant \left(\sum_{i \in I} c_i\right)^r$ for arbitrary $c_i \geqslant 0$. Therefore,

$$\int_G (m(x))^r \cdot \sum_{i \in I} |f(x_i)|^r \cdot \chi_{x_i U^{-1}}(x)\, dx \leqslant \|\Theta_f\|_{L_{r,m}}^r \leqslant N^r \cdot \|F\|_{L_{r,m}}^r$$
$$\leqslant N^r \cdot \left(\|f\|_{L_{r,m}} + \|\operatorname{osc}_U^R f\|_{L_{r,m}}\right)^r$$
$$\leqslant N^r \cdot (1 + C)^r \cdot \|f\|_{L_{r,m}}^r .$$

Finally, if $\chi_{x_i U^{-1}}(x) \neq 0$, then $x = x_i u^{-1}$ for some $u \in U$, and therefore $m(x_i) = m(xu) \leqslant m(x) \cdot w(u) \leqslant C' \cdot m(x)$ for $C' := \sup_{u \in U} w(u)$, which is finite since w is continuous and U is compact.

Overall, we have thus shown

$$(C')^{-r} \cdot \sum_{i \in I} (m(x_i))^r \cdot |f(x_i)|^r \cdot |x_i U^{-1}| \leqslant \int_G (m(x))^r \cdot \sum_{i \in I} |f(x_i)|^r \cdot \chi_{x_i U^{-1}}(x)\, dx$$
$$\leqslant N^r \cdot (1 + C)^r \cdot \|f\|_{L_{r,m}}^r ,$$

which — because of $|x_i U^{-1}| = |U^{-1}|$ — shows

$$\|\operatorname{Samp}_{\mathcal{X}} f\|_{\ell_{r,m}} \leqslant C' N (1 + C) \cdot |U^{-1}|^{-1/r} \cdot \|f\|_{L_{r,m}} \quad \text{for all } f \in \mathcal{M}_{r,m} ,$$

which finally proves that $\operatorname{Samp}_{\mathcal{X}}$ is well-defined and bounded. $\qquad\square$

Next, we use the result above to show that a sufficiently fine sampling of the point evaluations yields a Banach frame for the reproducing kernel space $\mathcal{M}_{r,m}$.

Proposition 3.4.6. *Let Assumption 3.4.3 be satisfied and let U_0 be a compact unit neighborhood as in part (ii) of Lemma 3.4.4. Further let $U \subset U_0^{-1}$ be a compact unit neighborhood such that the constant $C_{U^{-1}}$ from (3.4.4) satisfies*

$$\|RC_K\|_{L_{r,m} \to L_{r,m}} \cdot C_{U^{-1}} < 1 . \tag{3.4.8}$$

Let $\mathcal{X} = (x_i)_{i \in I}$ be any relatively separated family in G for which there exists a U-BUPU $\Phi = (\varphi_i)_{i \in I}$ with localizing family $\mathcal{X}$. Then there is a bounded linear reconstruction map $R: \ell_{r,m}(I) \to \mathcal{M}_{r,m}$ which satisfies

$$R \circ \operatorname{Samp}_{\mathcal{X}} = \operatorname{id}_{\mathcal{M}_{r,m}}$$

for the sampling map $\operatorname{Samp}_{\mathcal{X}}$ from (3.4.7).

In other words, the family $(\delta_{x_i})_{i \in I}$ of point evaluations forms a Banach frame for $\mathcal{M}_{r,m}$ with coefficient space $\ell_{r,m}(I)$.

Proof. With the sampling map $\operatorname{Samp}_{\mathcal{X}}$ as in (3.4.7) and the synthesis operator $\operatorname{Synth}_{\mathcal{X},\Phi}$ as in Lemma 3.4.2, we define the map

$$B := \operatorname{Synth}_{\mathcal{X},\Phi} \circ \operatorname{Samp}_{\mathcal{X}} : \mathcal{M}_{r,m} \to L_{r,m}(G) .$$

Because of $f(x) = \sum_{i \in I} \varphi_i(x) f(x)$, we have

$$|f(x) - Bf(x)| \leqslant \sum_{i \in I} \varphi_i(x) \cdot |f(x) - f(x_i)| .$$

But if $\varphi_i(x) \neq 0$, then $x = x_i u \in x_i U$, so that $x_i = xu^{-1} \in xU^{-1}$, and hence $|f(x) - f(x_i)| = |f(x) - f(xu^{-1})| \leqslant \operatorname{osc}_{U^{-1}}^R f(x)$. Therefore,

$$|f(x) - Bf(x)| \leqslant \sum_{i \in I} \psi_i(x) \operatorname{osc}_{U^{-1}}^R f(x) = \operatorname{osc}_{U^{-1}}^R f(x).$$

By Proposition 3.2.7 the operator RC_K is a projection onto $\mathcal{M}_{r,m}$, therefore $RC_K f = f$ for $f \in \mathcal{M}_{r,m}$. Thus, the operator $A := RC_K \circ B \colon \mathcal{M}_{r,m} \to \mathcal{M}_{r,m}$ is well-defined and bounded, and we have

$$\begin{aligned}
\|f - Af\|_{L_{r,m}} &= \|RC_K(f - Bf)\|_{L_{r,m}} \\
&\leqslant \|RC_K\|_{L_{r,m} \to L_{r,m}} \cdot \|f - Bf\|_{L_{r,m}} \\
&\leqslant \|RC_K\|_{L_{r,m} \to L_{r,m}} \cdot \|\operatorname{osc}_{U^{-1}}^R f\|_{L_{r,m}} \\
&\leqslant \|RC_K\|_{L_{r,m} \to L_{r,m}} \cdot C_{U^{-1}} \cdot \|f\|_{L_{r,m}}
\end{aligned}$$

for all $f \in \mathcal{M}_{r,m}$.

In view of the assumption (3.4.8), a Neumann series argument, see [4, Sect. 5.7], shows that the bounded linear operator $R_0 := \sum_{n=0}^{\infty} (\operatorname{id}_{\mathcal{M}_{r,m}} - A)^n \colon \mathcal{M}_{r,m} \to \mathcal{M}_{r,m}$ satisfies

$$(R_0 \circ RC_K \circ \operatorname{Synth}_{\mathcal{X},\Psi}) \circ \operatorname{Samp}_{\mathcal{X}} = R_0 \circ A = \operatorname{id}_{\mathcal{M}_{r,m}}.$$

Thus, $R := R_0 \circ RC_K \circ \operatorname{Synth}_{\mathcal{X},\Psi} \colon \ell_{r,m}(I) \to \mathcal{M}_{r,m}$ is the desired reconstruction operator. Note that the action of this operator on a given sequence is independent of the choice of r, m, since the action of the operators RC_K, $\operatorname{Synth}_{\mathcal{X},\Phi}$ and $A = RC_K \circ \operatorname{Synth}_{\mathcal{X},\Phi} \circ \operatorname{Samp}_{\mathcal{X}}$ is independent of r, m, so that the same holds for $R_0 = \sum_{n=0}^{\infty} (\operatorname{id} - A)^n$. $\square$

Remark 3.4.7. The proof shows that the action of the reconstruction operator is *independent* of the choice of r, m.

In other words, if the condition (3.4.8) is satisfied for $L_{r_1,m_1}(G)$ and $L_{r_2,m_2}(G)$ and if both $R_1 \colon \ell_{r_1,m_1}(I) \to \mathcal{M}_{r_1,m_1}$ and $R_2 \colon \ell_{r_2,m_2}(I) \to \mathcal{M}_{r_2,m_2}$ denote the respective reconstruction operators, then $R_1 c = R_2 c$ for all $c \in \ell_{r_1,m_1}(I) \cap \ell_{r_2,m_2}(I)$.

The same holds true by the correspondance principle for the reconstruction operator in the following theorem.

We can finally use the correspondance principle to transfer these results from the reproducing kernel space $\mathcal{M}_{r,m}$ to the coorbit space $\operatorname{Co}(L_{r,m})$.

Theorem 3.4.8. *Under the same assumptions as in Proposition 3.4.6, the sampled family $(\pi(x_i)\psi)_{i \in I} \subset (\operatorname{Co}(L_{r,m}))'$ forms a Banach frame for the coorbit space $\operatorname{Co}(L_{r,m})$ with coefficient space $\ell_{r,m}(I)$.*

More precisely, the sampling operator

$$\operatorname{Samp}_{\mathcal{X},\operatorname{Co}} \colon \operatorname{Co}(L_{r,m}) \to \ell_{r,m}(I), \quad f \mapsto \left(V_{e,\psi}f(x_i)\right)_{i \in I} = \left(\langle f, \pi(x_i)\psi\rangle_{\mathcal{S}_w}\right)_{i \in I} \qquad (3.4.9)$$

is well-defined and bounded, and there exists a bounded linear reconstruction operator $R_{\operatorname{Co}} \colon \ell_{r,m}(I) \to \operatorname{Co}(L_{r,m})$ satisfying

$$R_{\operatorname{Co}} \circ \operatorname{Samp}_{\mathcal{X},\operatorname{Co}} = \operatorname{id}_{\operatorname{Co}(L_{r,m})}.$$

Finally, the action of the reconstruction operator R_{Co} is independent of the choice of r, m, that is, if the assumptions of the current theorem are satisfied for $L_{r_1,m_1}(G)$ and for $L_{r_2,m_2}(G)$ and if R_1, R_2 denote the reconstruction operators, then $R_1 c = R_2 c$ for all $c \in \ell_{r_1,m_1}(I) \cap \ell_{r_2,m_2}(I)$.

Proof. The correspondence principle, cf. Proposition 3.1.19, states that the extended voice transform $V_{e,\psi}\colon \mathrm{Co}(L_{r,m}) \to \mathcal{M}_{r,m}$ is an isometric isomorphism. Now, with the sampling map $\mathrm{Samp}_{\mathcal{X}}$ from Proposition 3.4.6, we have

$$(\mathrm{Samp}_{\mathcal{X}} \circ V_{e,\psi})f = \big(V_{e,\psi}f(x_i)\big)_{i\in I} = \big(\langle f, \pi(x_i)u\rangle_{\mathcal{S}_w}\big)_{i\in I} = \mathrm{Samp}_{\mathcal{X},\mathrm{Co}}\, f\,.$$

Thus, the sampling operator $\mathrm{Samp}_{\mathcal{X},\mathrm{Co}} = \mathrm{Samp}_{\mathcal{X}} \circ V_{e,\psi}\colon \mathrm{Co}(L_{r,m}) \to \ell_{r,m}(I)$ is indeed well-defined and bounded.

Now, with the reconstruction operator $R\colon \ell_{r,m}(I) \to \mathcal{M}_{r,m}$ from Proposition 3.4.6, we define $R_{\mathrm{Co}} := V_{e,\psi}^{-1} \circ R\colon \ell_{r,m}(I) \to \mathrm{Co}(L_{r,m})$. Then

$$R_{\mathrm{Co}} \circ \mathrm{Samp}_{\mathcal{X},\mathrm{Co}} = V_{e,\psi}^{-1} \circ R \circ \mathrm{Samp}_{\mathcal{X}} \circ V_e = V_{e,\psi}^{-1} \circ V_{e,\psi} = \mathrm{id}_{\mathrm{Co}(L_{r,m})},$$

as desired. Since the action of R is independent of the choice of r, m, so is the action of R_{Co}. $\square$

3.4.2 Atomic Decompositions

Finally, in this subsection we provide assumptions under which there exist atomic decompositions for coorbit spaces $\mathrm{Co}(L_{r,m})$. This main result will be stated in Theorem 3.4.15. From now on we assume, as usual, that $1 < r < \infty$ is fixed and that $m\colon G \to \mathbb{R}_+$ is a w-moderate weight.

The assumptions we impose are slightly different compared to the conditions for Banach frames in the previous subsection.

Assumption 3.4.9. *Assume that the kernel K_ψ from (3.1.2) satisfies Assumption 3.1.15, i.e., $K_\psi \in L_{p,w}(G)$ for all $1 < p < \infty$, and the right convolution operator $RC_K\colon L_{r,m}(G) \to L_{r,m}(G)$, $f \mapsto f * K_\psi$ is well-defined.*

Furthermore assume that there is a kernel $W\colon G \to \mathbb{C}$ with the following properties:

(i) W is continuous.

(ii) $W * K_\psi = K_\psi$.

(iii) $M_{U_0}^R W \in L_{1,w}(G) \cap L_{1,w\Delta^{-1}}(G)$ for some compact unit neighborhood $U_0 \subset G$, where

$$M_{U_0}^R W(x) = \|W\|_{L_\infty(U_0 x)}, \quad x \in G \tag{3.4.10}$$

is the right local maximum function defined in (2.1.23).

Remark 3.4.10. Comparing the assumptions we made for Banach frames and atomic decompositions, we see that the only difference is the order in part (ii) and (iii). To be more precise, in Assumption 3.4.3 we suppose that $K_\psi * W = K_\psi$ and pose conditions on the *left* local maximum function, whereas in Assumption 3.4.9 we have $W * K_\psi = K_\psi$ and pose conditions on the *right* local maximum function. Hence, the two assumptions are in a certain sense symmetrical.

Remark 3.4.11. We will use below that $\widetilde{M}_U^R W \in L_{1,w}(G) \cap L_{1,w\Delta^{-1}}(G)$ for *every* compact unit neighborhood $U \subset G$ if we assume $\widetilde{M}_{U_0}^R W \in L_{1,w}(G) \cap L_{1,w\Delta^{-1}}(G)$ for *some* unit neighborhood $U_0 \subset G$.

Indeed, by compactness of U, and since $U \subset \bigcup_{x\in G}(\mathrm{int}\, U_0)x$, there is a finite family $(x_i)_{i=1,\dots,n}$ in G with $U \subset \bigcup_{i=1}^n U_0 x_i$. Therefore, $Ux \subset \bigcup_{i=1}^n U_0 x_i x$, whence

$$\widetilde{M}_U^R W(x) = \|W\|_{L_\infty(Ux)} \leqslant \sum_{i=1}^n \|W\|_{L_\infty(U_0 x_i x)} = \sum_{i=1}^n (\widetilde{M}_{U_0}^R W)(x_i x)\,.$$

By solidity and (left) translation invariance of $L_{1,v}(G)$ for $v = w$ or $v = w\Delta^{-1}$, this implies

$$\|\widetilde{M}_U^R W\|_{L_{1,v}} \leqslant \sum_{i=1}^n \|\lambda(x_i^{-1})(\widetilde{M}_{U_0}^R W)\|_{L_{1,v}} < \infty.$$

Here, the left-translation invariance of $L_{1,v}(G)$ is a consequence of the submultiplicativity of v.

Before we are able to define a suitable reconstruction operator as in Definition 1.9.2, we define, as in the previous subsection, a suitable synthesis operator on $L_{r,m}(G)$. Similarly, we define an analysis operator that is the basis for a coefficient operator.

Lemma 3.4.12. *Let Assumption 3.4.9 be satisfied, and let* $\mathcal{X} = (x_i)_{i \in I}$ *be any relatively separated family in* G. *Then the following hold:*

(i) *If* $\Phi - (\varphi_i)_{i \in I}$ *is a* U-*BUPU with localizing family* $\mathcal{X}$, *then the analysis operator*

$$\mathrm{Ana}_{\mathcal{X},\Psi} : L_{r,m}(G) \to \ell_{r,m}(I), \quad f \mapsto (\langle f, \varphi_i \rangle_{L_2})_{i \in I}$$

is a well-defined bounded linear map.

(ii) *The synthesis map*

$$\mathrm{Synth}_{\mathcal{X},W} : \ell_{r,m}(I) \to L_{r,m}(G), \quad (c_i)_{i \in I} \mapsto \sum_{i \in I} c_i \cdot \lambda(x_i) W$$

is a well-defined bounded linear map, where the defining series is almost everywhere absolutely convergent.

Proof. For $x = x_i u \in x_i U$ we have $m(x_i) = m(xu^{-1}) \leqslant m(x)w(u^{-1}) \leqslant C \cdot m(x)$, where $C := \sup_{u \in U} w(u^{-1})$ is finite by continuity of w and compactness of U. Since we also have $\varphi_i \equiv 0$ on $G \backslash x_i U$, then we see by following the lines of the proof of Proposition 3.2.13 and using Jensen's inequality, see [36, Theorem 10.2.6]:

$$m(x_i)^r \cdot |\langle f, \varphi_i \rangle_{L_2}|^r \leqslant |x_i U|^r \cdot \left(\int_{x_i U} |f(x)| m(x_i) \varphi_i(x) \frac{dx}{|x_i U|} \right)^r$$

$$\leqslant |x_i U|^r \cdot \int_{x_i U} (|f(x)| \cdot C \cdot m(x) \varphi_i(x))^r \frac{dx}{|x_i U|}$$

$$\leqslant |U|^{r-1} \cdot C^r \cdot \int_G |(m \cdot f)(x)|^r \cdot \varphi_i(x) \, dx,$$

where the last step used the left invariance of the Haar measure and the estimate $(\varphi_i(x))^r \leqslant \varphi_i(x)$ which holds since $\varphi_i(x) \in [0,1]$ and $r \geqslant 1$.

Summing over $i \in I$ and applying the monotone convergence theorem, we thus get

$$\| \mathrm{Ana}_{X,\Phi} f \|_{\ell_{r,m}}^r = \sum_{i \in I} \big(m(x_i) \cdot |\langle f, \varphi_i \rangle_{L_2}| \big)^r$$

$$\leqslant |U|^{r-1} \cdot C^r \cdot \int_G |(m \cdot f)(x)|^r \cdot \sum_{i \in I} \varphi_i(x) \, dx$$

$$= |U|^{r-1} \cdot C^r \cdot \|f\|_{L_{r,m}}^r < \infty,$$

thereby proving (i), that is the boundedness and well-definedness of $\mathrm{Ana}_{X,\Psi}$.

We now consider the synthesis map $\mathrm{Synth}_{\mathcal{X},W}$. Let $V \subset G$ be any compact unit neighborhood, and set $Q := \mathrm{int}\, V$, so that $U := \overline{Q} \subset V$ is a compact unit neighborhood that satisfies $\overline{\mathrm{int}\, U} \supset$

$\overline{Q} = U$ and hence $U = \overline{\operatorname{int} U}$. As a consequence, as seen in the proof of Lemma 3.2.3, we have $\|W\|_{L_\infty(Ux)} = \sup_{y \in Ux} |W(y)|$ for all $x \in G$. Here we used that W is continuous.

Now, let $x \in G$ and $i \in I$ be arbitrary. For any $y = x_i u \in x_i U$ we then have $x_i^{-1} x = (yu^{-1})^{-1} x = uy^{-1}x \in Uy^{-1}x$, and thus

$$|W(x_i^{-1}x)| \leq \|W\|_{L_\infty(Uy^{-1}x)} = (\widetilde{M}_U^R W)(y^{-1}x) \quad \text{for all } x \in G \text{ and } y \in x_i U.$$

Writing $\Theta := \widetilde{M}_U^R W$, and averaging this estimate over $y \in x_i U$, we get

$$|\lambda(x_i)W(x)| \leq \frac{1}{|U|} \int_G \chi_{x_i U}(y) \cdot \Theta(y^{-1}x) \, dy \quad \text{for all } x \in G, \, i \in I. \tag{3.4.11}$$

Now, let $c = (c_i)_{i \in I} \in \ell_{r,m}(I)$ be arbitrary, and set $\Upsilon := \sum_{i \in I} |c_i| \cdot \chi_{x_i U}$. With the notation introduced in Lemma 3.4.2 we get $\Upsilon = \operatorname{Synth}_{\mathcal{X},U} |c|$ with $|c| = (|c_i|)_{i \in I}$. This easily implies $\Upsilon \in L_{r,m}(G)$ with

$$\|\Upsilon\|_{L_{r,m}} \leq C \cdot \|c\|_{\ell_{r,m}} \tag{3.4.12}$$

for a constant $C = C(m, \mathcal{X}, U, r)$ independent of c.

By weighting estimate (3.4.11) with $|c_i|$ and summing over $i \in I$, and by invoking the monotone convergence theorem, we see for all $x \in G$ that

$$\sum_{i \in I} |c_i| \cdot |(\lambda(x_i)W)(x)| \leq \frac{1}{|U|} \cdot \int_G \Upsilon(y) \cdot \Theta(y^{-1}x) \, dy = \frac{1}{|U|} \cdot (\Upsilon * \Theta)(x).$$

But since $\Theta = \widetilde{M}_U^R W \in L_{1,w}(G) \cap L_{1,w\Delta^{-1}}(G)$, see Assumption 3.4.9 and Remark 3.4.11, and since $\Upsilon \in L_{r,m}(G)$, Young's inequality, see Proposition 1.3.9, shows $\Upsilon * \Theta \in L_{r,m}(G)$. In particular, this implies $\Upsilon * \Theta(x) < \infty$ almost everywhere. Therefore, we already see that the series defining $\operatorname{Synth}_{\mathcal{X},W} c$ is almost everywhere absolutely convergent. Finally, we also see

$$\|\operatorname{Synth}_{\mathcal{X},W} c\|_{L_{r,m}} \leq \left\| \sum_{i \in I} |c_i| \cdot |\lambda(x_i)W| \right\|_{L_{r,m}} \leq \frac{1}{|U|} \cdot \|\Upsilon * \Theta\|_{L_{r,m}}$$

$$\leq \frac{1}{|U|} \cdot \max\{\|\Theta\|_{L_{1,w}}, \|\Theta\|_{L_{1,w\Delta^{-1}}}\} \cdot \|\Upsilon\|_{L_{r,m}}.$$

In view of (3.4.12), this proves the boundedness and well-definedness of $\operatorname{Synth}_{\mathcal{X},W}$. $\square$

With this at hand we can prove the existence of atomic decompositions for the reproducing kernel space $\mathcal{M}_{r,m}$.

Proposition 3.4.13. *Let Assumption 3.4.9 be satisfied. For each compact unit neighborhood $U \subset G$ write*

$$C_U := \max\{\|\operatorname{osc}_U W\|_{L_{1,w}}, \|\operatorname{osc}_U W\|_{L_{1,w\Delta^{-1}}}\} \tag{3.4.13}$$

and we assume that

$$C_U \cdot \|RC_K\|_{L_{r,m} \to L_{r,m}} < 1. \tag{3.4.14}$$

Finally, let $\mathcal{X} = (x_i)_{i \in I}$ be a relatively separated family for which there exists a U-BUPU $\Phi = (\varphi_i)_{i \in I}$ with localizing family $\mathcal{X}$. Then the family $(\lambda(x_i)K)_{i \in I}$ forms a family of atoms for $\mathcal{M}_{r,m}$ with associated sequence space $\ell_{r,m}(I)$. This means:

(i) The synthesis operator

$$\operatorname{Synth}_{\mathcal{X},K} : \ell_{r,m}(I) \to \mathcal{M}_{r,m}, \, (c_i)_{i \in I} \mapsto \sum_{i \in I} c_i \cdot \lambda(x_i)K_\psi \tag{3.4.15}$$

is well-defined and bounded, with unconditional convergence of the defining series. This even holds without assuming (3.4.14).

(ii) There is a bounded coefficient operator $C \colon \mathcal{M}_{r,m} \to \ell_{r,m}(I)$ *with*

$$\mathrm{Synth}_{\mathcal{X},K} \circ C = \mathrm{id}_{\mathcal{M}_{r,m}} .$$

Proof. **Step 1:** We start by showing the boundedness of the synthesis operator defined in (3.4.15). For this step we will *not* use condition (3.4.14). By Assumption 3.4.9, the right convolution operator $RC_K \colon L_{r,m}(G) \to L_{r,m}(G)$ is bounded, and we have $W * K_\psi = K$, which implies $(\lambda(x)W) * K_\psi = \lambda(x)(W * K_\psi) = \lambda(x)K_\psi$ for all $x \in G$.

Furthermore, Lemma 3.4.12 shows that the map

$$\mathrm{Synth}_{\mathcal{X},W} \colon \ell_{r,m_\mathcal{X}}(I) \to L_{r,m}(G), \quad (c_i)_{i \in I} \mapsto \sum_{i \in I} c_i \cdot \lambda(x_i)W$$

is well-defined and bounded. Because of $r < \infty$, each $c = (c_i)_{i \in I} \in \ell_{r,m}(I)$ satisfies $c = \sum_{i \in I} c_i \delta_i$ with unconditional convergence in $\ell_{r,m}(I)$, where $(\delta_i)_{i \in I}$ denotes the standard basis of $\ell_{r,m}(I)$. This implies that the series defining $\mathrm{Synth}_{\mathcal{X},W} c$ converges unconditionally in $L_{r,m}(G)$. Since bounded operators preserve unconditional convergence, we see

$$RC_K\big(\mathrm{Synth}_{\mathcal{X},W} c\big) = RC_K\left(\sum_{i \in I} c_i \, \lambda(x_i)W\right)$$

$$= \sum_{i \in I} c_i \left[(\lambda(x_i)W) * K_\psi\right] = \sum_{i \in I} c_i \, \lambda(x_i)K_\psi$$

with unconditional convergence of the series. We have thus shown that $\mathrm{Synth}_{\mathcal{X},K} = RC_K \circ \mathrm{Synth}_{\mathcal{X},W} \colon \ell_{r,m}(I) \to L_{r,m}(G)$ is well-defined and bounded, with unconditional convergence of the defining series.

Since $\lambda(x_i)K_\psi \in \mathcal{M}_{r,m}$ for all $i \in I$, we also see that the range of $\mathrm{Synth}_{\mathcal{X},K}$ is contained in the closed subspace $\mathcal{M}_{r,m} \subset L_{r,m}(G)$.

Step 2: In this step we will prove an alternative reproducing formular for the space $\mathcal{M}_{r,m}$, more precisely

$$f = (f * W) * K_\psi \qquad \text{for all } f \in \mathcal{M}_{r,m} . \tag{3.4.16}$$

This is almost trivial: For $f \in \mathcal{M}_{r,m}$, we have $f = f * K_\psi$ by definition of $\mathcal{M}_{r,m}$, and we have $K_\psi = W * K_\psi$ by Assumption 3.4.9. By combining these facts, we get $f = f * K_\psi = f * (W * K_\psi)$. Thus, all we need to verify is that the convolution is associative in the setting that we consider here.

In light of [18, Lemma 6.3] to prove this it remains to show $((|f| * |W|) * |K_\psi|)(x) < \infty$ for almost all $x \in G$. To this end, we first show $W \in L_{1,w}(G) \cap L_{1,w\Delta^{-1}}(G)$. In order to see this, let $V \subset G$ be any compact unit neighborhood, and set $Q := \mathrm{int}\, V$, so that $U := \overline{Q} \subset V$ is a compact unit neighborhood that satisfies $\overline{\mathrm{int}\, U} \supset \overline{Q} = U$ and hence $U = \overline{\mathrm{int}\, U}$. As a consequence of this and of the continuity of W, as seen in the proof of Lemma 3.2.3, we have

$$\widetilde{M}_U^R W(x) = \|W\|_{L_\infty(Ux)} = \sup_{y \in Ux} |W(y)| \geqslant |W(x)| \quad \text{for all } x \in G .$$

Since $\widetilde{M}_U^R W \in L_{1,w}(G) \cap L_{1,w\Delta^{-1}}(G)$, see Assumption 3.4.9 and Remark 3.4.11, we have $W \in L_{1,w}(G) \cap L_{1,w\Delta^{-1}}(G)$.

Now, fix some $s \in (r, \infty)$ and let $f \in \mathcal{M}_{r,m}$. Because of $W \in L_{1,w}(G) \cap L_{1,w\Delta^{-1}}(G)$, Proposition 1.3.11 shows $|f| * |W| \in L_{r,m}(G)$. Therefore, by the second part of Proposition 1.3.11, we see $(|f| * |W|) * |K_\psi| \in L_{s,m}(G)$, since $|K_\psi(x^{-1})| = |K_\psi(x)|$ and since $K_\psi \in L_{p,w}(G)$ for all $p \in (1, \infty)$. In particular, $((|f| * |W|) * |K_\psi|)(x) < \infty$ for almost all $x \in G$. By the considerations from above, we thus see that (3.4.16) holds.

Step 3: In this step we will discretize and approximate the convolution $f \mapsto f * W$. To this end let $\mathrm{Ana}_{\mathcal{X},\Phi} : L_{r,m}(G) \to \ell_{r,m}(I)$ and $\mathrm{Synth}_{\mathcal{X},W} : \ell_{r,m}(I) \to L_{r,m}(G)$ be as defined in Lemma 3.4.12, and define $A := \mathrm{Synth}_{\mathcal{X},W} \circ \mathrm{Ana}_{\mathcal{X},\Phi} : L_{r,m}(G) \to L_{r,m}(G)$. In this step, we will show

$$\|f * W - Af\|_{L_{r,m}} \leqslant C_U \cdot \|f\|_{L_{r,m}} \qquad \text{for all } f \in L_{r,m}(G), \tag{3.4.17}$$

with C_U as in (3.4.13).

To this end, recall from the previous step that $W \in L_{1,w}(G) \cap L_{1,w\Delta^{-1}}(G)$, so that Young's inequality, cf. Proposition 1.3.9, shows $|f| * |W| \in L_{r,m}(G)$ for $f \in L_{r,m}(G)$. In particular, this implies $|f| * |W|(x) < \infty$ for almost all $x \in G$. For each such $x \in G$, the dominated convergence theorem and the definition of the operators $\mathrm{Ana}_{\mathcal{X},\Psi}$, $\mathrm{Synth}_{\mathcal{X},W}$ and A shows

$$|f * W(x) - Af(x)| = \left| \sum_{i \in I} \int_G f(y)\varphi_i(y)W(y^{-1}x)\,dy - \sum_{i \in I} \langle f, \varphi_i \rangle_{L_2}(\lambda(x_i)W)(x) \right|$$

$$\leqslant \sum_{i \in I} \int_G \varphi_i(y) \cdot |f(y)| \cdot |W(y^{-1}x) - W(x_i^{-1}x)|\,dy.$$

Fix $i \in I$ for the moment. For $y \in G$ with $\varphi_i(y) \neq 0$, we have $y = x_i u \in x_i U$, and thus $x_i^{-1}x = uy^{-1}x \in Uy^{-1}x$. Therefore, $|W(y^{-1}x) - W(x_i^{-1}x)| \leqslant (\mathrm{osc}_U W)(y^{-1}x)$, by definition of the oscillation $\mathrm{osc}_U W$, see (3.2.8).

If we combine this with the estimate from above and with the monotone convergence theorem, we get

$$|f * W(x) - Af(x)| \leqslant \sum_{i \in I} \int_G \varphi_i(y) \cdot |f(y)| \cdot (\mathrm{osc}_U W)(y^{-1}x)\,dy$$

$$= \int_G |f(y)| \cdot (\mathrm{osc}_U W)(y^{-1}x)\,dy = (|f| * \mathrm{osc}_U W)(x).$$

In view of Young's inequality, see Proposition 1.3.9, and the definition of C_U, see (3.4.13), we see that (3.4.17) holds true.

Step 4: Finally, we use the preceding steps to conclude the proof. Recall that the operator $RC_K : L_{r,m}(G) \to \mathcal{M}_{r,m}$ is bounded by Assumption 3.4.9. Thus, the operator defined by $B := RC_K \circ A|_{\mathcal{M}_{r,m}} : \mathcal{M}_{r,m} \to \mathcal{M}_{r,m}$ is well-defined and bounded, with A as in the preceding step. Now, for arbitrary $f \in \mathcal{M}_{r,m}$ our results from Steps 2 and 3 show

$$\|f - Bf\|_{L_{r,m}} = \|RC_K(f * W - Af)\|_{L_{r,m}} \leqslant \|RC_K\|_{L_{r,m} \to L_{r,m}} \cdot C_U \cdot \|f\|_{L_{r,m}}.$$

In view of our assumption (3.4.14), a Neumann series argument, see [4, Sect. 5.7], shows that $C_0 := \sum_{n=0}^{\infty} (\mathrm{id}_{\mathcal{M}_{r,m}} - B)^n$ defines a bounded linear operator $C_0 : \mathcal{M}_{r,m} \to \mathcal{M}_{r,m}$ which satisfies $B \circ C_0 = \mathrm{id}_{\mathcal{M}_{r,m}}$.

But we saw in Step 1 that $\mathrm{Synth}_{\mathcal{X},K} = RC_K \circ \mathrm{Synth}_{\mathcal{X},W}$, so that

$$B = RC_K \circ A|_{\mathcal{M}_{r,m}} = RC_K \circ \mathrm{Synth}_{\mathcal{X},W} \circ \mathrm{Ana}_{\mathcal{X},\Psi}|_{\mathcal{M}_{r,m}}$$

$$= \mathrm{Synth}_{\mathcal{X},K} \circ \mathrm{Ana}_{\mathcal{X},\Psi}|_{\mathcal{M}_{r,m}}.$$

Thus, the operator $C := \mathrm{Ana}_{\mathcal{X},\Psi}|_{\mathcal{M}_{r,m}} \circ C_0 : \mathcal{M}_{r,m} \to \ell_{r,m}(I)$ satisfies

$$\mathrm{Synth}_{\mathcal{X},K} \circ C = B \circ C_0 = \mathrm{id}_{\mathcal{M}_{r,m}}.$$

It is not hard to see that the action of the coefficient operator C is independent of the choice of r, m. For more details see the end of the proof of Proposition 3.4.6, where a similar claim is shown. $\qquad \square$

Remark 3.4.14. (i) We note that condition (3.4.14) is always satisfied for U small enough, thanks to Lemma 3.2.3 and Assumption 3.4.9.

(ii) As in Proposition 3.4.6, the action of the coefficient operator C is independent of the choice of r, m, that is, if condition (3.4.14) is satisfied for $L_{r_1,m_1}(G)$ and $L_{r_2,m_2}(G)$ and if $C_1 \colon \mathcal{M}_{r_1,m_1}(I) \to \ell_{r_1,m_1}(I)$ and $C_2 \colon \mathcal{M}_{r_2,m_2} \to \ell_{r_2,m_2}(I)$ are the respective coefficient operators, then $C_1 f = C_2 f$ for all $f \in \mathcal{M}_{r_1,m_1} \cap \mathcal{M}_{r_2,m_2}$.

Now we can transfer the results from Proposition 3.4.13 from the reproducing kernel spaces $\mathcal{M}_{r,m}$ to the coorbit spaces $\mathrm{Co}(L_{r,m})$ using the corrseponding principle.

Theorem 3.4.15. *Under the same assumptions as in Proposition 3.4.13, the sampled family $(\pi(x_i)\psi)_{i\in I} \subset \mathrm{Co}(L_{r,m})$ forms a family of atoms for $\mathrm{Co}(L_{r,m})$ with coefficient space $\ell_{r,m}(I)$. More precisely, the synthesis operator*

$$\mathrm{Synth}_{\mathcal{X},\mathrm{Co}} \colon \ell_{r,m}(I) \to \mathrm{Co}(L_{r,m}), \quad (c_i)_{i\in I} \mapsto \sum_{i\in I} c_i \cdot \pi(x_i)\psi$$

is well-defined and bounded, and there is a bounded linear coefficient operator denoted by $C_{\mathrm{Co}} \colon \mathrm{Co}(L_{r,m}) \to \ell_{r,m}(I)$ satisfying $\mathrm{Synth}_{\mathcal{X},\mathrm{Co}} \circ C_{\mathrm{Co}} = \mathrm{id}_{\mathrm{Co}(L_{r,m})}$.

Finally, the action of the coefficient operator C_{Co} is independent of the choice of r, m. In other words, if the assumptions of the current theorem are satisfied for both $L_{r_1,m_1}(G)$ and for $L_{r_2,m_2}(G)$ and if C_1, C_2 denote the corresponding coefficient operators, then $C_1 f = C_2 f$ for all $f \in \mathrm{Co}(L_{r_1,m_1}) \cap \mathrm{Co}(L_{r_2,m_2})$.

Proof. The correspondence principle, cf. Proposition 3.1.19, states that the extended voice transform $V_{e,\psi} \colon \mathrm{Co}(L_{r,m}) \to \mathcal{M}_{r,m}$ is an isometric isomorphism. Furthermore,

$$(V_{e,\psi}\,\pi(x)\,u)(y) = \langle \pi(x)\,\psi, \pi(y)\,\psi\rangle_{\mathcal{S}_w} = \langle \pi(x)\,\psi, \pi(y)\,\psi\rangle_{\mathcal{H}}$$
$$= K_\psi(x^{-1}y) = (\lambda(x)K_\psi)(y)$$

for all $x, y \in G$. In other words, $V_{e,\psi}\,\pi(x)\,\psi = \lambda(x)K_\psi$ for all $x \in G$.

Now, since the bounded linear operator $V_{e,\psi}^{-1} \colon \mathcal{M}_{r,m} \to \mathrm{Co}(L_{r,m})$ preserves unconditional convergence of series, the synthesis operator $\mathrm{Synth}_{\mathcal{X},K}$ from Proposition 3.4.13 satisfies

$$(V_{e,\psi}^{-1} \circ \mathrm{Synth}_{\mathcal{X},K})\,(c_i)_{i\in I} = V_{e,\psi}^{-1}\left(\sum_{i\in I} c_i \cdot \lambda(x_i)K_\psi\right) = \sum_{i\in I} c_i \cdot V_{e,\psi}^{-1}(\lambda(x_i)K_\psi)$$
$$= \sum_{i\in I} c_i \cdot \pi(x_i)\psi = \mathrm{Synth}_{\mathcal{X},\mathrm{Co}}\,(c_i)_{i\in I},$$

for arbitrary $(c_i)_{i\in I} \in \ell_{r,m}(I)$, with unconditional convergence of all involved series. Thus, the operator $\mathrm{Synth}_{\mathcal{X},\mathrm{Co}} = V_{e,\psi}^{-1} \circ \mathrm{Synth}_{\mathcal{X},K} \colon \ell_{r,m}(I) \to \mathrm{Co}(L_{r,m})$ is indeed well-defined and bounded.

Now, with the coefficient operator $C \colon \mathcal{M}_{r,m} \to \ell_{r,m}(I)$ from Proposition 3.4.13, we define $C_{\mathrm{Co}} := C \circ V_{e,\psi} \colon \mathrm{Co}(L_{r,m}) \to \ell_{r,m}(I)$. Then

$$\mathrm{Synth}_{\mathcal{X},\mathrm{Co}} \circ C_{\mathrm{Co}} = V_{e,\psi}^{-1} \circ \mathrm{Synth}_{\mathcal{X},K} \circ C \circ V_{e,\psi} = V_{e,\psi}^{-1} \circ V_{e,\psi} = \mathrm{id}_{\mathrm{Co}(L_{r,m})},$$

as desired. Since the action of C is independent of the choice of r, m, so is the action of C_{Co}. $\square$

3.4.3   Revisiting Paley-Wiener Spaces II

In this subsection we revisit the Paley-Wiener spaces B_Ω^p introduced in Subsection 3.2.3 for the second time. We will not recall all properties of these spaces, we refer to Subsection 3.2.3 for a detailed describtion.

Let $\Omega \subset \mathbb{R}$ be a fixed set and $1 \leqslant p < \infty$, then the Paley-Wiener p-space B_Ω^p of functions with band in Ω is defined as

$$B_\Omega^p = \{f \in L_p(\mathbb{R}) : \mathrm{supp}(\mathcal{F}f) \subseteq \Omega\}. \tag{3.4.18}$$

These spaces can be understood as coorbit spaces with kernel $K = \mathcal{F}^{-1}\chi_\Omega$, at least if $\Omega = [-\omega, \omega]$ for some $\omega > 0$, so the kernel is clearly not in $L_1(\mathbb{R})$. The right convolution operator RC_K associated with K is a bounded operator on $L_r(\mathbb{R})$ for all $1 < r < \infty$. This implies that Assumption 3.2.4 is fulfilled, which is necessary for the discretization results in Subsection 3.2.2 as well as in the two preceding subsections.

However, in Subsection 3.3.2 we already revisited the Paley-Wiener spaces and showed in Proposition 3.3.5 that there also exists a kernel function $K = \mathcal{F}^{-1}\chi_C$ for a certain compact set $C \subset [0,1]$ whose associated right convolution operator does not act boundedly on $L_r(\mathbb{R})$ for any $r \neq 2$. In light of the results in Subsection 3.3.1 this means that no meaningful discretization of related Paley-Wiener spaces is possible.

In the present subsection we will apply the results of the preceding two subsections concerning Banach frames and atomic decompositions to the Paley-Wiener p-spaces B_Ω^p, thereby improving the discretization results we have obtained so far. Moreover, we point out sufficient conditions on the set Ω under which the discretization results hold. We present these condtions in the following assumption.

Assumption 3.4.16. *Let $\Omega \subset \mathbb{R}$ be measurable, and let $r \in (1, \infty)\backslash\{2\}$. Assume the following properties hold:*

   (i) Ω is bounded;

   (ii) the kernel $K := \mathcal{F}^{-1}\chi_\Omega$ satisfies $K \in \bigcap_{1<p<\infty} L_p(\mathbb{R})$;

   (iii) the convolution operator RC_K is well-defined on $L_r(\mathbb{R})$.

Remark 3.4.17. The last property means that the indicator function χ_Ω is an $L_r(\mathbb{R})$-Fourier multiplier, which implies that $\chi_{\Omega^c} = 1 - \chi_\Omega$ is an $L_r(\mathbb{R})$-Fourier multiplier as well. Therefore, [83, Theorem 1] shows that there is an open set $U \subset \mathbb{R}$ with $\chi_{\Omega^c} = \chi_U$ almost everywhere, and thus $\chi_{U^c} = \chi_\Omega$ almost everywhere. But since Ω is bounded, we have $\Omega \subset [-R, R]$ for some $R > 0$, and then $\chi_\Omega = \chi_\Omega\chi_{[-R,R]} = \chi_{U^c}\chi_{[-R,R]} = \chi_{[-R,R]\backslash U}$ almost everywhere. Thus, by modifying Ω on a null set, we will always assume that Ω is compact. This neither changes the kernel K, nor the underlying Hilbert space

$$\mathcal{H} := B_\Omega^2 := \{f \in L_2(\mathbb{R}) : \mathrm{supp}(\mathcal{F}f) \subseteq \Omega\}.$$

As seen in the discussion in Subsection 3.2.3, if we set $u := K = \mathcal{F}^{-1}\chi_\Omega$ as the admissible vector, then all standing assumptions from Subsection 3.1.2 are satisfied for $m = w \equiv 1$, so that the coorbit spaces $\mathrm{Co}(L_p)$ are well-defined for $1 < p < \infty$. Furthermore, we saw in (3.2.50) that the associated voice transform satisfies $V_u f = \mathcal{F}^{-1}(\hat{u}\hat{f}) = f$ for all $f \in \mathcal{H} = B_\Omega^2 \subset L_2(\mathbb{R})$. Using this identity, we can now show that the identy (3.4.18) holds true for the coorbit spaces $\mathrm{Co}(L_p)$ if $1 < p \leqslant 2$.

Lemma 3.4.18. *Setting* $\mathcal{T} := \bigcap_{1 < p < \infty} L_p(\mathbb{R})$ *as in (3.1.12), the space* $\mathcal{S}$ *from (3.1.16) satisfies*

$$\mathcal{S} = \{ f \in \mathcal{T} \,:\, f * K = f \} \,,$$

with topology generated by the family of norms $(\| \cdot \|_{L_p})_{1 < p < \infty}$.
Furthermore, with $\mathcal{M}_p = \{ f \in L_p(\mathbb{R}) \,:\, f * K = f \}$ *and* $\mathcal{M} := \bigcup_{1 < p < \infty} \mathcal{M}_p$, *the map*

$$\iota : \mathcal{M} \to \mathcal{S}', f \mapsto \Phi_f \quad \text{with} \quad \langle \Phi_f, g \rangle_{\mathcal{S} \times \mathcal{S}} := \langle f, g \rangle_{L_2} \quad \text{for all} \quad g \in \mathcal{S} \subset L_2(\mathbb{R}) \qquad (3.4.19)$$

is a bijection. If we identify each $\varphi \in \mathcal{S}'$ *with its inverse image* $\iota^{-1}\varphi \in \mathcal{M}$ *under this map, then the extended voice transform is the identity map, that is* $V_e \varphi = \iota^{-1}\varphi$.
According to the general result, the coorbit spaces $\mathrm{Co}(L_p)$ *are given by*

$$\mathrm{Co}(L_p) = \iota(\mathcal{M}_p) \qquad \text{for all } p \in (1, \infty) \,, \qquad (3.4.20)$$

which means that if we identify φ *with* $\iota^{-1}\varphi$, *then* $\mathrm{Co}(L_p) = \mathcal{M}_p$.
Finally for $1 < p \leqslant 2$ *we have*

$$\mathcal{M}_p = B_\Omega^p := \left\{ f \in L_p(\mathbb{R}) \,:\, \mathrm{supp}(\mathcal{F}f) \subseteq \Omega \right\}.$$

Therefore, up to canonical identifications, the coorbit spaces $\mathrm{Co}(L_p)$ *coincide with the Paley-Wiener spaces* B_Ω^p.

Proof. The following proof is similar to the proof of [18, Proposition 4.8] with some significant improvements and generalizations.

The first property is an immediate consequence of the definitions, combined with $Vf = f$ for $f \in \mathcal{H}$.

The map ι is indeed well-defined, since if $f \in \mathcal{M}_p$ for some $p \in (1, \infty)$, then $|\langle f, g \rangle_{L_2}| \leqslant \|f\|_{L_p} \cdot \|g\|_{L_{p'}}$, where $\| \cdot \|_{L_{p'}}$ is a continuous norm on $\mathcal{S}$.

To prove the surjectivity of ι, we first show that $\mathcal{M}$ is a complex vector space. Since each $\mathcal{M}_p$ is closed under multiplication with complex numbers, we only need to show that $\mathcal{M}$ is closed under addition. To this end, note for $f \in \mathcal{M}_p$ because of $K \in L_{p'}(\mathbb{R})$ that

$$|f(x)| = |(f * K)(x)| = |\langle f, \lambda(x)K \rangle_{L_2}| \leqslant \|f\|_{L_p} \cdot \|\lambda(x)K\|_{L_{p'}} \leqslant C_p \cdot \|f\|_{L_p}$$

for all $x \in \mathbb{R}$, and thus $\mathcal{M}_p \hookrightarrow L_\infty$. This embedding implies $\mathcal{M}_p \subset \mathcal{M}_q$ for $p \leqslant q$, and thus $\mathcal{M}_p + \mathcal{M}_q \subset \mathcal{M}_q + \mathcal{M}_q = \mathcal{M}_q \subset \mathcal{M}$. From this, we easily see that $\mathcal{M}$ is a vector space.

With $\mathcal{M}$ being a vector space, we see $\mathcal{M} = \mathrm{span}\bigcup_{1 < p < \infty} \mathcal{M}_p$. With notation as in (3.1.21), this means $\mathcal{M} = \mathcal{M}^\mathcal{U}$. Hence, Theorem 3.1.15 shows for arbitrary $\varphi \in \mathcal{S}'$ that $f := V_e \varphi \in \mathcal{M}^\mathcal{U} = \mathcal{M}$, and (3.1.20) shows because of $Vg = g$ for $g \in \mathcal{S} \subset \mathcal{H}$ that

$$\langle \Phi_f, g \rangle_\mathcal{S} = \langle f, g \rangle_{L_2} = \langle V_e \varphi, Vg \rangle_{L_2} = \langle \varphi, g \rangle_\mathcal{S} \,.$$

Hence, $\varphi = \Phi_f = \iota f = \iota V_e \varphi$. On the one hand, this shows that ι is surjective, and on the other hand—once we know that ι is bijective—it proves that the inverse of ι is given as a map $\iota^{-1} = V_e : \mathcal{S}' \to \mathcal{M}$.

In order to prove that ι is injective, note $\lambda(x)K \in \mathcal{S}$ for all $x \in \mathbb{R}$ and recall $\overline{K(x)} = K(-x)$. Hence,

$$\langle \Phi_f, \lambda(x)K \rangle_\mathcal{S} = \langle f, \lambda(x)K \rangle_{L_2} = (f * K)(x) = f(x) \quad \text{for} \quad f \in \mathcal{M} \,.$$

Therefore, if $\Phi_f = 0$, then $f = 0$ as well. Since the domain $\mathcal{M}$ of ι is a vector space, this shows that ι is injective.

Equation (3.4.20) is seen to be true by combining the identity $V_e = \iota^{-1}$ with the correspondence principle, see Proposition 3.1.19, which states that

$$V_e \colon \mathrm{Co}(L_p) \to \{f \in L_p(\mathbb{R}) \mid f * K = f\} = \mathcal{M}_p$$

is an isomorphism.

To prove $\mathcal{M}_p = B_\Omega^p$ for $p \in (1, 2]$, first note $\mathcal{F}(f * g) = \widehat{f} \cdot \widehat{g}$ for arbitrary $f, g \in L_2$, see e.g. [95, p. 270]. Therefore, for $f \in \mathcal{M}_p \hookrightarrow \mathcal{M}_2$ (here we used that $p \leqslant 2$) we see that $\widehat{f} = \widehat{f * K} = \widehat{f} \cdot \widehat{K} = \widehat{f} \cdot \chi_\Omega$, where the equality holds in the sense of tempered distributions. But since both sides are $L_2(\mathbb{R})$ functions, this implies $\widehat{f} = \widehat{f} \cdot \chi_\Omega$ almost everywhere, and thus $f \in B_\Omega^p$.

Conversely, let $f \in B_\Omega^p$ be arbitrary. Because of $p \leqslant 2$, [100, Theorem in Sect. 1.4.1] shows $f \in L_2(\mathbb{R})$. Furthermore, since $\widehat{f} \equiv 0$ almost everywhere on $\mathbb{R} \backslash \Omega$, we have $\mathcal{F}(f * K) = \widehat{f} \cdot \widehat{K} = \widehat{f} \cdot \chi_\Omega = \widehat{f}$, and thus $f = f * K$, so that $f \in \mathcal{M}_p$. $\qquad\square$

Remark 3.4.19. We do not know if in general the identity $\mathcal{M}_p = B_\Omega^p$ with

$$B_\Omega^p = \big\{f \in L_p(\mathbb{R}) : \text{the tempered distribution } \widehat{f} \text{ has } \mathrm{supp}(\widehat{f}) \subset \Omega\big\}$$

also holds for $p \in (2, \infty)$. In case of $\Omega = [-\omega, \omega]$, it was shown in [18, Proposition 4.8] that this is true, as we recalled in Subsection 3.2.3. Using this, one can show $\mathcal{M}_p = B_\Omega^p$ even if Ω is a finite disjoint union of compact intervals. For more general sets Ω, however, we do not know whether $\mathcal{M}_p = B_\Omega^p$ for $2 < p < \infty$.

Since the results in Lemma 3.4.18 have shown that Paley-Wiener spaces B_Ω^p can be interpreted as coorbit spaces $\mathrm{Co}(L_r)$ we can apply the discretization machinery from the previous subsections. Moreover, because the reproducing kernel spaces $\mathcal{M}_p$ coincide with the Paley-Wiener spaces, it suffices to find discretizations for $\mathcal{M}_p$.

In the next proposition we show the existence of Banach frames and atomic decompositions for $\mathcal{M}_p$. Additionally, in Subsections 3.4.1 and 3.4.2 we have shown that for these discretizations a sufficiently dense family of sampling points $(x_i)_{i \in I}$ in G is necessary. For the case of Paley-Wiener spaces, one can state quite precisely how dense the sampling points need to be.

Proposition 3.4.20. *Suppose that Assumption 3.4.16 is satisfied, and choose $R > 0$ and $\xi_0 \in \mathbb{R}$ with $\Omega \subset [\xi_0 - R, \xi_0 + R]$.*

Then the family

$$\left(K\left(\cdot - \frac{k}{2R}\right)\right)_{k \in \mathbb{Z}} = \left(\mathcal{F}^{-1}\chi_\Omega\left(\cdot - \frac{k}{2R}\right)\right)_{k \in \mathbb{Z}} \tag{3.4.21}$$

forms a Banach frame and an atomic decomposition for the reproducing kernel space $\mathcal{M}_r$ with coefficient space $\ell_r(\mathbb{Z})$. More precisely, the operators

$$\mathrm{Samp} \colon \mathcal{M}_r \to \ell_r(\mathbb{Z}), f \mapsto \left(f\left(\frac{k}{2R}\right)\right)_{k \in \mathbb{Z}} = \left(\left\langle f, K\left(\cdot - \frac{k}{2R}\right)\right\rangle_{L_2}\right)_{k \in \mathbb{Z}} \tag{3.4.22}$$

and

$$\mathrm{Synth} \colon \ell_r(\mathbb{Z}) \to \mathcal{M}_r, (c_k)_{k \in \mathbb{Z}} \mapsto \sum_{k \in \mathbb{Z}} c_k \cdot K\left(\cdot - \frac{k}{2R}\right) \tag{3.4.23}$$

are well-defined and bounded with $\mathrm{Synth} \circ \mathrm{Samp} = (2R)^{-1} \cdot \mathrm{id}_{\mathcal{M}_r}$.

Proof. Since $\Omega \subset \mathbb{R}$ is bounded, there is a Schwartz function $\psi \in \mathcal{S}(\mathbb{R})$ with $\psi \equiv 1$ on Ω. We then have $W := \mathcal{F}^{-1}\psi \in \mathcal{S}(\mathbb{R})$, so that W is continuous. Furthermore,

$$\widehat{W * K} = \widehat{K * W} = \hat{K} \cdot \widehat{W} = \chi_\Omega \cdot \psi = \chi_\Omega = \hat{K} \,,$$

and hence $W * K = K * W = K$. Since W is a Schwartz function, there is some $C > 0$ with $|W(x)| \leqslant C \cdot (1 + |x|)^{-2}$ for all $x \in \mathbb{R}$. Because of

$$1 + |x| \leqslant 2 + |x - y| \leqslant 2 \cdot (1 + |x - y|)$$

for any $y \in Q := U_0 := [-1, 1]$, this shows $|W(x+y)| \leqslant 4C \cdot (1+|x|)^{-2}$, and hence $\widetilde{M_Q^R} W \in L_1(\mathbb{R})$ and $M_{U_0}^L W \in L_1(\mathbb{R})$. Overall, noting that the modular function Δ of the abelian group $G = \mathbb{R}$ satisfies $\Delta \equiv 1$, we see using Lemma 3.4.4 that Assumptions 3.4.3 and 3.4.9 are both satisfied for $w = m \equiv 1$.

Now, define $I := \mathbb{Z}$ and $x_k := k/(2R)$ for $k \in \mathbb{Z}$. It is not hard to see that the family $(x_k)_{k \in \mathbb{Z}}$ is relatively separated in $G = \mathbb{R}$. Therefore, Lemma 3.4.5 and Proposition 3.4.13 show that the two operators from the statement of the current proposition are well-defined and bounded. It remains to show $\mathrm{Synth} \circ \mathrm{Samp} = (2R)^{-1} \mathrm{id}_{\mathcal{M}_r}$.

For this, it suffices to show $\mathrm{Synth}(\mathrm{Samp}\, f) = (2R)^{-1} \cdot f$ for $f \in \mathcal{M}_r \cap L_2(\mathbb{R})$, since Lemma 3.2.9 shows that $\mathrm{span}\{\lambda(x)K\}_{x \in \mathbb{R}} \subset \mathcal{M}_r \cap L_2(\mathbb{R})$ is dense in $\mathcal{M}_r$. But it is well-known that the family $(e_k)_{k \in \mathbb{Z}} = \big((2R)^{-1/2} \cdot e^{2\pi i \frac{k}{2R} \cdot}\big)_{k \in \mathbb{Z}}$ forms an orthonormal basis of $L_2(\Omega_0)$ where $\Omega_0 := \xi_0 + [-R, R]$. To make use of this orthonormal basis, first note for $f \in \mathcal{M}_r \cap L_2(\mathbb{R})$ that $\hat{f} = \widehat{f * K} = \hat{f} \cdot \hat{K} = \chi_\Omega \cdot \hat{f}$. Because of $\hat{f} = \hat{f} \cdot \chi_\Omega$, we get $\hat{f} \equiv 0$ almost everywhere on $\mathbb{R} \backslash \Omega \supset \mathbb{R} \backslash \Omega_0$.

Overall, since $\mathcal{F}(\lambda(k/(2R))K) = e^{-2\pi i \frac{k}{2R} \cdot} \chi_\Omega = (2R)^{1/2} \cdot e_{-k} \cdot \chi_\Omega$, we see

$$\hat{f} = \chi_\Omega \cdot \hat{f} = \chi_\Omega \cdot \sum_{k \in \mathbb{Z}} \langle \hat{f}, e_k \rangle_{L_2} e_k$$

$$= (2R)^{-1} \cdot \sum_{k \in \mathbb{Z}} \Big\langle \hat{f}, \mathcal{F}\big(\lambda(-k/(2R))K\big) \Big\rangle_{L_2} \cdot \mathcal{F}(\lambda(-k/(2R))\, K)$$

$$= (2R)^{-1} \cdot \mathcal{F}\left(\sum_{\ell \in \mathbb{Z}} \langle f, \lambda(\ell/(2R))K \rangle_{L_2} \cdot \lambda(\ell/(2R))\, K \right) \qquad (3.4.24)$$

$$= (2R)^{-1} \cdot \mathcal{F}(\mathrm{Synth}(\mathrm{Samp}\, f)) \,,$$

which implies $f = (2R)^{-1} \cdot (\mathrm{Synth} \circ \mathrm{Samp})f$ for all $f \in L_2(\mathbb{R}) \cap \mathcal{M}_r$, as desired. $\qquad \square$

Remark 3.4.21. The result from Proposition 3.4.20, precisely $f = (2R)^{-1} \cdot (\mathrm{Synth} \circ \mathrm{Samp})f$, can be restated for a symmetric interval $\Omega = [-R, R]$ using (3.2.51) as

$$f(x) = \sum_{n \in \mathbb{Z}} f\left(\frac{n}{2R}\right) \cdot \frac{\sin(2R\pi x - \pi n)}{2R\pi x - \pi n} \quad \text{for all } f \in B_\Omega^p, \, x \in \mathbb{R}\,. \qquad (3.4.25)$$

For $p = 2$ this is exactly the well-known Whittaker-Kotelnikov-Shannon sampling theorem [96], a famous result in sampling theory, which has been proven for $1 \leqslant p \leqslant \infty$ by Gensun [63]. The result of Proposition 3.4.20 can therefore be seen as an extension of this theorem, beacause in addition we have individually proven the boundedness of the sampling and synthesis operators.

Chapter 4

Generalized Coorbit Theory

In this chapter we present a generalization of the classic coorbit theory developed in Chapter 2. These results are due to Fornasier and Rauhut [53] and later extended with Ullrich [102]. The main idea behind this generalization is that a group structure is not necessary for defining coorbit spaces. Of course this implies that we have to find suitable substitutes for those objects in Chapter 2 that rely on the group structure. Let us briefly discuss this approach.

By choosing an *admissible vector* ψ in the classic coorbit theory, Proposition 2.1.5 yields that the voice transform is an isometry $V_\psi\colon \mathcal{H} \to L_2(G)$. This is the same as saying that the family $(\pi(x)\psi)_{x\in G}$ is a *Parseval frame* for $\mathcal{H}$ in the sense of Definition 1.7.1 indexed by the group G. Let us generalize this to assuming that instead we have some frame $\mathfrak{F} = (\psi_x)_{x\in X}$ for $\mathcal{H}$ indexed by a *measure space* X. This, however, means that we no longer have a convolution at hand, therefore *reproducing identities* like (2.1.6) do not make sense. The substitute for this are the *kernel operators* defined in Section 1.8. According to that section we define the reproducing kernel to be $K(x,y) = \langle \psi_y, \psi_x \rangle_{\mathcal{H}}$, which also serves as an operator on function spaces on X. By (1.7.9) this kernel operator is indeed reproducing. Another main ingredient in Chapter 2 was the integrability of the kernel. A substitute for this are the *kernel spaces* defined in Definition 1.8.2 and we will see that a condition for the coorbit spaces in this chapter to be well-defined will be $K \in \mathcal{A}_{1,w}$ for certain weights w.

Note that we mainly focus on target spaces $Y = L_{p,m}(G)$ as everywhere in this thesis. However, the same theory is also applicable to mixed norm spaces [53,102], quasi-Banach spaces [76] and function spaces with variable smoothness and integrability [79], just to name a few.

This chapter is structured as follows. In Section 4.1 we recall the coorbit theory developed in [53] and include some comparative notes where we consider it appropriate. In Section 4.2 we present discretization results for these coorbit spaces—in particular atomic decompositions and Banach frames—and give sufficient conditions for these to exist. Finally, in Section 4.3 we use results of [102] and apply this machinery to a special frame, namely an inhomogeneous wavelet frame, which yields inhomogeneous Besov spaces $B^s_{p,p}(\mathbb{R})$.

4.1 Coorbit Spaces

Assume $\mathcal{H}$ to be a separable Hilbert space and X a locally compact and σ-compact Hausdorff space endowed with a positive Radon measure μ with $\operatorname{supp}\mu = X$. Furthermore, let $\mathfrak{F} = (\psi_x)_{x\in X}$ be a *Parseval frame* for $\mathcal{H}$ indexed by X, i.e., for all $f \in \mathcal{H}$ the map $x \mapsto \langle f, \psi_x \rangle_{\mathcal{H}}$ is measurable and it holds

$$\|f\|_{\mathcal{H}}^2 = \int_X |\langle f, \psi_x \rangle_{\mathcal{H}}|^2 \, d\mu(x) \quad \text{for all } f \in \mathcal{H}, \tag{4.1.1}$$

see Definition 1.7.1. Let us additionally assume the map $x \to \psi_x$ to be weakly continuous for the sake of simplicity. The frame $\mathfrak{F}$ will serve as a substitute for the vectors $\pi(x)\psi$ in the previous two chapters.

Associated to the frame $\mathfrak{F}$ we define a voice transform as in (1.7.5).

Definition 4.1.1. The *voice transform* associated with $\mathfrak{F} = (\psi_x)_{x \in X}$ is defined as the mapping

$$V_{\mathfrak{F}} \colon \mathcal{H} \to L_2(X, \mu), \quad V_{\mathfrak{F}} f(x) := \langle f, \psi_x \rangle_{\mathcal{H}}. \tag{4.1.2}$$

We denote with

$$K_{\mathfrak{F}} \colon X \times X \to \mathbb{C}, \quad K_{\mathfrak{F}}(x, y) := V_{\mathfrak{F}} \psi_y(x) = \langle \psi_y, \psi_x \rangle_{\mathcal{H}} \tag{4.1.3}$$

the *kernel function* associated with $\mathfrak{F}$.

It is clear that the voice transform is well-defined due to (4.1.1). By the results of Section 1.7, specifically (1.7.9), the kernel $K_{\mathfrak{F}}$ suffices the *reproducing formula*

$$V_{\mathfrak{F}} f(x) = \int_X V_{\mathfrak{F}} f(y) K_{\mathfrak{F}}(x, y) \, d\mu(y) = K_{\mathfrak{F}}(V_{\mathfrak{F}} f)(x), \quad \text{for all } f \in \mathcal{H}, \tag{4.1.4}$$

where $K_{\mathfrak{F}}$ also denotes the kernel operator acting on functions by the action described above. As in the theories described in the preceding two chapters, this reproducing formula will be fundamental.

Let us futher note that the kernel function $K_{\mathfrak{F}}$ is continuous in both arguments by our assumption on the frame. Additionally, the frame equation (4.1.1) can be rewritten using the voice transform as

$$\|f\|_{\mathcal{H}} = \|V_{\mathfrak{F}} f\|_{L_2} \quad \text{for all } f \in \mathcal{H}. \tag{4.1.5}$$

In order for the theory in this chapter to work, we need some basic assumptions. For this let $w \colon X \times X \to \mathbb{R}_{\geqslant 0}$ be a *moderate weight* in the sense of Definition 1.8.1, that is

$$w(x, y) \leqslant w(x, z) \cdot w(z, y), \tag{4.1.6}$$
$$w(x, y) = w(y, x), \tag{4.1.7}$$
$$w(x, x) \leqslant C_w < \infty \tag{4.1.8}$$

for all $x, y, z \in X$.

The following assumption is crucial for the definition of suitable coorbit spaces.

Assumption 4.1.2. *We assume the following hold true.*

(i) $\|\psi_x\|_{\mathcal{H}} \leqslant C_{\mathfrak{F}} < \infty$ *for all* $x \in X$.

(ii) $K_{\mathfrak{F}} \in \mathcal{A}_{1,w}$, *that is,* $K_{\mathfrak{F}}$ *is measurable and*

$$\sup_{x \in X} \int_X |\langle \psi_x, \psi_y \rangle_{\mathcal{H}} \cdot w(x, y)| \, d\mu(y) < \infty, \tag{4.1.9}$$

where $\mathcal{A}_{1,w}$ *is the kernel space defined in Definition 1.8.2.*

Remark 4.1.3. (i) Equation (4.1.9) is a consequence of the definition of the norm $\| \cdot \|_{\mathcal{A}_{1,w}}$ in (1.8.5), where we used $K_{\mathfrak{F}}(x, y) = \overline{K_{\mathfrak{F}}(y, x)}$ and the symmetry of the weight w.

(ii) The condition (i) in Assumption 4.1.2 is not very restrictive, the second assumption however is more restrictive. The reader might compare this integrability condition to the definition of analyzing vectors in (2.1.13), which also requires some sort of integrability on the kernel. Therefore, this assumption is a substitute for analyzing vectors in Chapter 2. As we noted in part (i) of Remark 2.1.9, this condition is also restrictive in this general setting and in Chapter 5 we discuss a theory that does *not* rely on this integrability condition.

(iii) As an easy consequence of (i) the kernel $K_{\mathfrak{F}}$ is a bounded function on $X \times X$.

Next, we will define test spaces denoted with $\mathcal{H}_{1,v}$ for some weight v on X, whose dual space will serve as a reservoir for the coorbit spaces. For this let $v \colon X \to \mathbb{R}_+$ be a *w-moderate weight* on X that satisfies $v \geqslant 1$ and

$$v(x) \leqslant w(x,y) \cdot v(y) \quad \text{for all } x, y \in X. \tag{4.1.10}$$

A possible choice for v is $v(x) = w(x,z)$ for some fixed $z \in X$. With this weight at hand we define the following test spaces.

Proposition 4.1.4. *Let Assumption 4.1.2 be fulfilled, then the space*

$$\mathcal{H}_{1,v} := \{f \in \mathcal{H} \ : \ V_{\mathfrak{F}}f = \langle f, \psi. \rangle_{\mathcal{H}} \in L_{1,v}(X,\mu)\} \tag{4.1.11}$$

endowed with the natural norm $\|f\|_{\mathcal{H}_{1,v}} := \|V_{\mathfrak{F}}f\|_{L_{1,v}}$ fulfills the following:

(i) $\mathcal{H}_{1,v}$ is a Banach space;

(ii) the frame $\mathfrak{F}$ is contained in $\mathcal{H}_{1,v}$, that is, $(\psi_x)_{x \in X} \subset \mathcal{H}_{1,v}$;

(iii) $\mathcal{H}_{1,v}$ is densely embedded into $\mathcal{H}$.

Proof. The norm $\|\cdot\|_{\mathcal{H}_{1,v}}$ is indeed a norm, since $\|f\|_{\mathcal{H}_{1,v}} = 0$ implies $\langle f, \psi_x \rangle_{\mathcal{H}} = 0$ for almost all $x \in X$. But since $\mathfrak{F}$ is a frame this implies $f = 0$ almost everywhere. Let $(f_n)_{n \in \mathbb{N}}$ be a Cauchy sequence in $\mathcal{H}_{1,v}$, which means that $(g_n)_{n \in \mathbb{N}} := (V_{\mathfrak{F}}f_n)_{n \in \mathbb{N}}$ is a Cauchy sequence in $L_{1,v}(X,\mu)$. By the completeness of $L_{1,v}(X,\mu)$ there exists a unique $g \in L_{1,v}(X,\mu)$ with $g_n \to g$. We note that $|K_{\mathfrak{F}}(x,y)| \leqslant C_{\mathfrak{F}}^2$ where $C_{\mathfrak{F}}$ is the constant from Assumption 4.1.2 (i). Using this and $v \geqslant 1$ it holds that

$$|K_{\mathfrak{F}}(g)(x)| \leqslant \int_X |K_{\mathfrak{F}}(x,y)g(y)v(y)| \, d\mu(y) \leqslant C_{\mathfrak{F}}^2 \cdot \|g\|_{L_{1,v}}$$

for every $x \in X$, thus $K_{\mathfrak{F}}(g) \in L_\infty(X,\mu)$. Furthermore, by the reproducing formula (4.1.4) we have $K_{\mathfrak{F}}(g_n) = g_n$ for all $n \in \mathbb{N}$, so that by the continuity of $K_{\mathfrak{F}}$ it holds that $K_{\mathfrak{F}}(g) = g$ as well. Thus, $g \in L_\infty(X,\mu)$ and since $L_\infty \cap L_{1,v} \subset L_2$ it follows $g \in L_2$. Since the application of K_ψ is an orthogonal projection from L_2 onto the image of $V_{\mathfrak{F}}$ there exists $f \in \mathcal{H}$ such that $g = V_{\mathfrak{F}}f$. Moreover, $V_{\mathfrak{F}}f \in L_{1,v}$ means $f \in \mathcal{H}_{1,v}$ and $f_n \to f \in \mathcal{H}_{1,v}$.

Item (ii) is a direct consequence of $K_{\mathfrak{F}} \in \mathcal{A}_{1,w}$, since

$$\|\psi_x\|_{\mathcal{H}_{1,v}} = \int_X |V_{\mathfrak{F}}\psi_x(y) \cdot v(y)| \, d\mu(y) = \int_X |K_{\mathfrak{F}}(x,y) \cdot w(y,z)| \, d\mu(y)$$

$$\leqslant v(x) \int_X |K_{\mathfrak{F}}(x,y) \cdot w(x,y)| \, d\mu(y) = v(x) \cdot \|K_{\mathfrak{F}}\|_{\mathcal{A}_{1,w}} < \infty \tag{4.1.12}$$

for all $x \in X$.

It remains to prove part (iii). We first note that $\mathcal{H}_{1,v}$ is a dense subset of $\mathcal{H}$, since the frame $\mathfrak{F}$ of $\mathcal{H}$ is contained in $\mathcal{H}_{1,v}$ by part (ii). Next, $|K_{\mathfrak{F}}(x,y)| \leq C_{\mathfrak{F}}^2$ implies

$$\|K_{\mathfrak{F}}(g)\|_{L_{\infty,v}} \leq \operatorname*{ess\,sup}_{x \in X} \int_X |K(x,z)w(x,z) \cdot g(z)v(z)| \, d\mu(z)$$

$$\leq \operatorname*{ess\,sup}_{x,y \in X} |K_{\mathfrak{F}}(x,y)w(x,y)| \cdot \int_X |g(z)v(z)| \, d\mu(z) \leq C_{\mathfrak{F}}^2 C_w \cdot \|g\|_{L_{1,v}}$$

for all suitable g. Hence, by the frame property (4.1.5) an the reproducing formula (4.1.4), we obtain

$$\|f\|_{\mathcal{H}}^2 = \|V_{\mathfrak{F}}f\|_{L_2}^2 \leq \|V_{\mathfrak{F}}f\|_{L_{1,v}} \cdot \|V_{\mathfrak{F}}f\|_{L_{\infty,v}} = \|V_{\mathfrak{F}}f\|_{L_{1,v}} \cdot \|K_{\mathfrak{F}}(V_{\mathfrak{F}}f)\|_{L_{\infty,v}} \leq C_{\mathfrak{F}}^2 C_w \cdot \|V_{\mathfrak{F}}f\|_{L_{1,v}}^2$$

for all $f \in \mathcal{H}_{1,v}$. $\qquad\qquad\qquad\qquad\qquad\qquad\qquad\qquad\qquad\qquad\qquad\qquad\qquad\qquad\quad\square$

By the proposition above, this set of test functions leads to the Gelfand triple setting of dense embeddings

$$\mathcal{H}_{1,v} \hookrightarrow \mathcal{H} \cong \mathcal{H}' \hookrightarrow \mathcal{H}_{1,v}'$$

with $\mathcal{H}_{1,v}'$ being the canonical dual space of $\mathcal{H}_{1,v}$, i.e., the space of all conjugate linear continuous functionals on $\mathcal{H}_{1,v}$. This way it is possible to extend the voice transform in a canonical way to elements $f \in \mathcal{H}_{1,v}'$ by setting the *extended voice transform* $V_{e,\mathfrak{F}}$ as

$$V_{e,\mathfrak{F}}f(x) = f(\psi_x) = \langle f, \psi_x \rangle_{\mathcal{H}_{1,v}' \times \mathcal{H}_{1,v}}. \tag{4.1.13}$$

As in the previous chapters, the space $\mathcal{H}_{1,v}'$ will serve as the reservoir for the coorbit spaces.

Let us collect some properties of the extended voice transform in the following lemma.

Lemma 4.1.5 (Lemma 3 of [53]). *If Assumption 4.1.2 is fulfilled, then*

(i) *the map $V_{e,\mathfrak{F}} : \mathcal{H}_{1,v}' \to L_{\infty,v^{-1}}(X,\mu)$ is well-defined and injective;*

(ii) *the reproducing formula extends to $\mathcal{H}_{1,v}'$, i.e.,*

$$V_{e,\mathfrak{F}}f = K_{\mathfrak{F}}(V_{e,\mathfrak{F}}f) \quad \text{for all } f \in \mathcal{H}_{1,v}'; \tag{4.1.14}$$

(iii) *conversely, if $F \in L_{\infty,v^{-1}}(X,\mu)$ satisfies the reproducing formula $F = K_{\mathfrak{F}}(F)$, then there exists $f \in \mathcal{H}_{1,v}'$ such that $F = V_{e,\mathfrak{F}}f$.*

A direct consequence of the lemma above is that the extended voice transform $V_{e,\mathfrak{F}}$ induces an isomorphism between the dual space $\mathcal{H}_{1,v}'$ and the reproducing kernel space

$$\left\{ F \in L_{\infty,v^{-1}}(X,\mu) \; : \; K_{\mathfrak{F}}(F) = F \right\}.$$

As in the previous two chapters, the coorbit spaces are intended to be defined via the decay properties of images under the extended voice transform. The decay is measured by function spaces Y on X. As usual, these function spaces need to be admissible in a certain sense, which is made precise in the following.

Definition 4.1.6. We say Y is a *target space*, if it fulfills the following conditions:

(i) $(Y, \|\cdot\|_Y)$ is a non-trivial solid Banach space of functions on X, which is continuously embedded into $L_{1,\text{loc}}(X,\mu)$;

(ii) we have $\mathcal{A}_{1,w}(Y) \subset Y$ and

$$\|K(F)\|_Y \leqslant \|K\|_{\mathcal{A}_{1,w}} \cdot \|F\|_Y \quad \text{for all } K \in \mathcal{A}_{1,w},\ F \in Y, \tag{4.1.15}$$

where $\mathcal{A}_{1,w}$ is the weighted kernel space from Definition 1.8.2.

It is possible to relax condition (i) and also allow quasi-Banach spaces as target spaces [79]. Now we are ready to define the coorbit spaces $\mathrm{Co}(Y)$.

Definition 4.1.7. The *coorbit space* $\mathrm{Co}(Y)$ associated with a target space Y with respect to the frame $\mathfrak{F} = (\psi_x)_{x \in X}$ is defined as

$$\mathrm{Co}(Y) := \left\{ T \in \mathcal{H}'_{1,v} : V_{e,\mathfrak{F}} T \in Y \right\} \tag{4.1.16}$$

with natural norm $\|T\|_{\mathrm{Co}(Y)} = \|V_{e,\mathfrak{F}} T\|_Y$.

Of course we want these spaces to be Banach spaces that, as usual, fulfill some reproducing identity. This is shown in the following proposition.

Proposition 4.1.8 (Proposition 2 of [53]). *Suppose that Assumption 4.1.2 is fulfilled and $K_{\mathfrak{F}}(Y) \subset L_{\infty,v^{-1}}(X, \mu)$. Then,*

(i) $\mathrm{Co}(Y)$ *is a Banach space with its natural norm;*

(ii) *the extended voice transform $V_{e,\mathfrak{F}}$ induces an isometric isomorphism between $\mathrm{Co}(Y)$ and the reproducing kernel space*

$$\mathcal{M}_Y := \{ F \in Y : K_{\mathfrak{F}}(F) = F \}.$$

This implies that the reproducing identity

$$K_{\mathfrak{F}}(V_{e,\mathfrak{F}} T) = V_{e,\mathfrak{F}} T, \quad T \in \mathrm{Co}(Y),$$

holds true.

Remark 4.1.9. The condition $K_{\mathfrak{F}}(Y) \subset L_{\infty,v^{-1}}(X, \mu)$ above is a substitute for the condition in Theorem 2.1.16, which states that the kernel is contained in a certain Wiener-type space. We will show in Lemma 4.1.10 that this condition is fulfilled in the relevant cases. In general it is true as well under certain assumptions on the oscillation of the kernel, cf. [53, Corollary 5].

It has been shown in [53, Corollary 2], that we can express the basic spaces as coorbit spaces. To be precise, we have $\mathcal{H} = \mathrm{Co}(L_2)$ and $\mathcal{H}'_{1,v} = \mathrm{Co}(L_{\infty,v^{-1}})$.

As in the previous chapters we are particularly interested in target spaces $Y = L_{p,m}(X, \mu)$ for suitable weights m. For this, let $m \colon X \to \mathbb{R}_+$ be a *w-moderate weight* on X, i.e.,

$$m(x) \leqslant w(x,y) \cdot m(y) \quad \text{for all } x, y \in X. \tag{4.1.17}$$

Note that this is the same condition as the one for the weight v in (4.1.10), so that we can choose $v = m$. Then, a weighted version of Schur's test, see Corollary 1.8.5, shows that $L_{p,m}(X, \mu)$, $1 \leqslant p \leqslant \infty$, is a target space as defined in Definition 4.1.6. We can therefore define coorbit spaces

$$\mathrm{Co}(L_{p,m}) := \left\{ T \in \mathcal{H}'_{1,m} : V_{e,\mathfrak{F}} T \in L_{p,m}(X, \mu) \right\}, \quad 1 \leqslant p \leqslant \infty, \tag{4.1.18}$$

for suitable frames $\mathfrak{F}$. Following Proposition 4.1.8, which we will justify in a moment, this space is isometrically isomorphic to the closed subspace

$$\mathcal{M}_{p,m} := \mathcal{M}_{L_{p,m}} = \{ F \in L_{p,m}(X, \mu) : K_{\mathfrak{F}}(F) = F \}. \tag{4.1.19}$$

This space is a reproducing kernel Banach space, just like in the classical case.

The following Lemma justifies the application of Proposition 4.1.8.

Lemma 4.1.10. *For $1 \leqslant p \leqslant \infty$ and a w-moderate weight m we have the bounded embedding $K_{\mathfrak{F}}(L_{p,m}(X,\mu)) \subset L_{\infty,m^{-1}}(X,\mu)$, hence Proposition 4.1.8 is applicable.*

Proof. We first note that $K_{\mathfrak{F}} \in \mathcal{A}_{1,w}$ implies $K_{\mathfrak{F}} \in \mathcal{A}_{\infty,w}$. By symmetry of the kernel the condition $K_{\mathfrak{F}} \in \mathcal{A}_{1,w}$ is equivalent to $\sup_{x \in X} \|K_{\mathfrak{F}}(x,\cdot)w(x,\cdot)\|_{L_1} < \infty$. Since $K_{\mathfrak{F}}$ and w are continuous, this means $\sup_{x,y \in X} |K_{\mathfrak{F}}(x,y)w(x,y)| < \infty$, or equivalently $K_{\mathfrak{F}} \in \mathcal{A}_{\infty,w}$. Therefore, for any $1 < q < \infty$ we have

$$\|K_{\mathfrak{F}}\|_{\mathcal{A}_{q,w}} = \sup_{x \in X} \int_X |K(x,y)|^q w(x,y)^q \, d\mu(y) \leqslant \|K\|_{\mathcal{A}_{1,w}} \cdot \|K\|_{\mathcal{A}_{\infty,w}}^{q-1}.$$

By Lemma 1.8.7 the kernel operator associated to $K_{\mathfrak{F}}$ is a bounded operator from $L_{p,m}(X,\mu)$ to $L_{\infty,m}(X,\mu) \subset L_{\infty,m^{-1}}(X,\mu)$, which proves the claim. $\square$

We close this section by collecting some basic facts about the coorbit spaces $\mathrm{Co}(L_{p,m})$; we omit the proof since it is almost identical to the proof of Lemma 2.1.18.

Lemma 4.1.11. *Let m, m_1, m_2 be w-moderate weights on X. Then:*

(i) *The spaces $\mathrm{Co}(L_{p,m})$ are monotonically increasing with $1 \leqslant p \leqslant \infty$.*

(ii) *Let $1 \leqslant p_2 < p_1 \leqslant \infty$ and assume the quotient m_1/m_2 belongs to $L_r(X,\mu)$ for some $1 \leqslant r < \infty$ fulfilling $r^{-1} \leqslant p_2^{-1} - p_1^{-1}$, then $\mathrm{Co}(L_{p_1,m_1}) \subset \mathrm{Co}(L_{p_2,m_2})$.*

(iii) *$\mathrm{Co}(L_{p_1,m_1}) = \mathrm{Co}(L_{p_2,m_2})$ if and only if $p_1 = p_2$ and there are two constants $0 < C_1 \leqslant C_2$ such that $C_1 m_1(x) \leqslant m_2(x) \leqslant C_2 m_1(x)$ for all $x \in X$.*

4.2 Discretization

As usual, we are interested in discretizations of the coorbit spaces defined in the preceding section. In particular, we intend to find atomic decompositions and Banach frames and investigate conditions for the existence of such discretizations. We will keep this exhibition short and only state the main results, a thorough discussion and the proofs can be found in Section 5 of [53]. We restrict ourselves to the coorbit spaces $\mathrm{Co}(L_{p,m})$ associated to weighted Lebesgue spaces.

The main idea behind the discretization is, as in the classical setting in Chapter 2, to measure the oscillation of the kernel with respect to some suitable covering of X. For this we need an *admissible covering* of X, see also Definition 1.5.5, i.e., a family $\mathcal{U} = (U_i)_{i \in I}$ of subsets of X, such that each set U_i, $i \in I$, is relatively compact and has non-void interior. The family $\mathcal{U}$ covers X, that is, $X = \bigcup_{i \in I} U_i$, and at every point in X at most $N \in \mathbb{N}$ sets intersect, in other words $\sum_{i \in I} \chi_{U_i}(x) \leqslant N$ for all $x \in X$. Finally, we assume $\mathcal{U}$ to be *moderate*, which means that there exist constants $C, D > 0$ with $\mu(U_i) \geqslant C$ for all $i \in I$ and $\mu(U_i) \leqslant D\,\mu(U_j)$ for all $i, j \in I$ with $U_i \cap U_j \neq \varnothing$. Since we assumed X to be σ-compact, the index set I is countable.

Associated to an admissible covering $\mathcal{U}$ of X we define the $\mathcal{U}$-oscillation of a function F on X as

$$\mathrm{osc}_{\mathcal{U}} F(x,y) := \sup_{z \in Q_y} |F(x,y) - F(x,z)|, \tag{4.2.1}$$

where $Q_y := \{U_i : y \in U_i\}$. This definition is analogous to (2.2.1) in the classical setting. Note that we only measure the oscillation of F in the second argument.

For the discretization results to hold, we introduce an additional condition on w, and therefore on the target spaces Y. Let us assume there exists a constant $C_{w,\mathcal{U}} > 0$ such that

$$\sup_{x,y \in U_i} w(x,y) \leqslant C_{w,\mathcal{U}} \quad \text{for all } i \in I. \tag{4.2.2}$$

Obviously, this condition is automatically fulfilled in the group setting, but in the general case this is not clear. However, for the trivial weight $w \equiv 1$ (4.2.2) is fulfilled and in general it can be easily checked.

Finally, recall that the sequence spaces associated with the weighted Lebesue spaces $L_{p,m}(X, \mu)$ for a suitable weight m on X are given as

$$\ell_{p,m}(I) := \left\{ c = (c_i)_{i \in I} \in \mathbb{C}^I \ : \ \|c\|_{\ell_{p,m}}^p = \sum_{i \in I} |c_i|^p \mu(U_i) \sup_{x \in U_i} m(x)^p < \infty \right\}. \tag{4.2.3}$$

Note that (4.2.2) implies

$$m(x) \leqslant w(x, y) \cdot m(y) \leqslant C_{w, \mathcal{U}} \cdot m(y) \quad \text{for all } x, y \in U_i, \ i \in I$$

for any w-moderate weight m. By invoking Lemma 1.6.8 we can also write $m(x_i)$ instead of $\sup_{x \in U_i} m(x)$ for any family of points $(x_i)_{i \in I} \subset X$ with $x_i \in U_i$ for all $i \in I$ in the definition of the norm in (4.2.3).

4.2.1 Atomic Decompositions

In this subsection we present conditions under which there exist atomic decompositions for the coorbit spaces $\mathrm{Co}(L_{p,m})$ introduced in Section 4.1. The following theorem is a reduced form of Theorem 5 using Remark 4 of [53].

Theorem 4.2.1. *Let $1 \leqslant p < \infty$ and m be a w-moderate weight where w fulfills (4.2.2). Let $\mathfrak{F}$ fulfill Assumption 4.1.2, i.e., $K_{\mathfrak{F}} \in \mathcal{A}_{1,w}$, and assume there is an admissible covering $\mathcal{U}$ of X such that*

$$\|\mathrm{osc}_{\mathcal{U}} K_{\mathfrak{F}}\|_{\mathcal{A}_{1,w}} < 1. \tag{4.2.4}$$

Furthermore, choose points $(x_i)_{i \in I} \subset X$ such that $x_i \in U_i$ for all $i \in I$.

Then the family $(\psi_{x_i})_{i \in I}$ forms a family of atoms for the coorbit space $\mathrm{Co}(L_{p,m})$ with associated sequence space $\ell_{p,m}(I)$. This means:

(i) The synthesis operator

$$\mathrm{Synth} \colon \ell_{p,m}(I) \to \mathrm{Co}(L_{p,m}), \quad (c_i)_{i \in I} \mapsto \sum_{i \in I} c_i \cdot \psi_{x_i}$$

is well-defined and bounded, with unconditional convergence of the defining series. This even holds without assuming (4.2.4).

(ii) There is a bounded reconstruction operator

$$\mathrm{Coeff} \colon \mathrm{Co}(L_{p,m}) \to \ell_{p,m}(I) \quad \text{with} \quad \mathrm{Synth} \circ \mathrm{Coeff} = \mathrm{id}_{\mathrm{Co}(L_{p,m})}.$$

Theorem 5 in [53] is more thorough than the preceding theorem since it not only additionally includes Banach frames—which we will take a look at in the following subsection—, but also introduces a second family dual to $(\psi_{x_i})_{i \in I}$, and allows other target spaces than $L_{p,m}(X, \mu)$. Let us briefly compare the result above to the atomic decomposition in Subsection 2.2.1 and discuss the idea of the proof.

Comparing Theorem 2.2.1 with Theorem 4.2.1, we see that the prerequisits are very similar. As usual, we need some sort of covering and points in these sets. While in the group setting the covering was induced by some points $(x_i)_{i \in I}$, we have some freedom in choosing these points in the present setting. Nevertheless, the oscillation of the kernel with respect to this covering needs to satisfy some integrability condition in both cases.

The proof of Theorem 4.2.1 is also similar to the classical setting. The main idea is to discretize the reproducing identity $F = K_{\mathfrak{F}}(F)$ for all $\mathcal{M}_{p,m}$ and transfer this to the coorbit spaces $\mathrm{Co}(L_{p,m})$ via the correspondance principle, cf. Proposition 4.1.8. To show the invertibility of the discretization operator, a Neumann series argument is invoked. Here, the boundedness of the map $F \mapsto \mathrm{osc}_{\mathcal{U}} K_{\mathfrak{F}}(F)$ as an operator on $L_{p,m}(X,\mu)$ is being used, which is a direct consequence of condition (4.2.4) and a weighted version of Schur's test, cf. Corollary 1.8.6.

Note that the discretization operator U_Φ used in Theorem 5 of [53] is following the notation of Section 4.2 of [68], which is a composition of two operators S and T. Simply using T provides us with condition (4.2.4).

Altogether we see that the integrability of $K_{\mathfrak{F}}$—more precisely $K_{\mathfrak{F}}, \mathrm{osc}_{\mathcal{U}} K_{\mathfrak{F}} \in \mathcal{A}_{1,w}$—is crucial for obtaining atomic decompositions as in this setting.

4.2.2   Banach Frames

Dual to atomic decompositions, Theorem 5 of [53] also provides us with Banach frames for coorbit spaces $\mathrm{Co}(L_{p,m})$.

Theorem 4.2.2. *Let $1 \leqslant p < \infty$ and m be a w-moderate weight where w fulfills (4.2.2). Let $\mathfrak{F}$ fulfill Assumption 4.1.2, i.e., $K_{\mathfrak{F}} \in \mathcal{A}_{1,w}$, and assume there is an admissible covering $\mathcal{U}$ of X for some $\delta > 0$ such that*

$$\|\mathrm{osc}_{\mathcal{U}} K_{\mathfrak{F}}\|_{\mathcal{A}_{1,w}} \leqslant \delta, \qquad \|K_{\mathfrak{F}}\|_{\mathcal{A}_{1,w}} < \delta^{-1} \tag{4.2.5}$$

Furthermore choose points $(x_i)_{i \in I} \subset X$ such that $x_i \in U_i$ for all $i \in I$. Then the family of functionals $(\psi_{x_i})_{i \in I} \subset \mathrm{Co}(L_{p,m})'$ forms a Banach frame for $\mathrm{Co}(L_{p,m})$ with associated coefficient space $\ell_{p,m}(I)$. This means:

(i) The sampling operator

$$\mathrm{Samp} \colon \mathrm{Co}(L_{p,m}) \to \ell_{p,m}(I), \quad T \mapsto (V_{e,\mathfrak{F}} T(x_i))_{i \in I} = (\langle T, \psi_{x_i}\rangle_{\mathcal{H}'_{1,m} \times \mathcal{H}_{1,m}})_{i \in I}$$

is well-defined and bounded.

(ii) There exists a bounded linear reconstruction operator

$$\mathrm{Reco} \colon \ell_{p,m}(I) \to \mathrm{Co}(L_{p,m}) \quad \text{with} \quad \mathrm{Reco} \circ \mathrm{Samp} = \mathrm{id}_{\mathrm{Co}(L_{p,m})}.$$

Comparing this with Theorem 2.2.3 concerning the classic coorbit theory, we see that the condition (4.2.5) is very similar to (2.2.5). In fact, both proofs rely on the invertibility of a similar discretization operator, denoted with S in Section 4.2 of [68], for which a Neumann series argument is invoked. Again, this relies on the boundedness of the operators $F \mapsto K_{\mathfrak{F}}(F)$ and $F \mapsto \mathrm{osc}_{\mathcal{U}} K_{\mathfrak{F}}(F)$ as operators on weighted Lebesgue spaces, which is a consequence of the integrability of $K_{\mathfrak{F}}$ with respect to the kernel space $\mathcal{A}_{1,w}$ by a weighted version of Schur's test, cf. Corollary 1.8.6.

This shows that in order to obtain Banach frames for coorbit spaces $\mathrm{Co}(L_{p,m})$, again, integrability conditions on the kernel are indispensable.

4.3 An Example: Inhomogeneous Besov Spaces

In Subsection 2.3.1 we showed that the homogeneous Besov spaces $\dot{B}_{p,p}^{s-1/2+1/p}(\mathbb{R})$ for any integrability parameter $1 \leqslant p < \infty$ and smoothness parameter $s \in \mathbb{R}$ can be described using the classic coorbit theory developed in Chapter 2. Obviously, the group structure of the affine group was exploited in order to apply the classic coorbit theory. This, however, is not possible for the inhomogeneous Besov spaces $B_{p,p}^s(\mathbb{R})$. To bypass this obstacle we will invoke the generalized coorbit theory laid out in this chapter and see that it is indeed suitable for describing inhomogeneous Besov spaces as generalized coorbit spaces.

In this section we will first properly define the inhomogeneous Besov spaces, then define a suitable index set X and frame $\mathfrak{F}$, which we will then use to define the coorbit spaces after checking the necessary conditions. Finally, we give a discretization of these spaces according to the preceding section.

Note that the describtion of inhomogeneous Besov spaces via coorbit theory was first remarked by Fornasier and Rauhut [53] and later generalized for inhomogeneous function spaces of Besov-Lizorkin-Triebel type [102].

Let us first give a definition of the inhomogeneous Besov spaces taken from [54], again a detailed discussion on these spaces can be found in the books by Triebel [98–100].

Definition 4.3.1. Let $d \in \mathbb{N}$ and $\{\varphi_j\}_{j\in\mathbb{N}_0}$ be a family of functions satisfying (2.3.1)–(2.3.4) and let $\Phi \in \mathcal{S}(\mathbb{R}^d)$ satisfy

$$\operatorname{supp}\widehat{\Phi} \subset \left\{\xi \in \mathbb{R}^d \, : \, \|\xi\|_2 \leqslant 1\right\}, \tag{4.3.1}$$

$$\widehat{\Phi}(\xi) \geqslant c > 0 \quad \text{if} \quad \|\xi\|_2 \leqslant \frac{5}{6}. \tag{4.3.2}$$

Then the *inhomogeneous Besov space* $B_{p,q}^s(\mathbb{R}^d)$ for $s \in \mathbb{R}$ and $0 < p,q \leqslant \infty$ is defined as

$$B_{p,q}^s(\mathbb{R}^d) := \left\{ f \in \mathcal{S}'(\mathbb{R}^d) \, : \, \|f\|_{B_{p,q}^s} := \|\Phi * f\|_{L_p} + \left(\sum_{j=0}^{\infty} 2^{jsq} \|\varphi_j * f\|_{L_p}^q \right)^{1/q} < \infty \right\}, \tag{4.3.3}$$

with the usual change of notation if $q = \infty$.

The natural connection between homogeneous and inhomogeneous Besov spaces is given by

$$B_{p,q}^s(\mathbb{R}^d) = L_p(\mathbb{R}^d) \cap \dot{B}_{p,q}^s(\mathbb{R}^d),$$

see [97, Section 5.2.3].

Analogous to the homogeneous case, there are numerous different characterizations of inhomogeneous Besov spaces. We recall the following, which is a restatement of [98, Section 2.5.1] for the case $p = q$ and $d = 1$.

Corollary 4.3.2. *Let $0 < p \leqslant \infty$ and $s \in \mathbb{R}$. Let φ such that $\widehat{\Phi}$ and $\widehat{\varphi}$ be infinitely differentiable complex-valued function on $\mathbb{R}$ and $\mathbb{R}\backslash\{0\}$, respectively, satisfying the technical conditions (3)–(6) in [98, Section 2.5.1]. Then,*

$$\|f\|_{\widetilde{B}_{p,p}^s} := \left(\int_{\mathbb{R}} |\langle \Phi(\cdot - t), f \rangle_{L_2}|^p \, \mathrm{d}t \right)^{1/p} + \left(\int_{\mathbb{R}} \int_0^1 |\langle \varphi(a^{-1}(\cdot - t)), f \rangle_{L_2}|^p |a|^{-sp} \frac{\mathrm{d}a}{|a|^2} \, \mathrm{d}t \right)^{1/p} \tag{4.3.4}$$

is an equivalent (quasi-)norm on $B_{p,p}^s(\mathbb{R})$.

The structure of the equivalent norm will serve as a blueprint for the definition of the index space X. The inner products already indicate the nature of the voice transform as well as the frame that we will construct in the following.

Inhomogeneous Besov Frame

Let $\mathcal{H} = L_2(\mathbb{R})$ be the Hilbert space of functions and recall the *translation* and *dilation* operators acting on functions $f \in L_2(\mathbb{R})$ by

$$T_b f(x) = f(x - b) \quad \text{and} \quad D_a f(x) = |a|^{-1/2} f\left(\frac{x}{a}\right),$$

for $a \in \mathbb{R}^*$ and $b \in \mathbb{R}$. Now let $\psi \in \mathcal{S}(\mathbb{R})$ be a function satisfying the *Calderón condition* (2.3.13) given by

$$\int_{\mathbb{R}} \frac{|\widehat{\psi}(x)|^2}{|x|} \, dx =: C_\psi = 1, \tag{4.3.5}$$

i.e., ψ is an admissible vector for the affine group and the continuous wavelet transform, also simply called *wavelet*. Additionally, we assume $\operatorname{supp} \widehat{\psi} \subset \{c \in \mathbb{R} : \alpha^{-1} \leqslant |x| \leqslant \alpha\}$ for some $\alpha > 1$. Now let $\Phi \in \mathcal{S}(\mathbb{R})$ be a second function, such that

$$|\widehat{\Phi}(\xi)|^2 + \int_{-1}^{1} \frac{|\widehat{\psi}(y \cdot \xi)|^2}{|y|} \, dy = 1 \quad \text{for all } \xi \in \mathbb{R}, \tag{4.3.6}$$

or, equivalently, $|\widehat{\Phi}(\xi)|^2 = 1 - \int_{-\xi}^{\xi} |\widehat{\psi}(y)|^2/|y| \, dy$ for all $\xi \in \mathbb{R}$. By the Calderón condition (4.3.5) this implies $0 \leqslant |\widehat{\Phi}(\xi)| \leqslant 1$ for all $\xi \in \mathbb{R}$. Moreover, by the support property of $\widehat{\psi}$, the function $|\widehat{\Phi}|$ is a bump function with $|\widehat{\Phi}(\xi)| = 1$ for all $|\xi| \leqslant \alpha^{-1}$ and $\widehat{\Phi}(\xi) = 0$ for all $|\xi| \geqslant \alpha$.

These functions are the basis for our frame. Let us first define the index set

$$X = (\mathbb{R} \times \{\infty\}) \cup (\mathbb{R} \times [-1, 1]^*), \tag{4.3.7}$$

where ∞ denotes an isolated point and $[-1, 1]^* = [-1, 1] \backslash \{0\}$. The right side above is a subspace of the affine group, see Subsection 2.3.1. Let us define a measure on X, which is obviously based on the Haar measure of the affine group defined in (2.3.8).

Definition 4.3.3. On the set X a measure μ is defined via

$$\int_X F(x) \, d\mu(x) := \int_{\mathbb{R}} F(\infty, t) \, dt + \int_{\mathbb{R}} \int_{-1}^{1} F(a, t) \frac{da}{|a|^2} \, dt \tag{4.3.8}$$

for every complex valued function F on X which is measurable with respect to the Borel σ-algebra.

Now we define the family of functions $\mathfrak{F} = (\psi_x)_{x \in X}$ using the translation and dilation operators:

$$\psi_{(\infty, t)}(x) = T_t \Phi(x) = \Phi(x - t), \tag{4.3.9}$$

$$\psi_{(a, t)}(x) = T_t D_a \psi(x) = |a|^{-1/2} \psi\left(\frac{x - t}{a}\right). \tag{4.3.10}$$

This indeed constitutes a frame for $\mathcal{H}$, as the following corollary states.

Corollary 4.3.4. *The family* $\mathfrak{F} = (\psi_x)_{x \in X}$ *defined in* (4.3.9) *and* (4.3.10) *is a Parseval frame for* $L_2(\mathbb{R})$, *i.e.,*

$$\int_X |\langle f, \psi_x \rangle_{L_2}|^2 \, dx = \|f\|_{L_2}^2 \quad \text{for all } f \in L_2(\mathbb{R}).$$

Proof. We first note that the map $x \mapsto |\langle f, \psi_x \rangle_{L_2}|^2$ is measurable for all $f \in L_2(\mathbb{R})$. Then, similarly to (2.3.11), we have

$$\langle f, \psi_{(a,t)} \rangle_{L_2} = (f * \psi^*_{(a,0)})(t) = |a|^{1/2} \mathcal{F}^{-1}(\hat{f}(\cdot)\overline{\hat{\psi}(a\cdot)})(t), \qquad (4.3.11)$$

with an analogous equality for $\psi_{(\infty,t)}$. Using this as well as Fubini's and Plancherel's theorem we have

$$\int_X |\langle f, \psi_x \rangle_{L_2}|^2 \, d\mu(x) = \int_{\mathbb{R}} |\mathcal{F}^{-1}(\hat{f} \cdot \overline{\hat{\Phi}})(t)|^2 \, dt + \int_{-1}^{1} \int_{\mathbb{R}} |\mathcal{F}^{-1}(\hat{f}(\cdot)\overline{\hat{\psi}(a\cdot)})(t)|^2 \, dt \, \frac{da}{|a|}$$

$$= \int_{\mathbb{R}} |\hat{f}(t) \cdot \overline{\hat{\Phi}(t)}|^2 \, dt + \int_{-1}^{1} \int_{\mathbb{R}} |\hat{f}(t)\overline{\hat{\phi}(ta)}|^2 \, dt \, \frac{da}{|a|}$$

$$= \int_{\mathbb{R}} |f(\xi)|^2 \cdot \left(|\hat{\Phi}(\xi)|^2 + \int_{-1}^{1} \frac{|\hat{\psi}(y \cdot \xi)|^2}{|y|} \, dy \right) d\xi = \|f\|_{L_2}^2$$

by equation (4.3.6). $\qquad \square$

Following the lines of Section 4.1, we now define the voice transform $V_{\mathfrak{F}} \colon L_2(\mathbb{R}) \to L_2(X, \mu)$ as the map

$$V_{\mathfrak{F}} f(x) = \langle f, \psi_x \rangle_{L_2}$$

and the kernel $K_{\mathfrak{F}} \colon X \times X \to \mathbb{C}$ via $K_{\mathfrak{F}}(x, y) = V_{\mathfrak{F}} \psi_y(x) = \langle \psi_y, \psi_x \rangle_{L_2}$.

We are interested in coorbit spaces with respect to weighted Lebesgue spaces $L_{p,m}(X, \mu)$, where we choose weights m_s for $s \in \mathbb{R}$ of the type

$$m_s(a, t) = m_s(a) := |a|^s, \quad m_s(\infty, t) = 1.$$

For this weight to w-moderate for some suitable weight $w = w_s$, let

$$w_s((a_1, t_1), (a_2, t_2)) = w_s(a_1, a_2) := \left(\max \left\{ \frac{a_1}{a_2}, \frac{a_2}{a_1} \right\} \right)^{|s|},$$

with the obvious modification if $a_1 = \infty$ or $a_2 = \infty$. It is straightforward to show that w_s fulfils conditions (4.1.6)–(4.1.8) as well as (4.1.17) with $v = m$. Before defining the corresponding coorbit spaces, it remains to show that Assumption 4.1.2 is fulfilled. Part (i) is a direct consequence of the definition of $\mathfrak{F}$ with $C_{\mathfrak{F}} = \max\{\|\Phi\|_{L_2}, \|\psi\|_{L_2}\} < \infty$ and part (ii) is shown in the following theorem.

Theorem 4.3.5. *We have $K_{\mathfrak{F}} \in \mathcal{A}_{1,w_s}$ for all $s \in \mathbb{R}$.*

Proof. We restrict ourselves to the unweighted case $s = 0$, the weighted case can be shown analogously.

Let $a_1 \in [-1, 1]^*$ and $t_1 \in \mathbb{R}$, then by the support property of $\hat{\psi}$ we get $\hat{\psi}(a_1 \cdot)\hat{\psi}(a\cdot) \equiv 0$ if $|a| \leqslant |a_1|/\alpha^2$. The same then holds for $\mathcal{F}^{-1}(\hat{\psi}(a_1 \cdot)\hat{\psi}(a\cdot))$. Similarly, for any $|a| \leqslant \alpha^2$ it holds that $\mathcal{F}^{-1}(\hat{\Phi}(\cdot)\hat{\psi}(a\cdot)) \equiv 0$. Then, for fixed $x = (a_1, t_1)$ by applying Fubini's theorem and exploiting (4.3.11) we obtain

$$\int_X |\langle \psi_y, \psi_x \rangle_{L_2}| \, d\mu(y)$$

$$= \int_{\mathbb{R}} |\langle \Phi(\cdot - t), \psi(a_1^{-1}(\cdot - t)) \rangle_{L_2}| \, dt + \int_{\mathbb{R}} \int_{-1}^{1} |\langle \psi(a^{-1}(\cdot - t)), \psi(a_1^{-1}(\cdot - t_1)) \rangle_{L_2}| \frac{da}{|a|^2} \, dt$$

$$= |a_1|^{1/2} \int_{\mathbb{R}} |\mathcal{F}^{-1}(\hat{\Phi}(\cdot)\overline{\hat{\psi}(a_1 \cdot)})(t)| \, dt + |a_1|^{1/2} \int_{\mathbb{R}} \int_{-1}^{1} |\mathcal{F}^{-1}(\hat{\psi}(a\cdot)\overline{\hat{\psi}(a_1 \cdot)})(t)| \frac{da}{|a|^{3/2}} \, dt$$

$$\leqslant \|\mathcal{F}^{-1}(\hat{\Phi}(\cdot)\hat{\psi}(a_1 \cdot))\|_{L_1} + \int_{|a_1|/\alpha^2 \leqslant |a| \leqslant 1} \|\mathcal{F}^{-1}(\hat{\psi}(a\cdot)\hat{\psi}(a_1 \cdot))\|_{L_1} \frac{da}{|a|^{3/2}}.$$

We have $\|\mathcal{F}^{-1}(\hat{\Phi}(\cdot)\hat{\psi}(a_1\cdot))\|_{L_1} \leqslant C_1$ uniformly in a_1, so the first summand is finite. Similarly, we get $\|\mathcal{F}^{-1}(\hat{\psi}(a\cdot)\hat{\psi}(a_1\cdot))\|_{L_1} \leqslant C_2$ uniformly in a_1, so that the second term is finite as well.

If $a_1 = \infty$ and $t_1 \in \mathbb{R}$, using the same arguments as above we have

$$\int_X |\langle \psi_y, \psi_x \rangle_{L_2}| \, d\mu(y) = \|\hat{\Phi}\|_{L_2}^2 + \int_{\alpha^{-2} \leqslant |a| \leqslant 1} \|\mathcal{F}^{-1}(\hat{\psi}(a\cdot)\hat{\Phi}(\cdot))\|_{L_1} \frac{da}{|a|^{3/2}},$$

which is finite as well and independent of a_1.

By the symmetry property $K_{\mathfrak{F}}(x,y) = \overline{K_{\mathfrak{F}}(y,x)}$ we then obtain

$$\|K_{\mathfrak{F}}\|_{\mathcal{A}_{1,w_s}} = \sup_{x \in X} \int_X |K_{\mathfrak{F}}(x,y)| \, d\mu(y) < \infty.$$

The weighted case is treated exactly the same way using $w_s \geqslant 1$. $\qquad\square$

Finally, following the lines of Definition 4.1.7, the coorbit space associated to the frame $\mathfrak{F}$ is defined via

$$\mathcal{GWC}_p^s(\mathbb{R}) := \mathrm{Co}(L_{p,m_s}) = \{T \in \mathcal{H}'_{1,m_s} : V_{e,\mathfrak{F}}T \in L_{p,m_s}(X,\mu)\}, \quad 1 \leqslant p \leqslant \infty, s \in \mathbb{R}.$$

The condition $V_{e,\mathfrak{F}}T \in L_{p,m_s}(X,\mu)$ is very similar to the characterization of inhomogeneous Besov spaces in Corollary 4.3.2, more precisely, these spaces can be identified with the coorbit spaces above via $\mathcal{GWC}_p^s = B_{p,p}^{s-1/2+1/p}$. This is analogous to the identification of the homogeneous Besov spaces given in Subsection 2.3.1. For a more detailed discussion on the technicalities we refer to [102].

Discretization

Finally, we intend to discretize the coorbit spaces defined above to obtain atomic decompositions and Banach frames as in Section 4.2. In order to obtain this the first step is to design an *admissible covering* of X. We use the ideas of Subsection 2.3.1 and set for $\alpha > 1$ and $\tau > 0$ the family $\mathcal{U}_{\alpha,\tau} = (U_{\infty,m})_{m \in \mathbb{Z}} \cup (U_{\varepsilon,j,m})_{\varepsilon \in \{-1,1\}, j \in \mathbb{N}, m \in \mathbb{Z}}$ as

$$U_{\infty,m} := \{\infty\} \times \tau\left(m + \left[-\frac{1}{2}, \frac{1}{2}\right)\right),$$

$$U_{\varepsilon,j,m} := \varepsilon[\alpha^{-j}, \alpha^{-j+1}) \times \tau\left(m + \alpha^{-j}\left[-\frac{1}{2}, \frac{1}{2}\right)\right).$$

It is easy to see that this a partition of X, i.e., the sets defined above are pairwise disjoint and $X = \bigcup_{m \in \mathbb{Z}} U_{\infty,m} \cup \bigcup_{\varepsilon \in \{-1,1\}, j \in \mathbb{N}, m \in \mathbb{Z}} U_{\varepsilon,j,m}$. To show the moderateness of $\mathcal{U}_{\alpha,\tau}$, note that we have

$$\mu(U_{\infty,m}) = \int_{\tau(m-1/2)}^{\tau(m+1/2)} dt = \tau,$$

and similarly

$$\mu(U_{\varepsilon,j,m}) = \int_{\alpha^{-j}}^{\alpha^{-j+1}} \frac{da}{a^2} \cdot \int_{\tau(m-\alpha^{-j}/2)}^{\tau(m+\alpha^{-j}/2)} dt = (\alpha^j - \alpha^{j-1}) \cdot \alpha^{-j}\tau = (1 - \alpha^{-1})\tau.$$

This also shows that the covering can be made arbitrarily small with respect to its measure when $\alpha \searrow 1$ and $\tau \searrow 0$.

In [102, Proposition 4.22] it has been shown that in this setting the oscillation $\mathrm{osc}_{\mathcal{U}_{\alpha,\tau}} K_{\mathfrak{F}}$ defined in (4.2.1) is contained in $\mathcal{A}_{1,w_s}$ for all $s \in \mathbb{R}$. Additionally, the norm $\|\mathrm{osc}_{\mathcal{U}_{\alpha,\tau}} K_{\mathfrak{F}}\|_{\mathcal{A}_{1,w_s}}$

can be made arbitrarily small if $\alpha \searrow 1$ and $\tau \searrow 0$. Thus, Theorem 4.2.1 and Theorem 4.2.2 are applicable and we obtain atomic decompositions and Banach frames for $\mathcal{GWC}_p^s$. This means that every $f \in \mathcal{GWC}_p^s$ can be decomposed into

$$f = \sum_{m \in \mathbb{Z}} c(f)_{\infty,m} \cdot \psi_{(\infty,\tau(m-1/2))} + \sum_{\varepsilon \in \{-1,1\}} \sum_{j \in \mathbb{N}} \sum_{m \in \mathbb{Z}} c(f)_{\varepsilon,j,m} \cdot \psi_{(\varepsilon\alpha^{-j},\tau(m-\alpha^{-j}/2))},$$

for a sequence $c(f) \in \ell_{p,m_s}(\{\infty\} \times \mathbb{Z} \cup \{-1,1\} \times \mathbb{N} \times \mathbb{Z})$ depending on f and satisfying

$$\left(\sum_{m \in \mathbb{Z}} |c_{\infty,m}|^p + \sum_{\varepsilon \in \{-1,1\}} \sum_{j \in \mathbb{N}} \sum_{m \in \mathbb{Z}} |c_{\varepsilon,j,m}|^{1/p} |\alpha|^{jsp} \right)^p = \|c(f)\|_{\ell_{p,m_s}} \lesssim \|f\|_{\mathcal{GWC}_p^s}.$$

Conversely, if $d \in \ell_{p,m_s}(\{\infty\} \times \mathbb{Z} \cup \{-1,1\} \times \mathbb{N} \times \mathbb{Z})$ is given, then

$$f = \sum_{m \in \mathbb{Z}} d_{\infty,m} \cdot \psi_{(\infty,\tau(m-1/2))} + \sum_{\varepsilon \in \{-1,1\}} \sum_{j \in \mathbb{N}} \sum_{m \in \mathbb{Z}} d_{\varepsilon,j,m} \cdot \psi_{(\varepsilon\alpha^{-j},\tau(m-\alpha^{-j}/2))} \in \mathcal{GWC}_p^s,$$

and we have the inequality

$$\|f\|_{\mathcal{GWC}_p^s} \lesssim \|c\|_{\ell_{p,m_s}}.$$

Chapter 5

Generalized Coorbit Theory with Non-Integrable Kernel

The theory developed in the previous chapter crucially depended on Assumption 4.1.2, which assumed the reproducing kernel $K_{\mathfrak{F}}$ to be contained in the kernel space $\mathcal{A}_{1,w}$. This assumption laid the foundation for the definition of the space $\mathcal{H}_{1,v}$, whose dual space served as a reservoir for the coorbit spaces, for the well-definedness of the coorbit spaces itself and for the discretization of generalized coorbit spaces. As in Chapter 3 we are faced with the challenge of what is further possible. Therefore the motivation for the present chapter is the question: What happens if the kernel is not contained in $\mathcal{A}_{1,w}$?

The first assumption we suppose instead is the kernel $K_{\mathfrak{F}}$ being contained in some kernel space $\mathcal{A}_{p,w}$ for $1 < p$. This way we can define test spaces $\mathcal{H}_{\tau,v}$, $1 < \tau \leqslant 2$, whose intersection defines the Fréchet space $\mathcal{H}_v$ in (5.1.15), which then serves as a substitute for $\mathcal{H}_{1,v}$. This allows us to define coorbit spaces $\mathrm{Co}(L_{p,m})$ as subsets of $\mathcal{H}'_v$ and these spaces fulfill all desired properties as Proposition 5.1.9 shows.

We are left with the problem of discretizing these coorbit spaces and this is where the present chapter is inspired by the theory developed in Chapter 3. We use very similar ideas to show first, but unsatisfactory, discretization results in Theorem 5.2.17. In addition these results rely on the assumption that the space $\mathrm{span}\{\psi_x\}_{x \in X}$ is dense in $\mathrm{Co}(L_{p,m})$, which needs to be checked individually. As Theorem 5.3.1 shows this assumption is indeed necessary to obtain discretization results similar to the ones of the previous chapter.

However, if we pose an additional assumption we do have Banach frames and atomic decompositions as Theorem 5.4.7 and Theorem 5.4.11 show. For this we assume there exists a second kernel W which is sufficiently smooth and leaves the reproducing kernel invariant under kernel multiplication from the left or right. In a way this chapter can be seen as a fusion of Chapters 3 and 4, in the sense that we are looking at generalized coorbit theory, i.e., coorbit spaces without an underlying group structure, without an integrable reproducing kernel. Since some of the proofs are up to a change of notation almost the same as the proofs in Chapter 3, we omit those proofs.

The developed theory also has an application, namely the inhomogeneous shearlet coorbit spaces. These spaces are the inhomogeneous counterpart of the homogeneous shearlet coorbit spaces defined in Subsection 2.3.2, where the idea of inhomogeneity is inspired by the inhomogeneous Besov spaces in Section 4.3. These spaces are well-defined and we give a reconstruction result using the developed theory in Lemma 5.5.23.

This chapter is organized as follows. In Section 5.1 we define the generalized coorbit spaces with non-integrable reproducing kernel. Then, in Section 5.2, we present first discretization results. The results in Section 5.3 show that we cannot hope to find stable decompositions of the coorbit spaces. With the aid of an additional kernel W we present satisfactory discretization results in Section 5.4. Finally, in Section 5.5, we introduce inhomogeneous shearlet coorbit spaces and show that certain assumptions on the kernel are fulfilled.

5.1 Coorbit Spaces

We start very similar to Chapter 4 in the following fashion. Let $\mathcal{H}$ be a separable Hilbert space and X a locally compact and σ-finite Hausdorff space endowed with a positive Radon measure μ with $\operatorname{supp}\mu = X$. We start with a family of functions $\mathfrak{F} = (\psi_x)_{x\in X}$ indexed by the parameter space X which forms a *Parseval frame* for $\mathcal{H}$. That is, for all $f \in \mathcal{H}$ the map $x \mapsto \langle f, \psi_x \rangle_{\mathcal{H}}$ is measurable and we have

$$\|f\|_{\mathcal{H}} = \int_X |\langle f, \psi_x \rangle_{\mathcal{H}}|^2 \, d\mu(x) \quad \text{for all } f \in \mathcal{H}, \tag{5.1.1}$$

see Definition 1.7.1. Additionally, for the sake of simplicity, we assume the map $x \mapsto \psi_x$ to be weakly continuous. This frame is the foundation for the voice transform $V_{\mathfrak{F}}$ and reproducing kernel $K_{\mathfrak{F}}$ similar to Definition 4.1.1.

Definition 5.1.1. The *voice transform* associated with $\mathfrak{F} = (\psi_x)_{x\in X}$ is defined as the mapping

$$V_{\mathfrak{F}}\colon \mathcal{H} \to L_2(X,\mu), \quad V_{\mathfrak{F}}f(x) := \langle f, \psi_x \rangle_{\mathcal{H}}. \tag{5.1.2}$$

We denote with

$$K_{\mathfrak{F}}\colon X \times X \to \mathbb{C}, \quad K_{\mathfrak{F}}(x,y) := V_{\mathfrak{F}}\psi_y(x) = \langle \psi_y, \psi_x \rangle_{\mathcal{H}} \tag{5.1.3}$$

the *kernel function* associated with $\mathfrak{F}$.

Recall that the operator associated with a kernel $K : X \times X \to \mathbb{C}$ acting on a function f on X is defined by

$$K(f)(x) = \int_X K(x,y)f(y)\,d\mu(y) = \langle f, \overline{K(x,\cdot)} \rangle_{L_2}, \quad x \in X. \tag{5.1.4}$$

Due to (5.1.1), the voice transform $V_{\mathfrak{F}}$ is well-defined and by invoking (1.7.9) we have the *reproducing identity*

$$V_{\mathfrak{F}}f(x) = \int_X V_{\mathfrak{F}}f(y)K_{\mathfrak{F}}(x,y)\,d\mu(y) = K_{\mathfrak{F}}(V_{\mathfrak{F}}f)(x) \quad \text{for all } f \in \mathcal{H}, \tag{5.1.5}$$

where $K_{\mathfrak{F}}$ also denotes the kernel operator acting on functions by the action described in (5.1.4). Note further that the kernel function $K_{\mathfrak{F}}$ is continuous in both arguments by our assumption on the frame. Additionally, the frame equation (4.1.1) can be rewritten using the voice transform as

$$\|f\|_{\mathcal{H}} = \|V_{\mathfrak{F}}f\|_{L_2} \quad \text{for all } f \in \mathcal{H}. \tag{5.1.6}$$

As in the previous chapter, we introduce a weight $w\colon X \times X \to \mathbb{R}_{\geqslant 0}$ on $X \times X$ in the following fashion. Let w be an *admissible weight* in the sense of Definition 1.8.1, that is, w is positive and continuous and satisfies

$$w(x,y) \leqslant w(x,z) \cdot w(z,y), \tag{5.1.7}$$
$$w(x,y) = w(y,x), \tag{5.1.8}$$
$$w(x,x) \leqslant C_w < \infty \tag{5.1.9}$$

for all $x, y, z \in X$.

The main idea of this chapter is to avoid the fundamental Assumption 4.1.2 of the preceding chapter and use a weaker assumption. More precisely, we allow kernels K_ψ that are *not* contained in the kernel space $\mathcal{A}_{1,w}$.

Assumption 5.1.2. *We have $K_\mathfrak{F} \in \mathcal{A}_{q,w}$ for all $1 < q < \infty$, that is, $K_\mathfrak{F}$ is measurable and*

$$\sup_{x \in X} \int_X |\langle \psi_x, \psi_y \rangle_\mathcal{H}|^q \cdot w(x,y)^q \, d\mu(y) < \infty \quad \text{for all } 1 < q < \infty, \tag{5.1.10}$$

where $\mathcal{A}_{q,w}$ is the kernel space defined in Definition 1.8.2.

We add some remarks.

Remark 5.1.3.   (i) The assumption above is analogous to Assumption 3.1.12 in the group setting, which assumed the kernel K_ψ to be contained in all weighted Lebesgue spaces $L_{p,w}(G)$, $1 < p < \infty$.

 (ii) If we assume $w^{-1} \in \mathcal{A}_p$ for some $1 < p < \infty$, then Hölder's inequality shows

$$\|K\|_{\mathcal{A}_1} \leqslant \|K\|_{\mathcal{A}_{p',w}} \cdot \|w^{-1}\|_{\mathcal{A}_p},$$

so $K \in \mathcal{A}_1$ and we are in the setting of Chapter 4. However, in many interesting examples this is not true because w is independent of at least one variable, so that $w^{-1} \notin \mathcal{A}_p$ for all $1 < q < \infty$.

(iii) Assumption 5.1.2 is indeed weaker than the assumption $K_\mathfrak{F} \in \mathcal{A}_{1,w}$ we made in the previous chapter, which can be seen as follows. Since $K_\mathfrak{F} \cdot w$ is continuous, if $K_\mathfrak{F} \cdot w \in \mathcal{A}_1$ then it is bounded as well, so $K_\mathfrak{F} \cdot w \in \mathcal{A}_\infty$. But this obviously implies $K_\mathfrak{F} \in \mathcal{A}_{q,w}$ for all $1 < q < \infty$.

(iv) In stark contrast to Chapter 4, a weighted version of Schur's test, see Corollary 1.8.6, is no longer applicable to the kernel $K_\mathfrak{F}$. Instead we may only apply the weaker version given in Corollary 1.8.9, which assures the boundedness of the kernel operator associated to $K_\mathfrak{F}$ from $L_{p,m}(X,\mu)$ to $L_{r,m}(X,\mu)$ for all $1 \leqslant p < r \leqslant \infty$ and w-moderate weights m. We will exploit this numerous times in the present chapter.

The next step is to define suitable spaces $\mathcal{H}_{\tau,v}$ for weights v on X, which serves as a first step towards a test space on X. For this let $v \colon X \to \mathbb{R}_+$ be a *w-moderate weight* on X that satisfies $v \geqslant 1$ and

$$v(x) \leqslant w(x,y) \cdot v(y) \quad \text{for all } x, y \in X. \tag{5.1.11}$$

Note that if v is a w-moderate weight, then so is v^{-1}, a fact we will use time and time again. Then, for $1 \leqslant \tau \leqslant 2$ consider the spaces

$$\mathcal{H}_{\tau,v} := \{ f \in \mathcal{H} : V_\mathfrak{F} f = \langle f, \psi. \rangle_\mathcal{H} \in L_{\tau,v}(X,\mu) \} \tag{5.1.12}$$

equipped with the natural norm

$$\|f\|_{\mathcal{H}_{\tau,v}} := \|V_\mathfrak{F} f\|_{L_{\tau,v}}. \tag{5.1.13}$$

First we note that these spaces are non-empty; more precisely, the frame $\mathfrak{F}$ is contained in $\mathcal{H}_{\tau,v}$ and the following Lemma holds.

Lemma 5.1.4. *If* $1 \leqslant \tau \leqslant 2$ *and* $K_{\mathfrak{F}} \in \mathcal{A}_{\tau,w}$, *then*

$$\|\psi_x\|_{\mathcal{H}_{\tau,v}} \leqslant v(x) \cdot \|K_{\mathfrak{F}}\|_{\mathcal{A}_{\tau,w}},$$

hence $\psi_x \in \mathcal{H}_{\tau,v}$ *for all* $x \in X$.

Proof. It is

$$\|\psi_x\|_{\mathcal{H}_{\tau,v}}^{\tau} = \int_X |V_{\mathfrak{F}}\psi_x(y)|^{\tau} v(y)^{\tau} \, d\mu(y)$$

$$\leqslant v(x)^{\tau} \int_X |K_{\mathfrak{F}}(y,x)|^{\tau} w(y,x)^{\tau} \, d\mu(y)$$

$$\leqslant v(x)^{\tau} \cdot \|K_{\mathfrak{F}}\|_{\mathcal{A}_{\tau,w}}^{\tau},$$

for all $x \in X$, which proves the assertion.   □

Since $\mathfrak{F}$ constitutes a frame for $\mathcal{H}$ this means $\mathcal{H}_{\tau,v} \subset \mathcal{H}$ is dense. Moreover, the spaces $\mathcal{H}_{\tau,v}$ are Banach spaces, as the following lemma states.

Lemma 5.1.5. *If* $1 \leqslant \tau \leqslant 2$ *and* $K_{\mathfrak{F}} \in \mathcal{A}_{\tau',w}$, *then the space* $\mathcal{H}_{\tau,v}$ *is a Banach space and satisfies the continuous embedding* $\mathcal{H}_{\tau,v} \hookrightarrow \mathcal{H}$.

Proof. Let $(f_n)_{n\in\mathbb{N}} \subset \mathcal{H}_{\tau,v} \subset \mathcal{H}$ be a Cauchy sequence in $\mathcal{H}_{\tau,v}$, which means $(g_n)_{n\in\mathbb{N}} := (V_{\mathfrak{F}}f_n)_{n\in\mathbb{N}}$ is a Cauchy sequence in $L_{\tau,v}(X,\mu)$. By the completeness of $L_{\tau,v}(X,\mu)$ there exists a unique $g \in L_{\tau,v}(X,\mu)$ with $g_n \to g$. Then, by Hölder's inequality, for almost every $x \in X$ it holds that

$$|K_{\mathfrak{F}}(g)(x)| \leqslant \int_X |K_{\mathfrak{F}}(x,y)g(y)| \, d\mu(y)$$

$$\leqslant \|K_{\mathfrak{F}}(x,\cdot)\|_{L_{\tau',v^{-1}}} \cdot \|g\|_{L_{\tau,v}}$$

$$\leqslant v(x)^{-1} \cdot \|K_{\mathfrak{F}}\|_{\mathcal{A}_{\tau',w}} \cdot \|g\|_{L_{\tau,v}},$$

thus $K_{\mathfrak{F}}(g) \in L_{\infty}(X,\mu)$. Furthermore, by the reproducing formula 5.1.5 we have $K_{\mathfrak{F}}(g_n) = g_n$ for all $n \in \mathbb{N}$, and since $K_{\mathfrak{F}} : L_{\tau,v}(X,\mu) \to L_{\infty,v}(X,\mu)$ is continuous, see Lemma 1.8.7, this implies $K_{\mathfrak{F}}(g_n) \to K_{\mathfrak{F}}(g)$ in $L_{\infty,v}(X,\mu)$, yielding $K_{\mathfrak{F}}(g) = g$. Thus, $g \in L_{\infty}(X,\mu)$ and since $L_{\infty}(X,\mu) \cap L_{\tau,v}(X,\mu) \subset L_2(X,\mu)$ it follows $g \in L_2(X,\mu)$. Since the application of $K_{\mathfrak{F}}$ is the orthogonal projection from $L_2(X,\mu)$ onto the image of $V_{\mathfrak{F}}$ there exists $f \in \mathcal{H}$ such that $g = V_{\mathfrak{F}}f$. Moreover, $V_{\mathfrak{F}}f \in L_{\tau,v}(X,\mu)$ means $f \in \mathcal{H}_{\tau,v}$ and $f_n \to f \in \mathcal{H}_{\tau,v}$.

To show the continuity of the embedding $\mathcal{H}_{\tau,v} \subset \mathcal{H}$ for f and g as above it follows

$$\|g\|_{L_{\infty,v}} = \|K_{\mathfrak{F}}g\|_{L_{\infty,v}} \leqslant \|K_{\mathfrak{F}}\|_{\mathcal{A}_{\tau',w}} \cdot \|g\|_{L_{\tau,v}},$$

which implies

$$\|g\|_{L_2} \leqslant \|g\|_{L_{\tau}}^{\tau/2} \cdot \|g\|_{L_{\infty}}^{1-\tau/2} \leqslant \|g\|_{L_{\tau,v}}^{\tau/2} \cdot \|g\|_{L_{\infty,v}}^{1-\tau/2} \leqslant \|K_{\mathfrak{F}}\|_{\mathcal{A}_{\tau',w}}^{1-\tau/2} \cdot \|g\|_{L_{\tau,v}}.$$

Then,

$$\|f\|_{\mathcal{H}} = \|V_{\mathfrak{F}}f\|_{L_2} = \|g\|_{L_2} \leqslant \|K_{\mathfrak{F}}\|_{\mathcal{A}_{\tau',w}}^{1-\tau/2} \cdot \|g\|_{L_{\tau,v}} = \|K_{\mathfrak{F}}\|_{\mathcal{A}_{\tau',w}}^{1-\tau/2} \cdot \|f\|_{\mathcal{H}_{\tau,v}},$$

which concludes the proof.   □

Hence, these function spaces lead to the Gelfand triple setting of dense embeddings

$$\mathcal{H}_{\tau,v} \hookrightarrow \mathcal{H} \cong \mathcal{H}' \hookrightarrow \mathcal{H}'_{\tau,v}$$

where $\mathcal{H}'_{\tau,v}$ denotes the canonical anti-dual space of $\mathcal{H}_{\tau,v}$, i.e., the space of all conjugate linear, continuous functionals on $\mathcal{H}'_{\tau,v}$, and this space can be interpreted as a space of distributions. An element $h \in \mathcal{H}$ is hereby identified with the functional $\mathcal{H} \to \mathbb{C}$, $f \to \langle h, f \rangle_{\mathcal{H}}$. With these embeddings it is possible to extend the notion of the voice transform in a canonical way to elements $f \in \mathcal{H}'_{\tau,v}$ by setting

$$V_{\mathfrak{F},\tau} f(x) = f(\psi_x) = \langle f, \psi_x \rangle_{\mathcal{H}'_{\tau,v} \times \mathcal{H}'_{\tau,v}}. \tag{5.1.14}$$

With the assumption $K_{\mathfrak{F}} \in \mathcal{A}_{\tau,m_v}$ and Lemma 5.1.4 this is well-defined for all $x \in X$. Furthermore, the same lemma shows

$$|V_{\mathfrak{F},\tau} f(x)| \leqslant \|f\|_{\mathcal{H}'_{\tau,v}} \cdot \|\psi_x\|_{\mathcal{H}_{\tau,v}} \leqslant \|f\|_{\mathcal{H}'_{\tau,v}} \cdot \|K_{\mathfrak{F}}\|_{\mathcal{A}_{\tau,w}} \cdot v(x)$$

for all $x \in X$, i.e., $V_{\mathfrak{F},\tau} f \in L_{\infty,v^{-1}}(X,\mu)$.

Under particular assumptions on the reproducing kernel we can prove the following nesting properties.

Lemma 5.1.6. *Let Assumption 5.1.2 be fulfilled, i.e., $K_{\mathfrak{F}} \in \mathcal{A}_{q,w}$ for all $1 < q < \infty$, then we have the continuous embeddings $\mathcal{H}_{\sigma,v} \hookrightarrow \mathcal{H}_{\tau,v}$ and $\mathcal{H}'_{\tau,v} \hookrightarrow \mathcal{H}'_{\sigma,v}$ for all $1 \leqslant \sigma < \tau \leqslant 2$.*

Proof. Assume $f \in \mathcal{H}_{\sigma,v}$, which means $f \in \mathcal{H}$ with $V_{\mathfrak{F}} f \in L_{\sigma,v}(X,\mu)$. Since the reproducing identity (5.1.5) holds it follows $V_{\mathfrak{F}} f = K_{\mathfrak{F}}(V_{\mathfrak{F}} f) \in K_{\mathfrak{F}}(L_{\sigma,v}(X,\mu))$ and with Corollary 1.8.9 we derive $V_{\mathfrak{F}} f \in L_{\tau,v}(X,\mu)$, hence $f \in \mathcal{H}_{\tau,v}$ with

$$\|V_{\mathfrak{F}} f\|_{L_{\tau,v}} \leqslant \|K_{\mathfrak{F}}(V_{\mathfrak{F}} f)\|_{L_{\tau,v}} \leqslant \|K_{\mathfrak{F}}\|_{L_{\sigma,v} \to L_{\tau,v}} \cdot \|V_{\mathfrak{F}} f\|_{L_{\sigma,v}}.$$

The second assertion is immediate. $\qquad\square$

Now let us define the test spaces, whose dual will serve as reservoir for the coorbit spaces. Let $\mathcal{H}_v$ be the intersection of all $\mathcal{H}_{\tau,v}$, that is,

$$\mathcal{H}_v := \bigcap_{1 < \tau \leqslant 2} \mathcal{H}_{\tau,v}, \tag{5.1.15}$$

and endowed with the topology induced by the family of norms $\|\cdot\|_{\mathcal{H}_{\tau,v}}$ for $1 < \tau \leqslant 2$ this space is a Fréchet space, see Theorem 1.37 in [92]. With Lemma 5.1.4 we have $\psi_x \in \mathcal{H}_v$ for all $x \in X$. Furthermore, we have the continuous embeddings $\mathcal{H}_v \hookrightarrow \mathcal{H}_{\tau,v}$ for all $1 < \tau \leqslant 2$ by the definition of the topology.

Comparing $\mathcal{H}_v$ with the approach in Subsection 3.1.2, we see that this is a substitute for the space $\mathcal{S}_w$ defined in (3.1.16). In a similar fashion we can understand $\mathcal{H}_v$ as the space of functions $f \in \mathcal{H}$ with $V_{\mathfrak{F}} f \in L_{\tau,v}(X,\mu)$ for all $1 < \tau \leqslant 2$.

We denote with $\mathcal{H}'_v$ the anti-dual of $\mathcal{H}_v$ and observe $\mathcal{H}'_{\tau,v} \hookrightarrow \mathcal{H}'_v$ for all $1 < \tau \leqslant 2$. Now define $\mathcal{R}$ as the union of reproducing kernel spaces:

$$\mathcal{R} := \bigcup_{1 < q < \infty} \{F \in L_{q,v^{-1}}(X,\mu) : K_{\mathfrak{F}} F = F\}. \tag{5.1.16}$$

With this definition we propose the following lemma.

Lemma 5.1.7. *Let Assumption 5.1.2 be fulfilled, that is, $K_{\mathfrak{F}} \in A_{q,w}$ for all $1 < q < \infty$. The extended voice transform $V_{e,\mathfrak{F}} \colon \mathcal{H}'_v \to \mathcal{R}$ defined via*

$$V_{e,\mathfrak{F}} f(x) = f(\psi_x) = \langle f, \psi_x \rangle_{\mathcal{H}'_{\tau,v} \times \mathcal{H}_{\tau,v}}, \quad x \in X, \tag{5.1.17}$$

is a well-defined isomorphism, satisfying

$$f(\gamma) = \langle V_{e,\mathfrak{F}} f, V_{\mathfrak{F}} \gamma \rangle_{L_2} \tag{5.1.18}$$

for all $\gamma \in \mathcal{H}_v$ and $f \in \mathcal{H}'_v$.

Proof. We first show that $V_{e,\mathfrak{F}}(\mathcal{H}'_v) \subseteq K_{\mathfrak{F}}(L_{\tau',v^{-1}}(X, \mu))$. Note that from Lemma 5.1.6 it follows $\|\cdot\|_{\mathcal{H}_{\tau,v}} \lesssim \|\cdot\|_{\mathcal{H}_{\sigma,v}}$ for $1 < \sigma \leqslant \tau \leqslant 2$. From this and Theorem 1.37 in [92] it follows that the family $(U_{\tau,v})_{1 < \tau \leqslant 2, n \in \mathbb{N}}$ given by

$$U_{\tau,v} := \left\{ f \in \mathcal{H}_v \; : \; \|f\|_{\mathcal{H}_{\tau,v}} \leqslant \frac{1}{n} \right\}$$

is a local base at the origin in $\mathcal{H}_v$. With this at hand let $f \in \mathcal{H}'_v$ be arbitrary, then there is some $1 < \tau \leqslant 2$ and $C > 0$ satisfying $|f(\gamma)| \leqslant C \cdot \|\gamma\|_{\mathcal{H}_{\tau,v}}$ for all $\gamma \in \mathcal{H}_v$, see also Corollary 22.7 in [84]. We now define the antilinear functional

$$\varphi : V_{\mathfrak{F}}(\mathcal{H}_v) \to \mathbb{C}, \quad V_{\mathfrak{F}} \gamma \mapsto f(\gamma).$$

This functional is well-defined, since $V_{\mathfrak{F}} \colon \mathcal{H} \to L_2(X, \mu)$ is injective and $\mathcal{H}_v \subset \mathcal{H}$ by definition. Furthermore, by setting $\Gamma = V_{\mathfrak{F}} \gamma \in V_{\mathfrak{F}}(\mathcal{H}_v)$, we have

$$|\varphi(\Gamma)| = |f(\gamma)| \leqslant C \cdot \|\gamma\|_{\mathcal{H}_{\tau,v}} = C \cdot \|\Gamma\|_{L_{\tau,v}},$$

which means φ is bounded when considering $V_{\mathfrak{F}}(\mathcal{H}_v)$ as a subspace of $L_{\tau,v}(X, \mu)$. By the Hahn-Banach theorem, there is a continuous antilinear extension $\widetilde{\varphi} \colon L_{\tau,v}(X, \mu) \to \mathbb{C}$ of φ. By the Riesz representation theorem for the dual of L_p, there is some $G \in L_{\tau', \frac{1}{v}}(X, \mu)$ with

$$f(\gamma) = \varphi(V_{\mathfrak{F}} \gamma) = \widetilde{\varphi}(V_{\mathfrak{F}} \gamma) = \int_X G(x) \overline{V_{\mathfrak{F}} \gamma(x)} \, d\mu(x) \tag{5.1.19}$$

for all $\gamma \in \mathcal{H}_v$. Hence we observe

$$V_{e,\mathfrak{F}} f(x) = f(\psi_x) = \int_X G(y) \overline{V_{\mathfrak{F}} \psi_x(y)} \, d\mu(y) = \int_X G(y) K_{\mathfrak{F}}(x, y) \, d\mu(y) = K_{\mathfrak{F}} G(x)$$

for almost all $x \in X$, thus yielding $V_{e,\mathfrak{F}} f = K_{\mathfrak{F}} G$.

We will now show $V_{e,\mathfrak{F}}(\mathcal{H}'_v) \subset \mathcal{R}$. For this we note that X was chosen σ-finite, i.e., there exists a family of nested sets $(X_n)_{n \in \mathbb{N}}$ with $\mu(X_n) < \infty$, $X_n \subset X_{n+1}$ and $X = \bigcup_{n \in \mathbb{N}} X_n$. By setting

$$Y_n := \left\{ x \in X_n \; : \; \frac{1}{n} \leqslant v(x) \leqslant n \right\}$$

we observe $Y_n \subset Y_{n+1}$ and $X = \bigcup_{n \in \mathbb{N}} Y_n$. Now let $f \in \mathcal{H}'_v$ be arbitrary and $G \in L_{\tau',v^{-1}}(X, \mu)$ as above for some $1 < \tau \leqslant 2$ satisfy $V_{e,\mathfrak{F}} f = K_{\mathfrak{F}} G$. Set $G_n := \chi_{Y_n} G \in L_2(X, \mu)$, where χ_{Y_n} denotes the characteristic function of Y_n. It is obvious that $G_n \to G$ pointwise and $|G_n| \leqslant |G|$ for all $n \in \mathbb{N}$, hence $G_n \to G$ in $L_{\tau',v^{-1}}(X, \mu)$. By Corollary 1.8.9 the maps $K_{\mathfrak{F}} \colon L_{\tau',v^{-1}}(X, \mu) \to L_{q,v^{-1}}(X, \mu)$ and $K_{\mathfrak{F}} \colon L_{q,v^{-1}}(X, \mu) \to L_{r,v^{-1}}(X, \mu)$ are bounded for $2 \leqslant \tau' < q < r < \infty$, which means $K_{\mathfrak{F}}G_n \to K_{\mathfrak{F}}G$ in $L_{q,v^{-1}}(X,\mu)$ and $K_{\mathfrak{F}}K_{\mathfrak{F}}G_n \to K_{\mathfrak{F}}K_{\mathfrak{F}}G$ in $L_{r,v^{-1}}(X,\mu)$. Possibly after switching to a subsequence, this also yields convergence pointwise almost everywhere. Therefore, by the property $K_{\mathfrak{F}}K_{\mathfrak{F}} = K_{\mathfrak{F}}$ on $L_2(X,\mu)$, and since $G_n \in L_2(X,\mu)$,

$$K_{\mathfrak{F}}(V_{e,\mathfrak{F}}f) = K_{\mathfrak{F}}K_{\mathfrak{F}}G = \lim_{n\to\infty} K_{\mathfrak{F}}K_{\mathfrak{F}}G_n = \lim_{n\to\infty} K_{\mathfrak{F}}G_n = K_{\mathfrak{F}}G = V_{e,\mathfrak{F}}f,$$

and we have $V_{e,\mathfrak{F}}f \in L_{q,v^{-1}}(X,\mu)$. This yields $V_{e,\mathfrak{F}}f \in \mathcal{R}$.

We will now show equation (5.1.18). Let $f \in \mathcal{H}'_v$ and $G \in L_{\tau',v^{-1}}(X,\mu)$ with $V_{e,\mathfrak{F}}f = K_{\mathfrak{F}}G$ as above, as well as $G_n \in L_2(X,\mu)$ with $G_n \to G$ in $L_{\tau',v^{-1}}(X,\mu)$ and $K_{\mathfrak{F}}G_n \to K_{\mathfrak{F}}G$ in $L_{q,v^{-1}}(X,\mu)$ for $2 \leqslant \tau' < q < \infty$. By (5.1.19), the reproducing property (5.1.5) and the self-adjointness of $K_{\mathfrak{F}}$, it follows for arbitrary $\gamma \in \mathcal{H}_v \subset \mathcal{H}$ that

$$
\begin{aligned}
\langle V_{e,\mathfrak{F}}f, V_{\mathfrak{F}}\gamma \rangle_{L_2} &= \int_X K_{\mathfrak{F}}G(x)\overline{V_{\mathfrak{F}}f(x)}\,d\mu(x) \\
&= \lim_{n\to\infty} \int_X K_{\mathfrak{F}}G_n(x)\overline{V_{\mathfrak{F}}\gamma(x)}\,d\mu(x) \\
&= \lim_{n\to\infty} \int_X G_n(x)\overline{K_{\mathfrak{F}}V_{\mathfrak{F}}\gamma(x)}\,d\mu(x) \\
&= \int_X G(x)\overline{V_{\mathfrak{F}}\gamma(x)}\,d\mu(x) = f(\gamma),
\end{aligned}
$$

as claimed in (5.1.18).

With this at hand the injectivity of $V_{e,\mathfrak{F}}$ is immediate. Assume $V_{e,\mathfrak{F}}f \equiv 0$ for some $f \in \mathcal{H}'_v$, then by 5.1.18 we have $f(\gamma) = 0$ for all $\gamma \in \mathcal{H}_v$, hence $f \equiv 0$.

Finally we will show the surjectivity of $V_{e,\mathfrak{F}}$. To this end let $F \in L_{q,v^{-1}}(X,\mu)$ with $F = K_{\mathfrak{F}}F$ for some $1 < q < \infty$. Then, by Corollary 1.8.9, $F = K_{\mathfrak{F}}F \in L_{\tau,v^{-1}}(X,\mu)$ for every $q < \tau < \infty$, hence we can assume $2 \leqslant \tau < \infty$, or equivalently $1 < \tau' \leqslant 2$. Now we define

$$f: \mathcal{H}_v \to \mathbb{C}, \quad \gamma \mapsto \langle F, V_{\mathfrak{F}}\gamma \rangle_{L_2}.$$

Then, f is well-defined and antilinear with

$$|f(\gamma)| \leqslant \|F\|_{L_{\tau,v^{-1}}} \cdot \|V_{\mathfrak{F}}\gamma\|_{L_{\tau',v}} = \|F\|_{L_{\tau,v^{-1}}} \cdot \|\gamma\|_{\mathcal{H}_{\tau',v}},$$

i.e., $f \in \mathcal{H}'_{\tau',v} \hookrightarrow \mathcal{H}'_v$. Then,

$$
\begin{aligned}
V_{e,\mathfrak{F}}f(x) &= f(\psi_x) = \int_X F(y)\overline{V_{\mathfrak{F}}\psi_x(y)}\|\,d\mu(y) = \int_X K_{\mathfrak{F}}(x,y)F(y)\,d\mu(y) \\
&= K_{\mathfrak{F}}F(x) = F(x)
\end{aligned}
$$

for almost all $x \in X$, which shows surjectivity. $\qquad\square$

As in the entirety of this dissertation we are interested in coorbit spaces where the decay of the voice transform is measured in some weighted Lebesgue space $L_{p,m}(X,\mu)$ for suitable weights m. For this, let $m: X \to \mathbb{R}_+$ be a w-*moderate weight* on X, i.e.,

$$m(x) \leqslant w(x,y) \cdot m(y) \quad \text{for all } x,y \in X. \tag{5.1.20}$$

This is the same condition as for the weight v in (5.1.11), hence we can set $m = v$. Now we define the coorbit spaces as follows.

Definition 5.1.8. The coorbit spaces $\mathrm{Co}(L_{p,m})$ with respect to the frame $\mathfrak{F} = (\psi_x)_{x \in X}$ and target space $L_{p,m}(X, \mu)$ for $1 \leqslant p < \infty$ are defined as

$$\mathrm{Co}(L_{p,m}) := \left\{ f \in \mathcal{H}'_m : V_{e,\mathfrak{F}} f \in L_{p,m}(X, \mu) \right\}$$

endowed with the natural norms

$$\|f\|_{\mathrm{Co}(L_{p,m})} := \|V_{e,\mathfrak{F}} f\|_{L_{p,m}}.$$

We collect some basic facts in the following proposition.

Proposition 5.1.9. *Suppose $K_{\mathfrak{F}}(L_{p,m}(X, \mu)) \subset L_{r,m^{-1}}(X, \mu)$ for some $1 < r < \infty$. Then the following hold true.*

   (i) *A function $F \in L_{p,m}(X, \mu)$ is of the form $V_{e,\mathfrak{F}} f$ for some $f \in \mathrm{Co}_{\mathfrak{F}}(L_{p,m})$ if and only if $K_{\mathfrak{F}} F = F$.*

   (ii) *The space $(\mathrm{Co}(L_{p,m}), \|\cdot\|_{\mathrm{Co}_{\mathfrak{F}}(L_{p,m})})$ is a Banach space of functions.*

   (iii) *The map $V_{e,\mathfrak{F}} : \mathrm{Co}(L_{p,m}) \to L_{p,m}(X, \mu)$ induces an isometric isomorphism between the coorbit space $\mathrm{Co}(L_{p,m})$ and the reproducing kernel space*

$$\mathcal{M}_{p,m} := \{ F \in L_{p,m}(X, \mu) \, : \, K_{\mathfrak{F}} F = F \} \subset L_{p,m}(X, \mu). \tag{5.1.21}$$

Proof. To show (i) assume $f \in \mathrm{Co}(L_{p,m})$, then by definition $f \in \mathcal{H}'_m$ and the reproducing identity holds, thanks to Lemma 5.1.7. Conversely, if $F \in L_{p,m}(X, \mu)$ satisfies $K_{\mathfrak{F}} F = F$ we deduce by our assumption $F \in L_{r,m^{-1}}(X, \mu)$, hence $F \in \mathcal{R}$. By Lemma 5.1.7 there exists $f \in \mathcal{H}'_m$ such that $F = V_{e,\mathfrak{F}} f$, which shows the assertion.

To show (ii), we first note that $\|\cdot\|_{\mathrm{Co}(L_{p,m})}$ is indeed a well-defined norm. This follows from the properties of Lebesgue spaces and the injectivity of $V_{e,\mathfrak{F}}$ shown in Lemma 5.1.7. To show the completeness, suppose that $(f_n)_{n \in \mathbb{N}}$ is a Cauchy sequence in $\mathrm{Co}(L_{p,m})$, implying that $F_n := V_{e,\mathfrak{F}} f_n$ is a Cauchy sequence in $L_{p,m}(X, \mu)$. By the completeness of $L_{p,m}(X, \mu)$, this sequence convergences to an unique element $F \in L_{p,m}(X, \mu)$. By (i) and continuity of $K_{\mathfrak{F}} : L_{p,m}(X, \mu) \to L_{q,m}(X, \mu)$ for $p < q < \infty$ we get $K_{\mathfrak{F}} F_n = F_n$ for all $n \in \mathbb{N}$ and hence $K_{\mathfrak{F}} F = F$. Again by (i) there exists an $f \in \mathrm{Co}(L_{p,m})$ with $V_{\mathfrak{F},\tau} f = F$ and the completeness is shown.

The assertion (iii) follows from (i) and the injectivity of $V_{e,\mathfrak{F}}$. $\qquad\square$

Remark 5.1.10. The condition $K_{\mathfrak{F}}(L_{p,m}(X, \mu)) \subset L_{r,m^{-1}}(X, \mu)$ for some $1 < r < \infty$ in Proposition 5.1.9 may appear strange, but is naturally fulfilled for the following setting. Let Assumption 5.1.2 be fulfilled, i.e., $K_{\mathfrak{F}} \in \mathcal{A}_{q,w}$ for all $1 < q < \infty$, then by Corollary 1.8.9 we have $K_{\mathfrak{F}}(L_{p,m}(X, \mu)) \subset L_{r,m}(X, \mu) \subset L_{r,m^{-1}}(X, \mu)$ for all $1 \leqslant p < r \leqslant \infty$. This will be the setting we are interested in.

Further note that this condition is similar to the condition $K_{\mathfrak{F}}(Y) \subset L_{\infty,v^{-1}}(X, \mu)$ in Proposition 4.1.8 of the previous chapter.

We add some basic facts concerning the coorbit spaces $\mathrm{Co}(L_{p,m})$. We omit the proof since it is almost identical to the proof of Lemma 2.1.18.

Corollary 5.1.11. *Let m, m_1, m_2 be w-moderate weights on X. Then:*

   (i) *The spaces $\mathrm{Co}(L_{p,m})$ are monotonically increasing with $1 \leqslant p < \infty$.*

   (ii) *Let $1 \leqslant p_2 < p_1 \leqslant \infty$ and assume the quotient m_1/m_2 to belong to $L_r(X, \mu)$ for some $1 \leqslant r \leqslant \infty$ fulfilling $r^{-1} \leqslant p_2^{-1} - p_1^{-1}$, then $\mathrm{Co}(L_{p_1,m_1}) \subset \mathrm{Co}(L_{p_2,m_2})$.*

(iii) $\mathrm{Co}(L_{p_1,m_1}) = \mathrm{Co}(L_{p_2,m_2})$ *if and only if* $p_1 = p_2$ *and there are two constants* $0 < C_1 \leqslant C_2$ *such that* $C_1 m_1(x) \leqslant m_2(x) \leqslant C_2 m_1(x)$ *for all* $x \in X$.

Remark 5.1.12. Under the additional assumption $K_{\mathfrak{F}} \in \mathcal{A}_{1,w}$ the coorbit spaces from Chapter 4, see Definition 4.1.7, which we denote with $\widetilde{\mathrm{Co}}(L_{p,m})$ to avoid confusion, are well-defined by a weighted version of Schur's test, see Corollary 1.8.6. It can then be shown that the space $\widetilde{\mathrm{Co}}(L_{p,m})$ even coincides with $\mathrm{Co}(L_{p,v})$ up to canonical identifications.

This can be seen as follows. Recall the extended voice transform $V_{e,\mathfrak{F}}$ as defined in (4.1.13). Fix $1 \leqslant p < \infty$ and let $f \in \widetilde{\mathrm{Co}}(L_{p,m})$ be arbitrary, implying $V_{e,\mathfrak{F}} f \in L_{p,m}(X,\mu)$ and $K_{\mathfrak{F}} V_{e,\mathfrak{F}} f = V_{e,\mathfrak{F}} f$ by Proposition 4.1.8. Therefore, Corollary 1.8.9 shows $V_{e,\mathfrak{F}} f = K_{\mathfrak{F}} V_{e,\mathfrak{F}} f \in L_{q,m}(X,\mu)$ for all $q > p$. By Lemma 5.1.7 there exists some $g \in \mathcal{H}'_m$ with $V_{e,\mathfrak{F}} f = V_{e,\mathfrak{F}} g$. But since $V_{e,\mathfrak{F}} g \in L_{p,m}(X,\mu)$ this means $g \in \mathrm{Co}(L_{p,m})$.

By Lemma 5.1.6 we have $\mathcal{H}_{1,m} \hookrightarrow \mathcal{H}_m$, thus $\mathcal{H}'_m \hookrightarrow \mathcal{H}'_{1,m}$. This implies $V_{e,\mathfrak{F}} f = V_{e,\mathfrak{F}} g$, and since $V_{e,\mathfrak{F}}$ is injective we obtain $f = g$ as elements of $\mathcal{H}'_{1,m}$. Therefore $g \in \mathcal{H}'_m$ is an extension of $f \in \mathcal{H}'_{1,m}$. Overall, this shows that the map $\mathrm{Co}(L_{p,m}) \to \widetilde{\mathrm{Co}}(L_{p,m})$, $f \mapsto f|_{\mathcal{H}_{1,m}}$ is an isometric isomorphism.

This identification suggests that the presented generalization of the approach of Chapter 4 is to some extend natural.

Remark 5.1.13. The results in this section are less general than the coorbit spaces of the preceding chapters since we restrict ourselves to the target spaces $L_{p,m}(X,\mu)$. However, the coorbit spaces for a solid Banach spaces of functions Y continuously embedded into $L_{1,\mathrm{loc}}(X,\mu)$ and satisfying the continuous embedding $\mathcal{A}_{q,m}(Y) \subset Y$ for all $1 < q < \infty$, that is,

$$\|K(F)\|_Y \leqslant \|K\|_{\mathcal{A}_{q,w}} \cdot \|F\|_Y \quad \text{for all } K \in \mathcal{A}_{q,w}, \ F \in Y, \ 1 < q < \infty,$$

should also be well-defined. It is not clear if quasi-Banach spaces are suitable target spaces and if Assumption 5.1.2 can be replaced by the condition $K_{\mathfrak{F}} \in \mathcal{T}$, where $\mathcal{T}$ is a Fréchet space, as in Subsection 3.1.1. These problems are still left open for future publications.

5.2 Discretization

In this section we inted to establish a first discretization result for coorbit spaces $\mathrm{Co}(L_{p,m})$ described in Section 5.1. More precisely, we give a weak atomic decomposition for these spaces in Theorem 5.2.17. Before we start, let us briefly discuss how our apporach compares to Chapter 3 and Chapter 4.

The general idea of this section is heavily inspired by the results of Section 3.2, where we were interested in discretization results for coorbit spaces on groups with non-integrable kernel. Since in this chapter we are also concerned with kernels that are not integrable with respect to certain kernel spaces, this apporach is reasonable. The main differences obviously lie in the fact that we have no group structure available, which is why we have to adapt all the proofs to this circumstance. However, the final results are comparable.

As usual, the discretization of coorbit spaces is based on a suitable covering $\mathcal{U}$ of the index set X. This is why we will first recall the notion of admissible coverings. Then, the behaviour of the corresponding $\mathcal{U}$-oscillation $\mathrm{osc}_{\mathcal{U}} K_{\mathfrak{F}}$ of the kernel is a major ingredient. More precisely, we need $\mathrm{osc}_{\mathcal{U}} K_{\mathfrak{F}} \in \mathcal{A}_{r,m}$ for some suitable integrability parameter r. This is the analog of the integrability with respect to $\mathcal{A}_{1,m}$ of the $\mathcal{U}$-oscillation we assumed in Theorem 4.2.1 and Theorem 4.2.2, which was crucially supposed for the discretization results in Section 4.2 of the previous chapter to hold true. But since the kernel is not in $\mathcal{A}_{1,m}$ in our setting, the same will not be true for the $\mathcal{U}$-oscillation and we rely on different discretization techniques.

This section is organized as follows. First, we recall the main features of the previous section. Then, in Subsection 5.2.1 we present a basic assumption on the kernel $K_{\mathfrak{F}}$ that is sufficient for a weak form of atomic decomposition for generalized coorbit spaces with non-integrable kernel, which is presented in Subsection 5.2.2.

Let $\mathfrak{F} = (\psi_x)_{x \in X}$ be a *Parseval frame* for the Hilbert space $\mathcal{H}$. Furthermore we fix an *admissible weight* w on $X \times X$ as well as a w-moderate weight m and an integrablity parameter $1 \leqslant r < \infty$. We suppose from now on throughout this section that Assumption 5.1.2 from the previous section is fulfilled, i.e.,

$$K_{\mathfrak{F}} \in A_{q,w} \quad \text{for all } 1 < q < \infty, \tag{5.2.1}$$

where $K_{\mathfrak{F}}(x,y) = \langle \psi_y, \psi_x \rangle_{\mathcal{H}}$ is the *reproducing kernel*. By Proposition 5.1.9 the coorbit spaces with respect to $L_{r,m}(X, \mu)$ defined via

$$\mathrm{Co}(L_{r,m}) = \{T \in \mathcal{H}'_m \ : \ V_{e,\mathfrak{F}} T \in L_{r,m}(X, \mu)\} \tag{5.2.2}$$

are well-defined Banach spaces which are isometrically isomorphic to the reproducing kernel space

$$\mathcal{M}_{r,m} = \{F \in L_{r,m}(X, \mu) \ : \ K_{\mathfrak{F}} F = F\} \tag{5.2.3}$$

via the *extended voice transform* $V_{e,\mathfrak{F}}$. Let us characterize the anti-dual $\mathcal{M}'_{r,m}$ of the reproducing kernel space in the following lemma. We omit the proof since it is the same as the proof of Lemma 3.2.2.

Lemma 5.2.1. *The anti-dual $\mathcal{M}'_{r,m}$ of $\mathcal{M}_{r,m}$ is canonically isomorphic to $L_{r',m^{-1}}(X, \mu)/\mathcal{M}^{\perp}_{r,m}$, where*

$$\mathcal{M}^{\perp}_{r,m} = \left\{G \in L_{r',m^{-1}}(X, \mu) \ : \ \langle G, F \rangle_{L_2} = 0 \text{ for all } F \in \mathcal{M}_{r,m}\right\} \tag{5.2.4}$$

and r' denotes the dual exponent of r. Hence, for every $\Gamma \in \mathcal{M}'_{r,m}$ there is a $G \in L_{r,m^{-1}}(X, \mu)$ such that $\Gamma(G) = \langle G, F \rangle_{L_2}$ for all $\mathcal{M}_{r,m}$.

We recall the definition of the $\mathcal{U}$-oscillation of a kernel K. For this, let $\mathcal{U} = (U_i)_{i \in I}$ be any countable family of subsets of X which cover the whole index set, i.e., $\bigcup_{i \in I} U_i = X$. Then, the $\mathcal{U}$-oscillation of a kernel $K : X \times X \to \mathbb{C}$ is defined as

$$\mathrm{osc}_{\mathcal{U}} K(x,y) = \sup_{z \in Q_y} |K(x,y) - K(x,z)| \quad \text{for } x, y \in X, \tag{5.2.5}$$

where $Q_y = \bigcup_{y \in U_i \in \mathcal{U}} U_i$. Obviously the inequalities

$$|K(x,z)| \leqslant \mathrm{osc}_{\mathcal{U}} K(x,y) + |K(x,y)| \quad \text{for } z \in Q_y \tag{5.2.6}$$

and

$$\mathrm{osc}_{\mathcal{U}} K(x,y) \leqslant \sup_{z \in Q_y} |K(x,z)| + |K(x,y)| \tag{5.2.7}$$

hold true. The $\mathcal{U}$-oscillation of $K_{\mathfrak{F}}$ will play an important role for any discretization of coorbit spaces.

5.2.1   An Assumption on the Kernel

A major observation throughout this chapter is that Assumption 5.1.2 is not sufficient for a discretization of coorbit spaces $L_{p,m}(X,\mu)$. Instead, as in Subsection 3.2.1, we propose the following additional assumption on the kernel $K_{\mathfrak{F}}$.

Assumption 5.2.2. *The space* $\mathrm{span}\{K_{\mathfrak{F}}(\cdot,x)\}_{x\in G}$ *is dense in* $\mathcal{M}_{r,m}$.

Remark 5.2.3. The assumption above is crucial as can be seen as follows. By the isomorphism between $\mathcal{M}_{r,m}$ and $\mathrm{Co}(L_{r,m})$, see Proposition 5.1.9, the assumption above is equivalent to the density of the linear span of the frame $\mathfrak{F} = (\psi_x)_{x\in X}$ in $\mathrm{Co}(L_{r,m})$. In any discretization result, however, the goal is to extract a discrete frame for the coorbit space from the continuous frame $\mathfrak{F}$. This can only succeed, if the linear span of the frame is dense in the coorbit space, hence Assumption 5.2.2 makes sense.

Before we provide a natural condition on $K_{\mathfrak{F}}$ under which Assumption 5.2.2 is fulfilled in Lemma 5.2.7, we need a couple of auxiliary results. For this, let $K\colon X \times X \to \mathbb{C}$ be an arbitrary kernel fulfilling Assumption 5.1.2.

Proposition 5.2.4. *Assume for all* $f \in L_{r,m}(X,\mu)$ *we have* $K(x,\cdot)f \in L_1(X,\mu)$ *for almost all* $x \in X$ *and* $K(f) \in L_{r,m}(X,\mu)$, *then the integral operator associated with the kernel*

$$K\colon L_{r,m}(x,\mu) \to L_{r,m}(X,\mu), \quad K(f)(x) = \int_X K(x,y)f(y)\,d\mu(y)$$

is bounded.

Proof. We follow the lines of the proof of Proposition 3.2.6, which is based on [73, Proposition 3.10]. In this reference, the assertion is proven for $r = 2$ and $w \equiv 1$, but the basic idea can be applied to this case as well. For this we note that by the closed graph theorem it suffices to show that the integral operator is closed. To show this we take a sequence $(f_n)_{n\in\mathbb{N}}$ converging to $f \in L_{r,m}(X,\mu)$ such that the image sequence $(K(f_n))_{n\in\mathbb{N}}$ converges to $g \in L_{r,m}(X,\mu)$. By a sharp version of the Riesz-Fischer theorem, see [3, Theorem 13.6], there exists a positive function $h \in L_{r,m}(X,\mu)$ such that, possibly passing twice to a subsequence, for almost every $x,y \in X$ it holds that

$$|f_n(y)| \leqslant h(y),$$
$$\lim_{n\to\infty} f_n(y) = f(y),$$
$$\lim_{n\to\infty} K(f_n)(x) = g(x).$$

Furthermore, we get that for almost all $x \in X$ and all $n \in \mathbb{N}$ the mappings

$$y \mapsto K(x,y)f_n(y), \quad y \mapsto K(x,y)h(y)$$

are integrable. Then, for almost all $x,y \in X$ we have

$$|K(x,y)f_n(y)| \leqslant |K(x,y)h(y)|, \quad \lim_{n\to\infty} K(x,y)f_n(y) = K(x,y)f(y),$$

so by the dominated convergence theorem we obtain

$$g(x) = \lim_{n\to\infty} \int_X K(x,y)f_n(y)\,d\mu(y) = \int_X K(x,y)f(y)\,d\mu(y) = K(f)(x),$$

so the kernel operator K is indeed closed. $\square$

We recall that the duality pairing for $f \in L_{r,m}(X,\mu)$ and $g \in L_{r,m}(X,\mu)' = L_{r',m^{-1}}(X,\mu)$ is given by the form

$$\langle f, g \rangle_{L_2} = \int_X f(x)\overline{g(x)}\, d\mu(x).$$

Proposition 5.2.5. *Assume the integral operator*

$$K \colon L_{r,m}(x,\mu) \to L_{r,m}(X,\mu), \quad K(f)(x) = \int_X K(x,y)f(y)\, d\mu(y)$$

to be bounded, then

(i) *the integral operator K is bounded on $L_{r',m^{-1}}(X,\mu)$ and it coincides with the adjoint of K;*

(ii) *the integral operator K is a projection from $L_{r,m}(X,\mu)$ onto the reproducing kernel Banach space $\mathcal{M}_{r,m}^K = \{f \in L_{r,m}(X,\mu) : K(f) = f\}$.*

Proof. Let us show (i). Since the integral operator K is bounded on $L_{r,m}(X,\mu)$, the adjoint K^* is bounded on $L_{r',m^{-1}}(X,\mu) \simeq L_{r',m^{-1}}(X,\mu)'$. Let $g \in L_{r',m^{-1}}(X,\mu)$ and $f \in C_c(X) \subset L_{r,m}(X,\mu)$, then

$$\langle K^*(g), f \rangle_{L_2} = \langle g, K(f) \rangle_{L_2} = \int_X g(x) \int_X \overline{K(x,y)f(y)}\, d\mu(y)\, d\mu(x)$$

$$= \int_X \overline{f(y)} \int_X K(y,x)g(x)\, d\mu(x)\, d\mu(y) = \langle K(g), f \rangle_{L_2}.$$

Note that we can interchange the integrals by Fubini's theorem, since by Lemma 1.8.7 we have

$$\int_X \int_X |g(x)K(x,y)f(y)|\, d\mu(y)\, d\mu(x) \leqslant \|g\|_{L_{r',m^{-1}}} \cdot \| |K|(|f|) \|_{L_{r,m}}$$

$$\leqslant \|g\|_{L_{r',m^{-1}}} \cdot \|f\|_{L_{1,m}} \cdot \|R\|_{\mathcal{A}_{r,w}}$$

Using $f \in C_c(X) \subset L_{1,m}(X,\mu)$ and $r > 1$ this justifies the use of Fubini's theorem. Moreover, the function $K(g)$ is well-defined and almost everywhere finite. Since $C_c(X)$ is dense in $L_{r,m}(X,\mu)$ we get that $K^*(g) = K(g)$ and since K^* is bounded this implies $K(g) \in L_{r',m^{-1}}(X,\mu)$. Hence, the integral operator K is bounded on $L_{r',m^{-1}}(X,\mu)$ and it coincides with K^*.

It remains to prove (ii). For this let $f \in L_{r,m}(X,\mu)$, then by Lemma 1.8.7 it holds that

$$\int_X \int_X |K(x,y)K(y,z)f(z)|\, d\mu(y)\, d\mu(z) \leqslant \|K(x,\cdot)w(x,\cdot)\|_{L'_p} \cdot \| |K|(|f|) \|_{L_{p,m}} \cdot m(x)^{-1}$$

$$\leqslant \|K\|_{\mathcal{A}_{p',w}} \cdot \|K\|_{\mathcal{A}_{q,w}} \cdot \|f\|_{L_{r,m}} \cdot m(x)^{-1}$$

for all $x \in X$ and $p,q > 1$ such that $1 - 1/q = 1/p - 1/r$. By Fubini's theorem and Lemma 5.1.7 it follows that

$$K(K(f))(x) = \int_X f(z) \int_X \overline{K(y,x)}K(y,z)\, d\mu(y)\, d\mu(z) = \int_X f(z) \cdot \langle V_{\mathfrak{F}}\psi_x, V_{\mathfrak{F}}\psi_z \rangle_{L_2}\, d\mu(z)$$

$$= \int_X f(z) \cdot \langle \psi_x, \psi_z \rangle_{L_2}\, d\mu(z) = K(f)(x)$$

for all $x \in X$. Since K is bounded on $L_{r,m}(X,\mu)$ by assumption we get that $K^2 = K$, hence $\operatorname{Ran} K \subseteq \mathcal{M}_{r,m}^K$. The other inclusion is trivial. $\qquad \square$

The following corollary is a consequence of the result above applied to the reproducing kernel $K_{\mathfrak{F}}$.

Corollary 5.2.6. *Assume the integral operator $K_{\mathfrak{F}}$ to be bounded on $L_{r,m}(X,\mu)$. The sesquilinear pairing on $\mathrm{Co}(L_{r,m}) \times \mathrm{Co}(L_{r',m^{-1}})$ given by*

$$\langle T, T' \rangle_{\mathrm{Co}(L_{r,m})} = \langle V_{e,\mathfrak{F}}T, V_{e,\mathfrak{F}}T' \rangle_{L_2} \tag{5.2.8}$$

is such that the linear map

$$T' \mapsto \left(T \mapsto \overline{\langle T, T' \rangle_{\mathrm{Co}(L_{r,m})}}\right) \tag{5.2.9}$$

is an isomorphism of $\mathrm{Co}(L_{r',m^{-1}})$ onto the anti-linear dual of $\mathrm{Co}(L_{r,m})$.

Proof. We identify $\mathrm{Co}(L_{r,m})$ with $\mathcal{M}_{r,m}$ by the extended voice transform $V_{\mathfrak{F},e}$, see Proposition 5.1.9, so that the pairing becomes $\langle f, g \rangle_{L_2}$ for $f \in \mathcal{M}_{r,m}$ and $g \in \mathcal{M}_{r',m^{-1}}$. It therefore remains to prove that this pairing induces an isomorphism between $\mathcal{M}_{r',m^{-1}}$ and the dual of $\mathcal{M}_{r,m}$.

For fixed $g \in \mathcal{M}_{r',m^{-1}} \subset L_{r',m^{-1}}(X,\mu)$ the map $\Gamma_g: \mathcal{M}_{r,m} \to \mathbb{C}$, $\Gamma_g(f) = \overline{\langle f, g \rangle_{L_2}}$ is continuous and conjugate linear, and the operator norm can be estimated by

$$\|\Gamma_g\| = \sup\left\{|\langle f, g \rangle_{L_2}| : f \in \mathcal{M}_{r,m}, \|f\|_{L_{r,m}} \leqslant 1\right\}$$
$$\leqslant \sup\left\{|\langle h, g \rangle_{L_2}| : h \in L_{r,m}(X,\mu), \|h\|_{L_{r,m}} \leqslant 1\right\} = \|g\|_{L_{r',m^{-1}}}.$$

This implies that Γ_g is an element of the anti-dual of $\mathcal{M}_{r,m}$.

Next, since $L_{r,m}(X,\mu)$ is the dual of $L_{r',m^{-1}}(X,\mu)$, there is $h \in L_{r,m}(X,\mu)$ with $\|h\|_{L_{r,m}}$ such that $\|g\|_{L_{r,m}} = \langle h, g \rangle_{L_2}$. Let $c = \|K_{\mathfrak{F}}\|_{L_{r,m} \to L_{r,m}}$ and $f = c^{-1}K_{\mathfrak{F}}(h)$, we have $\|f\|_{L_{r,m}} \leqslant 1$ and by Proposition 2.3 we derive

$$\langle f, g \rangle_{L_2} = c^{-1}\langle R(f), g \rangle_{L_2} = c^{-1}\langle f, R^*(g) \rangle_{L_2} = c^{-1}\langle f, R(g) \rangle_{L_2} = c^{-1}\langle f, g \rangle_{L_2} = c^{-1}\|g\|_{L_{r',m^{-1}}},$$

hence $\|\Gamma_g\| \geqslant c^{-1}\|g\|_{L_{r,m}}$. This implies the injectivity of the map $g \mapsto \Gamma_g$.

It remains to prove the surjectivity of the map $g \mapsto \Gamma_g$. Take γ in the anti-linear dual of $\mathcal{M}_{r,m} \subset L_{r,m}(X,\mu)$. By an anti-linear version of the Riesz representation theorem there is $g' \in L_{r',m^{-1}}(X,\mu)$ such that $\Gamma(f) = \overline{\langle f, g' \rangle_{L_2}}$ for all $f \in \mathcal{M}_{r,m}$. Applying Proposition 5.2.5, we have $g = K_{\mathfrak{F}}(g') = K_{\mathfrak{F}}^*(g') \in \mathcal{M}_{r',m^{-1}}$, and hence

$$\Gamma(f) = \overline{\langle f, g' \rangle_{L_2}} = \overline{\langle K_{\mathfrak{F}}(f), g \rangle_{L_2}} = \overline{\langle f, K_{\mathfrak{F}}^*(g') \rangle_{L_2}} = \overline{\langle f, g \rangle_{L_2}} = \Gamma_g(f)$$

for all $f \in \mathcal{M}_{r,m}$, thus $\Gamma = \Gamma_g$. $\qquad\square$

Finally, the following lemma gives a sufficient condition for Assumption 5.2.2 to be fulfilled.

Lemma 5.2.7. *Assume the integral operator $K_{\mathfrak{F}}$ is bounded on $L_{r,m}(X,\mu)$. Then the sets* $\mathrm{span}\{\psi_x\}_{x \in X}$ *and* $\mathrm{span}\{K_{\mathfrak{F}}(\cdot, x)\}_{x \in X}$ *are dense in $\mathrm{Co}(L_{r,m})$ and $\mathcal{M}_{r,m}$, respectively. Thus, Assumption 5.2.2 is fulfilled.*

Proof. By Proposition 5.1.9 it is enough to show the second claim. For this let $\Gamma \in \mathcal{M}'_{r,m}$ such that for all $x \in X$ it holds that

$$\Gamma(K_{\mathfrak{F}}(\cdot, x)) = 0.$$

By Corollary 5.2.6 there exists $\overline{g} \in \mathcal{M}_{r',m^{-1}}$ such that $\Gamma(f) = \langle \overline{g}, f \rangle_{L_2}$ for all $f \in \mathcal{M}_{r,m}$. In particular,

$$0 = \Gamma(K_{\mathfrak{F}}(\cdot, x)) = \langle \overline{g}, K_{\mathfrak{F}}(\cdot, x) \rangle_{L_2} = \int_X K_{\mathfrak{F}}(x, y) g(y) \, d\mu(y) = K_{\mathfrak{F}}(g)(x)$$

for all $x \in X$, that is $K_{\mathfrak{F}}(g) = 0$. Since $\overline{g} \in \mathcal{M}_{r',m^{-1}}$ and $K_{\mathfrak{F}}$ is a projection onto $\mathcal{M}_{r',m^{-1}}$, this means $g = 0$, thus $\Gamma = 0$ as an element of $\mathcal{M}'_{r,m}$. Since Γ was chosen arbitrarily, this implies that $\mathrm{span}\{K_{\mathfrak{F}}(\cdot, x)\}_{x \in X}$ is dense in $\mathcal{M}_{r,m}$. $\qquad \square$

As in Chapter 3 we are interested in the setting where $K_{\mathfrak{F}} \notin \mathcal{A}_{1,w}$; but we still need the boundedness of the kernel operator K_ψ on $L_{r,m}(X, \mu)$. If we had $K_{\mathfrak{F}} \in \mathcal{A}_{1,w}$ then by invoking Corollary 1.8.6 the kernel operator was automatically bounded on $L_{r,m}(X, \mu)$ and we were in the setting of Chapter 4. The present setting is tricky but it is not clear how Assumption 5.2.2 can be ensured without assuming the boundedness condition.

5.2.2   Atomic Decomposition

In this subsection we will derive weak atomic decomposition results for coorbit spaces $\mathrm{Co}(L_{r,m})$ provided that Assumption 5.2.2 is fulfilled. The main result of this subsection will be stated in Theorem 5.2.17.

Suppose $\mathcal{U} = (U_i)_{i \in I}$ is an *admissible covering* of X for a countable index set I. By Definition 1.5.5 this means the following. First, each set $U_i \subset X$, $i \in I$, is relatively compact with non-void interior such that $X = \bigcup_{i \in I} U_i$. Moreover, each point $x \in X$ is contained in at most $N_{\mathcal{U}} \in \mathbb{N}$ sets, i.e.,

$$\sum_{i \in I} \chi_{U_i}(x) \leqslant N_{\mathcal{U}} \quad \text{for all } x \in X. \tag{5.2.10}$$

We replace the notion of moderateness by the following stronger assumption: Suppose there exist constants $C_{\mathcal{U}}, D_{\mathcal{U}} > 0$ such that

$$C_{\mathcal{U}} \leqslant \mu(U_i) \leqslant D_{\mathcal{U}} \quad \text{for all } i \in I. \tag{5.2.11}$$

It is easy to see that the condition above is stronger than part (iv) and (v) of Definition 1.5.5. However, we will see that we need the additional bound above in the following.

From now on assume for every $n \in \mathbb{N}$ there is an admissible covering $U_n = (U_{n,i})_{i \in I_n}$ satisfying (5.2.11) such that the covering numbers $N_{\mathcal{U}_n}$ in (5.2.10) are uniformly bounded from above:

$$N_{\mathcal{U}_n} \leqslant N \quad \text{for all } n \in \mathbb{N} \tag{5.2.12}$$

Now choose arbitrary points $x_{n,i} \in U_{n,i}$ for all $n \in \mathbb{N}$ and $i \in I_n$ such that the sets $Y_n := (x_{n,i})_{i \in I_n}$, $n \in \mathbb{N}$, fulfill

$$Y_n \subset Y_{n+1}, \tag{5.2.13}$$

$$\overline{\bigcup_{n \in \mathbb{N}} Y_n} = X. \tag{5.2.14}$$

Recall that the sequence spaces $\ell_{r,m}(I_n)$ for $1 \leqslant r < \infty$ and any weight m on X associated to a covering $\mathcal{U}_n$ are defined via

$$\ell_{r,m}(I_n) = \ell_{r,m}(I_n, \mathcal{U}_n) = \left\{ c = (c_{n,i})_{i \in I_n} \in \mathbb{C}^{I_n} \; : \; \|c\|_{\ell_{r,m}}^r = \sum_{i \in I_n} |c_{n,i}|^r \mu(U_{n,i}) \sup_{x \in U_{n,i}} m(x)^r < \infty \right\},$$

$$\tag{5.2.15}$$

see Lemma 1.6.7. If we additionally assume $m(x) \leqslant Cm(y)$ for all $x, y \in U_{n,i}$, $i \in I_n$, then we can replace $\sup_{x \in U_{n,i}} m(x)$ in (5.2.15) by the particular choice $m(x_{n,i})$ for all $i \in I_n$, i.e.,

$$\|c\|_{\ell_{r,m}}^r = \sum_{i \in I_n} |c_{n,i}|^r \mu(U_{n,i}) m(x_{n,i})^r. \tag{5.2.16}$$

We will assume this from now on.

For each $n \in \mathbb{N}$ let $\Phi_n = (\varphi_{n,i})_{i \in I_n}$ be a bounded admissible partition of unity subordinate to $\mathcal{U}_n$, see Definition 1.5.7. This means that each $\varphi_{n,i} \colon X \to [0,1]$ is a measurable function with $\operatorname{supp} \varphi_{n,i} \subset U_{n,i}$ and

$$\sum_{i \in I_n} \varphi \equiv 1 \quad \text{on } X.$$

By Lemma 1.5.8 such a $\mathcal{U}_n$-BAPU always exists. Let us additionally assume that each family Φ_n is linearly independent as almost everywhere defined functions, that is, if for any finitely supported sequence $(c_i)_{i \in I_n} \in \mathbb{C}^{I_n}$ the condition

$$\sum_{i \in I_n} c_i \varphi_{n,i}(x) = 0$$

is fulfilled for almost every $x \in X$, then $c = 0$.

Now consider for each $n \in \mathbb{N}$ a *finite* subset $X_n \subset Y_n$ that we will index by $X_n = (x_{i,n})_{i \in \mathcal{I}_n}$, $\#\mathcal{I}_n < \infty$, such that the conditions

$$X_n \subset X_{n+1}, \tag{5.2.17}$$

$$\bigcup_{n \in \mathbb{N}} X_n = X \tag{5.2.18}$$

are fulfilled. The sets X_n are proper subsets of Y_n, since otherwise Y_n had to be a finite set which is only possible if $\mu(X) < \infty$—a setting we are not interested in. By the conditions above it is clear that $\lim_{n \to \infty} \#\mathcal{I}_n = \infty$. Note that the family $(\varphi_{n,j})_{j \in \mathcal{I}_n}$ is similar to a partition of unity subordinate to the finite set $(U_{n,j})_{j \in \mathcal{I}_n}$. The sequence spaces $\ell_{r,m}(\mathcal{I}_n)$ are defined similar to (5.2.15).

As usual we now fix $1 < r < \infty$ as well as a w-moderate weight m on X and consider the following operator for any $n \in \mathbb{N}$:

$$T_n \colon L_{r,m}(X,\mu) \to \mathcal{M}_{r,m}, \quad T_n F := \sum_{i \in \mathcal{I}_n} \langle F, \varphi_{n,i} \rangle_{L_2} K_{\mathfrak{F}}(\cdot, x_{n,i}). \tag{5.2.19}$$

Lemma 5.2.8. *The operator T_n, $n \in \mathbb{N}$, is well-defined.*

Proof. We first note that the defining sum in (5.2.19) is finite by definition of $\mathcal{I}_n$. To show $T_n F \in L_{r,m}(X,\mu)$ we compute

$$\|T_n F\|_{L_{r,m}} \leqslant \sum_{i \in \mathcal{I}_n} |\langle F, \varphi_{n,i} \rangle_{L_2}| \cdot \|K_{\mathfrak{F}}(\cdot, x_{n,i})\|_{L_{r,m}}$$

$$\leqslant \sum_{i \in \mathcal{I}_n} \|F\|_{L_{r,m}} \cdot \|\varphi_{n,i}\|_{L_{r',m^{-1}}} \cdot \|K_{\mathfrak{F}}\|_{\mathcal{A}_{r,m}}.$$

Since each $\varphi_{n,i}$ is bounded with compact support, it is contained in all weighted Lebesgue spaces and hence the right hand side is finite. It remains to show the reproducing identity. We have

$$K_{\mathfrak{F}}(K_{\mathfrak{F}}(\cdot, x_{n,i}))(x) = K_{\mathfrak{F}}(V_{\mathfrak{F}} \psi_{x_{n,i}})(x) = V_{\mathfrak{F}} \psi_{x_{n,i}}(x) = K_{\mathfrak{F}}(x, x_{n,i})$$

for all $i \in \mathcal{I}_n$, which, by linearity of the kernel operator $K_{\mathfrak{F}}$, implies $K_{\mathfrak{F}}(T_n F) = T_n F$. $\qquad\square$

Let us define $V_n = \operatorname{Ran} T_n$, which is a finite-dimensional subspace of $\mathcal{M}_{r,m}$, as well as $\widetilde{V}_n = V_{e,\mathfrak{F}}^{-1}(V_n)$, which is a finite-dimensional subspace of $\operatorname{Co}(L_{r,m})$ by the correspondence principle, see Proposition 5.1.9. We collect some basic properties of V_n in the following lemma.

Lemma 5.2.9. *For every $n \in \mathbb{N}$ we have*

$$V_n = \operatorname{span}\{K_{\mathfrak{F}}(\cdot, x_{n,i})\}_{i \in \mathcal{I}_n}, \tag{5.2.20}$$

$$V_n \subset V_{n+1}. \tag{5.2.21}$$

If, in addition, the linear span of $\{K_{\mathfrak{F}}(\cdot, x)\}_{x \in X}$ is dense in $\mathcal{M}_{r,m}$, then

$$\overline{\bigcup_{n \in \mathbb{N}} V_n} = \mathcal{M}_{r,m}. \tag{5.2.22}$$

Proof. It is clear that $V_n \subset \operatorname{span}\{K_{\mathfrak{F}}(\cdot, x_{n,i})\}_{i \in \mathcal{I}_n}$. To show the other inclusion we must show that the map

$$F \mapsto (\langle F, \phi_{n,i}\rangle_{L_2})_{i \in \mathcal{I}_n}$$

is surjective from $L_{r,m}(X, \mu)$ to $\mathbb{C}^{\mathcal{I}_n}$. Assume this was not the case, then there would be a nonzero family $(\alpha_i)_{i \in \mathcal{I}_n} \in \mathbb{C}^{\mathcal{I}_n}$ such that $\sum_{i \in \mathcal{I}_n} \alpha_i \langle F, \phi_{n,i}\rangle_{L_2} = 0$. But by the linear independency of Φ_n this can only be the case if $\alpha = 0$, which is a contradiction. Therefore $\operatorname{span}\{R(\cdot, x_{n,i})\}_{i \in \mathcal{I}_n} \subset V_n$ holds true. The nesting property is straightforward.

It remains to show the second assertion. By definition, it holds $V_n \subset \mathcal{M}_{r,m}$ for all $n \in \mathbb{N}$ as well as the nesting property, hence $\overline{\bigcup_{n \in \mathbb{N}} V_n}$ is a closed linear subspace of $\mathcal{M}_{r,m}$. By the Hahn-Banach theorem the assertion is equivalent to the following condition: If $\Gamma \in \mathcal{M}'_{r,m}$ satisfies

$$\langle \Gamma, F\rangle_{\mathcal{M}'_{r,m} \times \mathcal{M}_{r,m}} = 0 \quad \text{for all } F \in V_n, n \in \mathbb{N},$$

then $\Gamma = 0$ as a functional in $\mathcal{M}'_{r,m}$. We can write $\langle \Gamma, F\rangle_{\mathcal{M}'_{r,m} \times \mathcal{M}_{r,m}} = \langle g, F\rangle_{L_2}$ for all $F \in \mathcal{M}_{r,m}$ for a suitable $g \in L_{r',m^{-1}}(X, \mu)$ (which is unique up to $\mathcal{M}_{r,m}^{\perp}$). Then, for any $f \in L_{r,m}(X, \mu)$, we can write

$$0 = \langle \Gamma, T_n f\rangle_{\mathcal{M}'_{r,m} \times \mathcal{M}_{r,m}} = \sum_{i \in \mathcal{I}_n} \langle f, \varphi_{n,i}\rangle_{L_2} \cdot \langle \Gamma, K_{\mathfrak{F}}(\cdot, x_{n,i})\rangle_{\mathcal{M}'_{r,m} \times \mathcal{M}_{r,m}}$$

$$= \sum_{i \in \mathcal{I}_n} \langle f, \varphi_{n,i}\rangle_{L_2} \cdot \langle g, K_{\mathfrak{F}}(x_{n,i}, \cdot)\rangle_{L_2} = \sum_{i \in \mathcal{I}_n} \langle f, \varphi_{n,i}\rangle_{L_2} \cdot K_{\mathfrak{F}}(g)(x_{n,i})$$

$$= \langle f, \sum_{i \in \mathcal{I}_n} K_{\mathfrak{F}}(g)(x_{n,i})\varphi_{n,i}\rangle_{L_2}.$$

But this holds true for any $f \in L_{r,m}(X, \mu)$, therefore we obtain $\sum_{i \in \mathcal{I}_n} K_{\mathfrak{F}}(g)(x_{n,i})\varphi_{n,i} = 0$ as a function in $L_{r',m^{-1}}(X, \mu)$. The family $\{\varphi_{n,i}\}_{i \in \mathcal{I}_n}$ is linearly independent as elements of $L_{r',m^{-1}}(X, \mu)$, therefore we get $K_{\mathfrak{F}}(g)(x_{n,i}) = 0$ for all $i \in \mathcal{I}_n$, $n \in \mathbb{N}$. By definition of X_n, the function $K_{\mathfrak{F}}(g)$ vanishes on a dense subset of X, i.e. $K_{\mathfrak{F}}(g) = 0$ almost everywhere. To show that it vanishes everywhere, we need to show the continuity of $K_{\mathfrak{F}}(g)$. To this extend we apply the dominated convergence theorem to the continuous function $x \mapsto K_{\mathfrak{F}}(x, y)g(y)$ for fixed $y \in X$. Note that the continuity of $K_{\mathfrak{F}}$ follows from the weak continuity of $x \mapsto \psi_x$. Since the function $y \mapsto K_{\mathfrak{F}}(x, y)g(y)$ for fixed $x \in X$ is integrable with respect to $L_1(X, \mu)$ by an application of Hölder's inequality:

$$\|K_{\mathfrak{F}}(x, \cdot)g(\cdot)\|_{L_1} \leqslant \|K_{\mathfrak{F}}(x, \cdot)\|_{L_{r,m}} \cdot \|g\|_{L_{r',m^{-1}}} \leqslant \|K_{\mathfrak{F}}\|_{\mathcal{A}_{r,m}} \cdot \|g\|_{L_{r',m^{-1}}},$$

the function $K_{\mathfrak{F}}(g)$ is continuous. Therefore $K_{\mathfrak{F}}(g)$ vanishes on X, which implies

$$0 = K_{\mathfrak{F}}(g)(x) = \langle g, K_{\mathfrak{F}}(\cdot, x) \rangle_{L_2} = \langle \Gamma, K_{\mathfrak{F}}(\cdot, x) \rangle_{\mathcal{M}'_{r,m} \times \mathcal{M}_{r,m}}$$

for all $x \in X$. But we assumed $\{K_{\mathfrak{F}}(\cdot, x)\}_{x \in X}$ to be dense in $\mathcal{M}_{r,m}$, hence, $\Gamma = 0$ as an element of $\mathcal{M}'_{r,m}$ and we are done. $\qquad\square$

Remark 5.2.10. As in Remark 3.2.11, analogous results to (5.2.20)–(5.2.22) can be obtained for $\widetilde{V}_n = V_{e,\mathfrak{F}}^{-1}(V_n) \subset \mathrm{Co}(L_{r,m})$.

Since the spaces V_n will serve as a finite dimensional substitute for $\mathcal{M}_{r,m}$, we are interested in projections onto V_n. To this end, let $\pi_n \colon \mathcal{M}_{r,m} \to V_n$ be the metric projection defined by

$$\pi_n(F) := \mathrm{argmin}_{g \in V_n} \|F - g\|_{L_{r,m}}. \tag{5.2.23}$$

This projection is well-defined and unique due to [64, Proposition 3.1]. Similarly, define the projection $\widetilde{\pi}_n \colon \mathrm{Co}(L_{r,m}) \to \widetilde{V}_n$ by $\widetilde{\pi}_n T = V_{e,\mathfrak{F}}^{-1} \pi_n V_{e,\mathfrak{F}} T$ for every $T \in \mathrm{Co}(L_{r,m})$.

The next result gives a first norm estimate for the metric projection above. We omit the proof since it is the same as the proof of Lemma 3.2.12.

Lemma 5.2.11. *Given any $\varepsilon > 0$ and $F \in \mathcal{M}_{r,m}$, there exists $n^* = N_{F,\varepsilon}^* \in \mathbb{N}$ such that for all $n \geqslant n^*$ it holds that*

$$\|F - \pi_n(F)\|_{L_{r,m}} \leqslant \varepsilon, \tag{5.2.24}$$

$$\|\pi_n(F)\|_{L_{r,m}} \leqslant (1 + \varepsilon)\|F\|_{L_{r,m}}. \tag{5.2.25}$$

This result shows that we can approximate $\mathcal{M}_{r,m}$ arbitrarily well with functions in the space $\pi_n(\mathcal{M}_{r,m}) = V_n$. Moreover, since π_n realizes the best approximation, for every $\varepsilon > 0$ and $F \in \mathcal{M}_{r,m}$ we have $\mathrm{dist}(F, V_{n*}) \leqslant \varepsilon$.

Now let us formulate a first upper estimate for the coefficients in the defining series of T_n in (5.2.19) with respect to the $\ell_{r,m}$-norm.

Proposition 5.2.12. *For any $F \in L_{r,m}(X, \mu)$ and $n \in \mathbb{N}$, let the coefficients $c_{n,i} \in \mathbb{C}$, $i \in \mathcal{I}_n$, be defined via*

$$c_{n,i} := \langle F, \varphi_{n,i} \rangle_{L_2},$$

then we have the inequality

$$\left(\sum_{i \in \mathcal{I}_n} |c_{n,i}|^r \mu(U_{n,i}) m(x_{n,i})^r \right)^{1/r} \leqslant \max_{i \in \mathcal{I}_n} \left[\mu(U_{n,i}) \cdot \sup_{z \in U_{n,i}} w(x_{n,i}, z) \right] \cdot \|F\|_{L_{r,m}}. \tag{5.2.26}$$

Proof. We first note that the coefficients $c_{n,i}$ are well-defined for all n and i, since $\varphi_{n,i}$ is compactly supported and bounded.

Then, we estimate

$$\sum_{i \in \mathcal{I}_n} |c_{n,i}|^r \mu(U_{n,i}) m(x_{n,i})^r$$

$$\leqslant \sum_{i \in \mathcal{I}_n} \mu(U_{n,i}) m(x_{n,i})^r \cdot \left[\int_X |F(y) \cdot \varphi_{n,i}(y)| \, d\mu(y) \right]^r$$

$$\leqslant \sum_{i \in \mathcal{I}_n} \mu(U_{n,i}) \left[\int_X |F(y) \cdot \varphi_{n,i}(y) \cdot w(x_{n,i}, y) \cdot m(y)| \, d\mu(y) \right]^r$$

$$\leqslant \sum_{i \in \mathcal{I}_n} \mu(U_{n,i}) \sup_{z \in U_{n,i}} w(x_{n,i}, z)^r \|\varphi_{n,i}\|_{L_1}^r \cdot \left[\int_X |(F \cdot m)(y)| \cdot \frac{\varphi_{n,i}(y)}{\|\varphi_{n,i}\|_{L_1}} \, d\mu(y) \right]^r.$$

We see that by the properties of Φ_n we have $\|\varphi_{n,i}\|_{L_1} = \mu(U_{n,i})$ for all $i \in I_n$ and by invoking Jensen's inequality similarly to the proof of Proposition 3.2.13 we obtain

$$\sum_{i \in \mathcal{I}_n} |c_{n,i}|^r m(x_{n,i})^r \leqslant \sum_{i \in \mathcal{I}_n} \sup_{z \in U_{n,i}} w(x_{n,i}, z)^r \mu(U_{n,i})^r \cdot \int_X |(F \cdot m)(y)|^r \varphi_{n,i}(y) \, d\mu(y)$$

$$\leqslant \max_{i \in \mathcal{I}_n} \left[\mu(U_{n,i})^r \cdot \sup_{z \in U_{n,i}} w(x_{n,i}, z)^r \right] \cdot \int_X |(F \cdot m)(y)|^r \sum_{j \in \mathcal{I}_n} \varphi_{n,j}(y) \, d\mu(y)$$

$$\leqslant \max_{i \in \mathcal{I}_n} \left[\mu(U_{n,i})^r \cdot \sup_{z \in U_{n,i}} w(x_{n,i}, z)^r \right] \cdot \|F\|_{L_{r,m}}^r,$$

which concludes the proof. $\qquad\qquad\square$

Using this result we can now state a first atomic decomposition of $V_n \subset \mathcal{M}_{r,m}$ for all $n \in \mathbb{N}$.

Lemma 5.2.13. *Given $n \in \mathbb{N}$, for all $F \in V_n$ the following atomic decomposition holds true:*

$$F = \sum_{i \in \mathcal{I}_n} c(F)_{n,i} K_{\mathfrak{F}}(\cdot, x_{n,i}), \qquad (5.2.27)$$

where the coefficients are of the form

$$c(F)_{n,i} = \langle S_n F, \varphi_{n,i} \rangle_{L_2} \qquad (5.2.28)$$

and S_n denotes any linear right inverse of T_n. In particular, the coefficients depend linearly on F and satisfy

$$\left(\sum_{i \in \mathcal{I}_n} |c(F)_{n,i}|^r \mu(U_{n,i}) m(x_{n,i})^r \right)^{1/r} \leqslant \max_{i \in \mathcal{I}_n} \left[\mu(U_{n,i}) \cdot \sup_{z \in U_{n,i}} w(x_{n,i}, z) \right] \cdot \|S_n\| \cdot \|F\|_{L_{r,m}}. \quad (5.2.29)$$

Proof. Let us first recall that T_n does indeed admit a right inverse $S_n : V_n \to L_{r,m}(X, \mu)$. This is an application of [8, Theorem 2.12] using that V_n is finite dimensional.

We fix any right inverse S_n of T_n and for any $F \in V_n$ we have

$$F = T_n S_n F = \sum_{i \in \mathcal{I}_n} \langle S_n F, \varphi_{n,i} \rangle_{L_2} K_{\mathfrak{F}}(\cdot, x_{n,i}),$$

which proves (5.2.27) and (5.2.28). Obviously the coefficients depend linearly on F. By invoking Proposition 5.2.12 we obtain the estimate

$$\left(\sum_{i \in \mathcal{I}_n} |c(F)_{n,i}|^r \mu(U_{n,i}) m(x_{n,i})^r \right)^{1/r} \leqslant \max_{i \in \mathcal{I}_n} \left[\mu(U_{n,i}) \cdot \sup_{z \in U_{n,i}} w(x_{n,i}, z) \right] \cdot \|S_n F\|_{L_{r,m}}$$

$$\leqslant \max_{i \in \mathcal{I}_n} \left[\mu(U_{n,i}) \cdot \sup_{z \in U_{n,i}} w(x_{n,i}, z) \right] \cdot \|S_n\| \cdot \|F\|_{L_{r,m}},$$

where $\|S_n\|$ denotes the operator norm of S_n as an operator from V_n to $L_{r,m}(X, \mu)$. $\qquad\square$

Remark 5.2.14. Similarly to the group setting in Subsection 3.2.2, we are interested in the boundedness of the sequence

$$\left(\max_{i \in \mathcal{I}_n} \left[\mu(U_{n,i}) \cdot \sup_{z \in U_{n,i}} w(x_{n,i}, z) \right] \cdot \|S_n\| \right)_{n \in \mathbb{N}}.$$

Observe that by (5.2.11) we have $C_{\mathcal{U}_n} \leqslant \max_{i \in \mathcal{I}_n} \mu(U_{n,i}) \leqslant D_{\mathcal{U}_n}$, where both $(C_n)_{n \in \mathbb{N}}$ and $(D_n)_{n \in \mathbb{N}}$ can be assumed to be decreasing sequences that tend to zero as n goes to infinity. Additionally, in practice we can assume $(\max_{i \in \mathcal{I}_n} \sup_{z \in U_{n,i}} w(x_{n,i}, z))_{n \in \mathbb{N}}$ to be a non-increasing sequence by assuming conditions like (4.2.2). Then, all we need to do is to check the behaviour of $\|S_n\|$. This, however, is already non-trivial in the group setting as we showed in Subsection 3.2.3. Recall that Lemma 3.2.21 gives a partial answer on how to characterize the opertor norm $\|S_n\|$.

The next auxiliary result is necessary for the main theorem of this subsection. Recall that $N \in \mathbb{N}$ is defined via (5.2.12).

Lemma 5.2.15. *Let $1 \leqslant p \leqslant \infty$, $n \in \mathbb{N}$, and $(d_i)_{i \in \mathcal{I}_n} \in \ell_{p,m}(\mathcal{I}_n)$ be a countable sequence, then*

$$\Big\| \sum_{i \in \mathcal{I}_n} |d_i| \chi_{U_{n,i}} \Big\|_{L_{p,m}} \leqslant \sup_{i \in \mathcal{I}_n} \sup_{x,y \in U_{n,i}} w(x,y) \cdot N^{1/p'} \cdot \|d_i\|_{\ell_{p,m}} \tag{5.2.30}$$

with the convention $\frac{1}{\infty} = 0$.

Proof. Fix $1 \leqslant p < \infty$ and $n \in \mathbb{N}$, the we have

$$\int_{U_{n,i}} m(y)^p \, d\mu(y) \leqslant m(x_{n,i})^p \int_{U_{n,i}} w(x_{n,i}, y)^p \, d\mu(y)$$
$$\leqslant m(x_{n,i})^p \cdot \sup_{x,y \in U_{n,i}} w(x,y)^p \cdot \mu(U_{n,i})$$

for all $i \in \mathcal{I}_n$. By property (5.2.10) of $\mathcal{U}_n$ we can find N disjoint subfamilies $\mathcal{V}_n^k = (U_{n,i_k})_{i_k \in I_n^k} \subset \mathcal{U}_n$, $1 \leqslant k \leqslant N$, with $\bigcup_{k=1}^N \mathcal{V}_n^k = \mathcal{U}_n$ and which are pairwise disjoint, i.e., $U_{n,i_k} \cap U_{n,j_l} = \varnothing$ for all $k \neq l$. Therefore, $\mathcal{I}_n = \bigcup_{k=1}^N I_n^k$. Then, using the disjointness we derive

$$\Big\| \sum_{i \in \mathcal{I}_n} |d_i| \chi_{U_{n,i}} \Big\|_{L_{p,m}} \leqslant \sum_{k=1}^N \Big\| \sum_{i_k \in I_n^k} |d_{i_k}| \chi_{U_{n,i_k}} \Big\|_{L_{p,m}}$$
$$= \sum_{k=1}^N \Big(\sum_{i_k \in I_n^k} |d_{i_k}|^p \int_{U_{n,i_k}} m(y)^p \, d\mu(y) \Big)^{1/p}$$
$$\leqslant \sup_{i \in \mathcal{I}_n} \sup_{x,y \in U_{n,i}} w(x,y) \cdot \sum_{k=1}^N \Big(\sum_{i_k \in I_n^k} |d_{i_k}|^p m(x_{n,i_k})^p \Big)^{1/p}$$
$$\leqslant \sup_{i \in \mathcal{I}_n} \sup_{x,y \in U_{n,i}} w(x,y) \cdot N^{1-\frac{1}{p}} \cdot \|(d_i)_{i \in \mathcal{I}_n}\|_{\ell_{p,m}}.$$

It remains to prove the case $p = \infty$. For this we operate as before and estimate

$$\Big\| \sum_{i \in \mathcal{I}_n} |d_i| \chi_{U_{n,i}} \Big\|_{L_{\infty,m}} \leqslant \sum_{k=1}^N \Big\| \sum_{i_k \in I_n^k} |d_{i_k}| \chi_{U_{n,i_k}} \Big\|_{L_{\infty,m}}$$
$$= \sum_{k=1}^N \sup_{i_k \in I_n^k} \Big(|d_{i_k}| \sup_{y \in U_{n,i_k}} m(y) \Big)$$
$$\leqslant N \cdot \Big(\sup_{i \in \mathcal{I}_n} |d_i| \cdot m(x_{n,i}) \Big) \cdot \sup_{i \in \mathcal{I}_n} \sup_{x,y \in U_{n,i}} w(x,y)$$
$$= \sup_{i \in \mathcal{I}_n} \sup_{x,y \in U_{n,i}} w(x,y) \cdot N \cdot \|(d_i)_{i \in \mathcal{I}_n}\|_{\ell_{\infty,m}},$$

which concludes the proof. $\qquad \square$

We add another technical lemma.

Lemma 5.2.16. *Denote with* $K \colon X \times X \to \mathbb{C}$ *an arbitrary kernel, let* $(d_i)_{i \in I}$ *be a countable sequence and* $\mathcal{U}$ *an admissible covering of* X; *then we have the following pointwise estimate:*

$$\left| \sum_{i \in I} d_i K(y, x_i) \right| \leqslant C_{\mathcal{U}}^{-1} \cdot [\mathrm{osc}_{\mathcal{U}} K + |K|] \left(\sum_{i \in I} |d_i| \chi_{U_i} \right)(y) \quad \textit{for all } y \in X, \tag{5.2.31}$$

where $C_{\mathcal{U}}$ *is the constant from* (5.2.11).

Proof. The proof is similar to the proof of [20, Chap. 3, p. 100]. Using (5.2.6), we estimate for any $y \in X$:

$$\left| \sum_{i \in I} d_i K(y, x_i) \right| = \left| \sum_{i \in I} d_i K(y, x_i) \mu(U_i)^{-1} \int_{U_i} 1 \, d\mu(z) \right|$$

$$\leqslant C_{\mathcal{U}}^{-1} \sum_{i \in I} |d_i| \int_{U_i} |K(y, x_i)| \, d\mu(z)$$

$$\leqslant C_{\mathcal{U}}^{-1} \sum_{i \in I} |d_i| \int_{U_i} \left(\mathrm{osc}_{\mathcal{U}} K(y, z) + |K(y, z)| \right) d\mu(z)$$

$$\leqslant C_{\mathcal{U}}^{-1} \int_X \left(\mathrm{osc}_{\mathcal{U}} K(y, z) + |K(y, z)| \right) \cdot \sum_{i \in I} |d_i| \chi_{U_i}(z) \, d\mu(z).$$

This can be written as (5.2.31) when viewing $[\mathrm{osc}_{\mathcal{U}} K + |K|]$ as a kernel operator. $\square$

Finally, we formulate the main result of this subsection.

Theorem 5.2.17. *Suppose Assumption 5.1.2 and Assumption 5.2.2 are fulfilled, i.e.,* $K_{\mathfrak{F}} \in \mathcal{A}_{q,w}$ *for all* $1 < q < \infty$ *and* $\mathrm{span}\{K_{\mathfrak{F}}(\cdot, x)\}_{x \in X}$ *is dense in* $\mathcal{M}_{r,m}$. *Assume further there is some* $p < r$ *with*

$$\mathrm{osc}_{\mathcal{U}_n} K_{\mathfrak{F}} \in \mathcal{A}_{p,w} \quad \textit{for all } n \in \mathbb{N}.$$

(i) *Fix* $\varepsilon > 0$; *then for any* $T \in \mathrm{Co}(L_{p,m})$ *there exists* $n^* = n^*_{T,\varepsilon} \in \mathbb{N}$, *such that for all* $n \geqslant n^*$

$$\left\| T - \sum_{i \in \mathcal{I}_n} c(T)_{n,i} \psi_{x_{n,i}} \right\|_{\mathrm{Co}(L_{p,m})} \leqslant \varepsilon, \tag{5.2.32}$$

where the family $(c(T)_{n,i})_{i \in I_n}$ *satisfies*

$$\|(c(T))_{n,i})_{i \in \mathcal{I}_n} \|_{\ell_{r,w}} \leqslant C_n (1 + \varepsilon) \cdot \|T\|_{\mathrm{Co}(L_{r,m})} \tag{5.2.33}$$

with

$$C_n = \max_{i \in \mathcal{I}_n} \left[\mu(U_{n,i}) \sup_{x,y \in U_{n,i}} w(x,y) \right] \cdot \|S_n\| \tag{5.2.34}$$

and where S_n *denotes any linear right inverse of the operator* $T_n \colon L_{r,m}(X, \mu) \to V_n$ *defined in* (5.2.19).

(ii) *Conversely, let* $n \in \mathbb{N}$ *and* $(d_{n,i})_{i \in I} \in \ell_{q,m}(I_n)$. *Then* $T = \sum_{i \in I_n} d_{n,i} \psi_{x_{n,i}} \in \mathrm{Co}(L_{r,m})$ *and the estimate*

$$\|T\|_{\mathrm{Co}(L_{r,m})} \leqslant D_n \cdot \|(d_i)_{i \in I}\|_{\ell_{q,m}} \tag{5.2.35}$$

holds true, where $1/q + 1/p = 1 + 1/r$, and

$$D_n = C_{\mathcal{U}_n}^{-1} \cdot \sup_{i \in I_n} \sup_{x,y \in U_{n,i}} w(x,y) \cdot N^{1/q'} \cdot \left(\|\mathrm{osc}_{\mathcal{U}_n} K_{\mathfrak{F}}\|_{\mathcal{A}_{p,w}} + \|K_{\mathfrak{F}}\|_{\mathcal{A}_{p,w}} \right), \tag{5.2.36}$$

where $C_{\mathcal{U}_n}$ is the constant in (5.2.11) and N the constant in (5.2.12). This even holds true without assuming Assumption 5.2.2.

Proof. To prove (i) we note that by the correspondence principle it suffices to prove the claim for $\mathcal{M}_{r,m}$ instead of $\mathrm{Co}(L_{r,m})$. Choose $n^* = n_{F,\varepsilon}^*$ as in Lemma 5.2.11 with $F = V_{e,\mathfrak{F}} T$. Then, following the lines of the proof of Theorem 3.2.17, the assertion is a consequence of Lemma 5.2.11, Proposition 5.2.12 and Lemma 5.2.13.

It remains to prove (ii). Again, we use the reproducing identity $V_{e,\mathfrak{F}} \psi_{x_{n_i}} = K_{\mathfrak{F}}(\cdot, x_{n,i})$. Then, by exploiting Lemma 5.2.16, Lemma 1.8.7 and Lemma 5.2.15 we obtain

$$\begin{aligned}
\|T\|_{\mathrm{Co}(L_{r,m})} &= \left\| \sum_{i \in I} d_{n,i} K_{\mathfrak{F}}(\cdot, x_{n,i}) \right\|_{L_{r,m}} \\
&\leqslant C_{\mathcal{U}_n}^{-1} \left\| [\mathrm{osc}_{\mathcal{U}} K_{\mathfrak{F}} + |K_{\mathfrak{F}}|] \left(\sum_{i \in I} |d_{n,i}| \chi_{U_i} \right) \right\|_{L_{r,m}} \\
&\leqslant C_{\mathcal{U}_n}^{-1} \|\mathrm{osc}_{\mathcal{U}} K_{\mathfrak{F}} + |K_{\mathfrak{F}}|\|_{\mathcal{A}_{p,w}} \cdot \left\| \sum_{i \in I} |d_{n,i}| \chi_{U_i} \right\|_{L_{q,m}} \\
&\leqslant C_{\mathcal{U}_n}^{-1} \cdot \left(\|\mathrm{osc}_{\mathcal{U}_n} K_{\mathfrak{F}}\|_{\mathcal{A}_{p,w}} + \|K_{\mathfrak{F}}\|_{\mathcal{A}_{p,w}} \right) \cdot \sup_{i \in I_n} \sup_{x,y \in U_{n,i}} w(x,y) \cdot N^{1/q'} \cdot \|(d_{n,i})_{i \in I}\|_{\ell_{q,m}}.
\end{aligned}$$

By the assumptions made the right hand side is finite. $\qquad \square$

Remark 5.2.18. (i) The coefficients $c(T)_{n,i}$, $i \in I_n$, in Theorem 5.2.17 (i) depend linearly on T if and only if the projection π_n in (5.2.23) is linear.

(ii) The proof of Theorem 5.2.17 (ii) shows that Assumption 5.2.2, i.e., the density of $\mathrm{span}\{K_{\mathfrak{F}}(\cdot, x)\}_{x \in X}$ in $\mathcal{M}_{r,m}$, is not used. The reconstruction of a function in $\mathrm{Co}(L_{r,m})$ therefore does not require the assumptions developed in Subsection 5.2.1.

If we assume an additional condition on the weight w similar to the condition in (4.2.2), i.e.,

$$\sup_{x,y \in U_{n,i}} w(x,y) \leqslant C_{w,\mathcal{U}_n} \quad \text{for all } i \in I_n,$$

then by invoking $\mu(U_{n,i}) \leqslant D_{\mathcal{U}_n}$ the constant in C_n in (5.2.34) can be bounded from above by

$$C_n \leqslant D_{\mathcal{U}_n} \cdot C_{w,\mathcal{U}_n} \cdot \|S_n\|. \tag{5.2.37}$$

Similarly, the constant D_n in (5.2.36) can be estimated by

$$D_n \leqslant C_{\mathcal{U}_n}^{-1} \cdot C_{w,\mathcal{U}_n} \cdot N^{1/q'} \cdot \left(\|\mathrm{osc}_{\mathcal{U}_n}(K_{\mathfrak{F}})\|_{\mathcal{A}_{p,w}} + \|K_{\mathfrak{F}}\|_{\mathcal{A}_{p,w}} \right). \tag{5.2.38}$$

Let us briefly discuss the constants above. For a more detailed discussion on Theorem 5.2.17 we refer to Subsection 3.2.2. First note that $C_{w,\mathcal{U}_n}$ is non-increasing and bounded, therefore negligible. We can reasonably assume $D_{\mathcal{U}_n}$ to be a decreasing sequence tending towards zero. Therefore the beviour of C_n is dominated by the behaviour of the operator norm $\|S_n\|$. However, as mentioned in Remark 5.2.14, we have no hope of understanding this operator norm.

Additionally, the second constant in (5.2.38) can be estimated from above by $D_n \lesssim C_{\mathcal{U}_n}^{-1} \cdot$ $\left(\|\mathrm{osc}_{\mathcal{U}_n}(K_{\mathfrak{F}})\|_{\mathcal{A}_{p,w}} + \|K_{\mathfrak{F}}\|_{\mathcal{A}_{p,w}} \right)$. This obviously implies that the upper bound is increasing as $C_{\mathcal{U}_n}$

decreases with $n \to \infty$. But if we are simply interested in a reconstruction result we only need part (ii) of Theorem 5.2.17 to hold true for one particular $n \in \mathbb{N}$ and the asymptotic behaviour of D_n is negligible.

The following proposition is a variation of Theorem 5.2.17.

Proposition 5.2.19. *Under the same assumptions as in Theorem 5.2.17 the following holds true: Fix $\varepsilon > 0$ and $T \in \mathrm{Co}(L_{r,m})$; then there exists $n^* = n^*_{T;\varepsilon} \in \mathbb{N}$ sucht that for all $n \geqslant n^*$*

$$\frac{1}{\tau_n(1+\varepsilon)} \cdot \|(c(T)_{n,i})_{i \in \mathcal{I}_n}\|_{\ell_{q,m}} \leqslant \|T\|_{\mathrm{Co}(L_{r,m})} \tag{5.2.39}$$

and

$$\|T\|_{\mathrm{Co}(L_{r,m})} \leqslant \varepsilon + D_n \cdot \|(c(T)_{n,i})_{i \in \mathcal{I}_n}\|_{\ell_{q,m}}, \tag{5.2.40}$$

with D_n as in (5.2.36) and C_n as in (5.2.34), $\tau_n := C_n \cdot |\mathcal{I}_n|^{\frac{1}{q}-\frac{1}{r}}$, $|\mathcal{I}_n|$ being the cardinality of $\mathcal{I}_n$ and $1/q + 1/p = 1 + 1/r$.

Proof. The proof is the same as the proof of Proposition 3.2.19 and a direct consequence of Theorem 5.2.17. □

We close this subsection with the following proposition, which presents a recipe for obtaining reconstruction operators with $q = r$. For this we additionally assume that both the oscillation of $K_{\mathfrak{F}}$ and the kernel itself define bounded kernel operators on $L_{r,m}(X,\mu)$. Note that by Lemma 5.2.7 this a sufficient condition for Assumption 5.2.2 to be fulfilled.

Proposition 5.2.20. *Suppose that Assumption 5.1.2 is fulfilled, i.e., we have $K_{\mathfrak{F}} \in \mathcal{A}_{q,w}$ for all $1 < q < \infty$, and that for some fixed $n \in \mathbb{N}$ the kernel operators associated with $\mathrm{osc}_{\mathcal{U}_n} K_{\mathfrak{F}}$ and $K_{\mathfrak{F}}$ are bounded on $L_{r,m}(X,\mu)$, that is,*

$$\|\mathrm{osc}_{\mathcal{U}_n} K_{\mathfrak{F}}(F)\|_{L_{r,m}} \leqslant C \cdot \|F\|_{L_{r,m}}, \tag{5.2.41}$$

$$\|K_{\mathfrak{F}}(F)\|_{L_{r,m}} \leqslant C \cdot \|F\|_{L_{r,m}} \tag{5.2.42}$$

for all $F \in L_{r,m}(X,\mu)$ and some $C > 0$.

Let $d = (d_{n,i})_{i \in I_n} \in \ell_{r,m}(I_n)$, then $T = \sum_{i \in I_n} d_{n,i} \psi_{n,i}$ is in $\mathrm{Co}(L_{r,m})$ with

$$\|T\|_{\mathrm{Co}(L_{r,m})} \leqslant 2C \cdot C_{\mathcal{U}_n}^{-1} \cdot \sup_{i \in I_n} \sup_{x,y \in U_{n,i}} w(x,y) \cdot N^{1/r'} \cdot \|(d_{n,i})_{i \in I}\|_{\ell_{r,m}}. \tag{5.2.43}$$

Proof. The proof is very similar to the proof of Theorem 5.2.17 (ii). Again, we invoke both Lemma 5.2.16 and Lemma 5.2.15 to obtain

$$\|T\|_{\mathrm{Co}(L_{r,m})} = \left\| \sum_{i \in I} d_i K_{\mathfrak{F}}(\cdot, x_{n,i}) \right\|_{L_{r,m}}$$

$$\leqslant C_{\mathcal{U}_n}^{-1} \left\| \mathrm{osc}_{\mathcal{U}} K_{\mathfrak{F}} \left(\sum_{i \in I} |d_i| \chi_{U_i} \right) \right\|_{L_{r,m}} + C_{\mathcal{U}_n}^{-1} \left\| K_{\mathfrak{F}} \left(\sum_{i \in I} |d_i| \chi_{U_i} \right) \right\|_{L_{r,m}}$$

$$\leqslant 2C \cdot C_{\mathcal{U}_n}^{-1} \cdot \left\| \sum_{i \in I} |d_i| \chi_{U_i} \right\|_{L_{r,m}}$$

$$\leqslant 2C \cdot C_{\mathcal{U}_n}^{-1} \cdot \sup_{i \in I_n} \sup_{x,y \in U_{n,i}} w(x,y) \cdot N^{1/r'} \cdot \|(d_{n,i})_{i \in I}\|_{\ell_{r,m}},$$

which concludes the proof. □

5.3 Obstruction to Discretization

In this section we show that a discretization similar to the results in Section 4.2 of generalized coorbit spaces $\mathrm{Co}(L_{r,m})$ with non-integrable kernel implies that the kernel operator associated to $K_{\mathfrak{F}}$ is a bounded operator on $L_{r,m}(X,\mu)$. This result is dual to the result of Section 3.3 for coorbit spaces with non-integrable kernel in the group setting. Since the ideas are lended from this section we will shorten some proofs.

The discretization results in Chapter 4 relied heavily on the assumption that the kernel $K_{\mathfrak{F}}(x,y) = V_{\mathfrak{F}}\psi_y(x) = \langle \psi_y, \psi_x \rangle_{\mathcal{H}}$ is in $\mathcal{A}_{1,w}$, which by virtue of a weighted version of Schur's test, see Corollary 1.8.6, implies that the corresponding kernel operator is bounded on all $L_{r,m}(X,\mu)$. By omitting this condition and replacing it by the weaker assumption $K_{\mathfrak{F}} \in \mathcal{A}_{q,w}$ for all $1 < q < \infty$ as we did in Section 5.1 we have no a priori knowledge if the kernel operator associated to $K_{\mathfrak{F}}$ is still bounded on $L_{r,m}(X,\mu)$. As we will show in Theorem 5.3.1 this is necessary for a meaningful discretization of the generalized coorbit spaces.

Note that if we assume the kernel operator to be bounded on $L_{r,m}(X,\mu)$, then Lemma 5.2.7 shows that Assumption 5.2.2 is fulfilled so that Theorem 5.2.17 and Proposition 5.2.20 imply that we obtain a weak atomic decomposition of the generalized coorbit spaces. Therefore the boundedness is sufficient for a weak atomic decomposition and in this section we show it to be also necessary for certain atomic decompositions and Banach frames.

Let us start by briefly recalling the discretization results of Section 4.2 for generalized coorbit spaces with kernel in $\mathcal{A}_{1,w}$. We assume the family $\mathcal{U} = (U_i)_{i \in I}$ of subsets of X to be a *moderate admissible covering* of X as defined in Definition 1.5.5 and choose arbitrary points $x_i \in U_i$ for all $i \in I$. Assume further the admissible weight w fulfills the condition

$$\sup_{x,y \in U_i} w(x,y) \leqslant C_{w,\mathcal{U}} \quad \text{for all } i \in I$$

Then, Theorem 4.2.1 shows that the synthesis operator

$$\mathrm{Synth}\colon \ell_{p,m}(I) \to \mathrm{Co}(L_{p,m}), \quad (c_i)_{i \in I} \mapsto \sum_{i \in I} c_i \cdot \psi_{x_i}$$

is well-defined and bounded for all $1 < p < \infty$ and w-moderate weights m. Precisely, we have shown that the family $(\psi_{x_i})_{i \in I}$ is a family of atoms and there is an atomic decomposition of generalized coorbit spaces.

Under similar assumptions we have shown in Theorem 4.2.2 that there also exist Banach frames. This means the sampling operator

$$\mathrm{Samp}\colon \mathrm{Co}(L_{p,m}) \to \ell_{p,m}(I), \quad T \mapsto (\langle T, \psi_{x_i} \rangle_{\mathcal{H}'_{1,m}, \mathcal{H}_{1,m}})_{i \in I}$$

is well-defined and bounded and admits a bounded linear left inverse that we call reconstruction operator. Therefore $(\psi_{x_i})_{i \in I}$ is a Banach frame for $\mathrm{Co}(L_{p,m})$ with coefficient space $\ell_{p,m}(I)$.

It is easy to see that if m is w-moderate weight, then so is m^{-1}. Therefore the statements above also hold true for $L_{r',m^{-1}}(X,\mu)$, $1 < r < \infty$. If we assume weaker versions of atomic decompositions and Banach frames for both $L_{r,m}(X,\mu)$ and $L_{r',m^{-1}}(X,\mu)$, then this implies that the kernel operator associated to $K_{\mathfrak{F}}$ is bounded as an operator on $L_{r,m}(X,\mu)$, as the following theorem states.

Theorem 5.3.1. *Let $1 < r < \infty$ be arbitrary. Suppose Assumption 5.1.2 is fulfilled, i.e., we have $K_{\mathfrak{F}} \in \mathcal{A}_{q,w}$ for all $1 < q < \infty$. Furthermore, assume that for some family $(x_i)_{i \in I}$ in X and for some weight $\theta = (\theta_i)_{i \in I}$ on the index set I the following hold:*

(i) *"Weak Banach frame condition for* $\mathrm{Co}(L_{r,m})$*"*: The analysis map

$$A \colon \mathrm{Co}(L_{r,m}) \to \ell_{r,\theta}(I), \quad f \mapsto (f(\psi_{x_i}))_{i \in I} \tag{5.3.1}$$

is well-defined and bounded, with

$$\|Af\|_{\ell_{r,\theta}} \asymp \|f\|_{\mathrm{Co}(L_{r,m})} \quad \text{for all } \varphi \in \mathrm{Co}(L_{r,m}). \tag{5.3.2}$$

(ii) *"Weak atomic decomposition condition for* $\mathrm{Co}(L_{r',m^{-1}})$*"*: The synthesis map

$$S \colon \ell_{r',\theta^{-1}}(I) \to \mathrm{Co}(L_{r',m^{-1}}), \quad (c_i)_{i \in I} \mapsto \sum_{i \in I} c_i \psi_{x_i} \tag{5.3.3}$$

is well-defined and bounded.

Then the integral operator associated to $K_{\mathfrak{F}}$ *defines a bounded linear operator on* $L_{r,m}(X,\mu)$.

We split the proof of the theorem in several part. Similar to the proof of Theorem 3.3.1, we proof the theorem via contradiction. First, we extend the voice transform to the dual space $[\mathrm{Co}(L_{r,m})]'$ by defining the *special voice transform* V_{sp} to be

$$V_{\mathrm{sp}}\varphi \colon X \to \mathbb{C}, \quad x \mapsto \varphi(\psi_x) = \langle \varphi, \psi_x \rangle_{[\mathrm{Co}(L_{r,m})]' \times \mathrm{Co}(L_{r,m})} \tag{5.3.4}$$

for every $\varphi \in [\mathrm{Co}(L_{r,m})]'$. This is well-defined since $\psi_x \in \mathrm{Co}(L_{r,m})$ with

$$\|\psi_x\|_{\mathrm{Co}(L_{r,m})} = \|K_{\mathfrak{F}}(x,\cdot)\|_{L_{r,m}} \leqslant m(x) \cdot \|K_{\mathfrak{F}}\|_{\mathcal{A}_{r,w}} \quad \text{for all } x \in X,$$

see Lemma 5.1.4 with $v = m$.

Using this, the following lemma shows that if the kernel operator associated with $K_{\mathfrak{F}}$ does *not* act boundedly on $L_{r',m^{-1}}(X,\mu)$, then there exists certain pathological functionals on $\mathrm{Co}(L_{r,m})$.

Lemma 5.3.2. *Let* $1 < r < \infty$ *be arbitrary and suppose Assumption 5.1.2 is fulfilled. If the integral operator associated to the kernel* $K_{\mathfrak{F}}$ *is not a well-defined and bounded operator on* $L_{r',m^{-1}}(X,\mu)$, *then there exists a conjugate linear continuous functional* $\varphi \in [\mathrm{Co}(L_{r,m})]'$ *satisfying* $V_{\mathrm{sp}}\varphi \notin L_{r',m^{-1}}(X,\mu)$.

Proof. First note that by Corollary 1.8.9 the integral operator $K_{\mathfrak{F}}$ is bounded as a map $K_{\mathfrak{F}} \colon L_{r',m^{-1}}(X,\mu) \to L_{q,m^{-1}}(X,\mu)$ for all $r' < q \leqslant \infty$. Hence an application of the closed graph theorem shows that there exists some $\Phi \in L_{r',m^{-1}}(X,\mu)$ such that $K_{\mathfrak{F}}(\Phi) \notin L_{r',m^{-1}}(X,\mu)$, see also the proof of Lemma 3.3.3.

Now we define the functional

$$\varphi \colon \mathrm{Co}(L_{r,m}) \to \mathbb{C}, \quad f \mapsto \int_X \Phi(y) \overline{V_{e,\mathfrak{F}} f(y)} \, d\mu(y).$$

This functional is well-defined and it holds that

$$|\varphi(f)| \leqslant \|\Phi\|_{L_{r',m^{-1}}} \cdot \|V_{e,\mathfrak{F}} f\|_{L_{r,m}} = \|\Phi\|_{L_{r',m^{-1}}} \cdot \|f\|_{\mathrm{Co}(L_{r,m})},$$

hence $\varphi \in [\mathrm{Co}(L_{r,m})]'$. Then, for all $x \in X$ we have

$$V_{\mathrm{sp}}\varphi(x) = \varphi(\psi_x) = \int_X \Phi(y) \overline{V_{e,\mathfrak{F}} \psi_x(y)} \, d\mu(y) = \int_X \Phi(y) \overline{K_{\mathfrak{F}}(y,x)} \, d\mu(y) = K_{\mathfrak{F}}(\Phi)(x).$$

By our choice of Φ we have $V_{\mathrm{sp}}\varphi \in L_{q,m^{-1}}(X,\mu)$ for all $r' < q \leqslant \infty$, but $V_{\mathrm{sp}}\varphi \notin L_{r',m^{-1}}(X,\mu)$ as desired. $\qquad\square$

The next lemma paves the way for the proof of Theorem 5.3.1. We omit the proof since it is analogous to the proof of Lemma 3.3.4. Simply replace $\pi(x)\psi$ by ψ_x as well as the occurring spaces.

Lemma 5.3.3. *Under the assumptions of Theorem 5.3.1 and with the notation as in Lemma 5.3.2, every conjugate linear continuous functional $\varphi \in [\mathrm{Co}(L_{r,m})]'$ satisfies the assertion $V_{\mathrm{sp}}\varphi \in L_{r',m^{-1}}(X,\mu)$.*

Now we can finally proof Theorem 5.3.1.

Proof of Theorem 5.3.1. We assume towards a contradiction that the kernel operator defined by $K_{\mathfrak{F}} \colon L_{r,m}(X,\mu) \to L_{r,m}(X,\mu)$ is not bounded. By Proposition 5.2.5 this implies that the operator $K_{\mathfrak{F}} \colon L_{r',m^{-1}}(X,\mu) \to L_{r',m^{-1}}(X,\mu)$ is also not bounded. Therefore Lemma 5.3.2 yields a conjugate linear continuous functional $\varphi \in [\mathrm{Co}(L_{r,m})]'$ with $V_{\mathrm{sp}}\varphi \notin L_{r',m^{-1}}(X,\mu)$. In view of Lemma 5.3.3 this gives the desired contradiction. Hence, Theorem 5.3.1 holds true. $\quad\square$

5.4 Improved Discretization Results Under Additional Assumptions

So far we have shown in Subsection 5.2.2 that under the Assumption that the kernel operator associated with $K_{\mathfrak{F}}$ acts boundedly on $L_{r,m}(X,\mu)$, we have a weak atomic decomposition of generalized coorbit spaces. On the other hand the preceding section showed that this condition is necessary for obtaining atomic decompositions and Banach frames. But it seems that we lack the conditions to show the existence of atomic decompositions and Banach frames.

In this section we present conditions which ensure these results. For this we use a similar approach to Section 3.4, which treats the group setting. To be more precise, we assume there exists a second kernel $W \colon X \times X \to \mathbb{C}$ being contained in $\mathcal{A}_{1,w}$ and fulfilling the reproducing identity $W \circ K_{\mathfrak{F}} = K_{\mathfrak{F}}$, where $\circ$ is the multiplication of kernels, i.e.,

$$W \circ K_{\mathfrak{F}}(x,y) = \int_X W(x,z)K_{\mathfrak{F}}(z,y)\,d\mu(z).$$

If this identity is fulfilled we are able to find atomic decompositions as well as Banach frames.

The main ideas of this section are borrowed from Section 3.4 and translated to the general setting. This section is structured as follows. We first recall the setting and basic notations. In Subsection 5.4.1 we show the existence of Banach frames under certain additional assumptions on W. Then, in Subsection 5.4.2, we similarly show the existence of atomic decompositions and specify the assumptions.

Before we start showing the discretization results, let us quickly recall some notions of the preceding sections. As usual, we denote with $\mathcal{U} = (U_i)_{i \in I}$ an *admissible covering* of X. This means that each set $U_i \subset X$, $i \in I$, is relatively compact and has non-void interior with $X = \bigcup_{i \in I} U_i$. Moreover, there exists a constant $N \in \mathbb{N}$ such that each point $x \in X$ is contained in at most N sets, i.e.,

$$\sum_{i \in I} \chi_{U_i}(x) \leqslant N \quad \text{for all } x \in X.$$

Instead of assuming the moderateness of $\mathcal{U}$, we suppose there are two constants $C_\mathcal{U}, D_\mathcal{U} > 0$ such that $C_\mathcal{U} \leqslant \mu(U_i) \leqslant D_\mathcal{U}$ for all $i \in I$.

Recall that the oscillation $\mathrm{osc}_\mathcal{U}K$ of a kernel $K \colon X \times X \to \mathbb{C}$ subordinate to $\mathcal{U}$ is defined as

$$\mathrm{osc}_\mathcal{U}K(x,y) = \sup_{z \in Q_y} |K(x,y) - K(x,z)|,$$

where $Q_y = \bigcup_{y \in U_i} U_i$.

Next, we say $\Phi = (\varphi_i)_{i \in I}$ is a bounded partition of unity subordinate to $\mathcal{U}$ in the following sense. Each $\varphi_i \colon X \to [0, 1]$ is measureable and suffices $\operatorname{supp} \varphi_i \subseteq U_i$ and we have $\sum_{i \in I} \varphi_i \equiv 1$. Thanks to Lemma 1.5.8 such a family of functions always exists. Finally, let us choose arbitrary but fixed points $x_i \in U_i$ for all $i \in I$.

From now on we will additionally assume the weight w to fulfill

$$\sup_{x,y \in U_i} w(x,y) \leqslant C_{w,\mathcal{U}} \quad \text{for all } i \in I. \tag{5.4.1}$$

and the w-moderate weight m to fulfill

$$m(x) \leqslant C \cdot m(y) \quad \text{for all } x, y \in U_i, \ i \in I,$$

for some $C, C_{w,\mathcal{U}} > 0$. Recall that the sequence spaces $\ell_{r,m}(I)$ for $1 \leqslant r < \infty$ are defined as

$$\ell_{r,m}(I) = \left\{ c = (c_i)_{i \in I} \in \mathbb{C}^I : \|c\|_{\ell_{r,m}}^r = \sum_{i \in I} |c_i|^r \mu(U_i)^r m(x_i)^r < \infty \right\} \tag{5.4.2}$$

in this setting. Then, the next preparing lemma holds true concerning two synthesis operators.

Lemma 5.4.1. *Let $1 \leqslant r < \infty$ and $\mathcal{U} = \{U_i\}_{i \in I}$ be a moderate admissible covering of X as well as $\Phi = \{\varphi_i\}_{i \in I}$ be a partition of unity subordinate to $\mathcal{U}$. Then*

(i) the synthesis operator

$$\operatorname{Synth}_{\mathcal{U}} \colon \ell_{r,m}(I) \to L_{r,m}(X, \mu), \quad c = (c_i)_{i \in I} \mapsto \sum_{i \in I} c_i \chi_{U_i} \tag{5.4.3}$$

is well-defined and bounded, with pointwise absolute convergence of the defining series;

(ii) the synthesis operator

$$\operatorname{Synth}_{\Phi} \colon \ell_{r,m}(I) \to L_{r,m}(X, \mu), \quad c = (c_i)_{i \in I} \mapsto \sum_{i \in I} c_i \varphi_i \tag{5.4.4}$$

is well-defined and bounded, with pointwise absolute convergence of the defining series.

Proof. We first note that (ii) is a consequence of (i). To see this we recall the properties of Φ, that is $0 \leqslant \varphi_i \leqslant 1$ and $\operatorname{supp} \varphi_i \subseteq U_i$ for all $i \in I$. Then,

$$|\operatorname{Synth}_{\Phi} c(x)| \leqslant \sum_{i \in I} |c_i| \phi_i(x) \leqslant \sum_{i \in i} |c_i| \chi_{U_i}(x) = \operatorname{Synth}_{\mathcal{U}} |c|(x)$$

for all $x \in X$ and $c \in \ell_{r,m}(I)$, where $|c| = (|c_i|)_{i \in I} \in \ell_{r,m}$, so that

$$\|\operatorname{Synth}_{\Phi} c\|_{L_{r,m}} \leqslant \|\operatorname{Synth}_{\mathcal{U}} |c|\|_{L_{r,m}} \lesssim \| |c| \|_{\ell_{r,m}} = \|c\|_{\ell_{r,m}}.$$

Thus, it remains to prove (i). By the definition of $\mathcal{U}$ we have $\sum_{i \in I} \chi_{U_i} \leqslant N \in \mathbb{N}$ for some $n \in \mathbb{N}$. This means that for every $x \in X$ only finitely many summands of the defining series of $\operatorname{Synth}_{\mathcal{U}}$ do not vanish. Hence, the series is pointwise absolute convergent. Moreover, for any $c \in \ell_{r,m}(I)$ we see

$$|\operatorname{Synth}_{\mathcal{U}} c(x)|^r \leqslant \left(\sum_{i \in I} |c_i| \chi_{U_i}(x) \chi_{U_i}(x) \right)^r \leqslant \left(\sup_{j \in I} |c_j| \chi_{U_j}(x) \cdot \sum_{i \in I} \chi_{U_i}(x) \right)^r$$

$$= N^r \cdot \sup_{j \in I} |c_j|^r \chi_{U_j}(x)^r \leqslant N^r \cdot \sum_{i \in I} |c_i|^r \chi_{U_i}(x)^r$$

for all $x \in X$, hence

$$\|\mathrm{Synth}_{\mathcal{U}} c\|_{L_{r,m}}^r \leqslant N^r \int_X \sum_{i \in I} |c_i|^r \chi_{U_i}(x) m(x)^r \, d\mu(x) = N^r \cdot \sum_{i \in I} |c_i|^r \int_{U_i} m(x)^r \, d\mu(x).$$

For any $x \in U_i$, $i \in I$, we have

$$m(x) \leqslant w(x_i, x) \cdot m(x_i) \leqslant m(x_i) \cdot \sup_{z \in U_i} w(x_i, z) \leqslant m(x_i) \cdot C_{w,\mathcal{U}},$$

which implies

$$\|\mathrm{Synth}_{\mathcal{U}} c\|_{L_{r,m}}^r \leqslant N^r \cdot \sum_{i \in I} |c_i| m(x_i)^r \cdot \mu(U_i) \cdot C_{w,\mathcal{U}}$$

$$\leqslant N^r \cdot D_{\mathcal{U}} \cdot C_{w,\mathcal{U}} \cdot \|c\|_{\ell_{r,m}}^r.$$

Since the constants above are finite, the operator $\mathrm{Synth}_{\mathcal{U}}$ is bounded. $\qquad\square$

5.4.1 Banach Frames

In this subsection we show the existence of Banach frames and specify the conditions on the kernel $K_{\mathfrak{F}}$. The main result is stated in Theorem 5.4.7. As usual, we fix $1 < r < \infty$ and a w-moderate weight m on X.

We suppose the following assumption is fulfilled.

Assumption 5.4.2. *Suppose Assumption 5.1.2 is fulfilled, i.e., $K_{\mathfrak{F}} \in \mathcal{A}_{q,w}$ for all $1 < q < \infty$, and the associated kernel operator is bounded on $L_{r,m}(X, \mu)$. Assume further there exists a kernel $W : X \times X \to \mathbb{C}$ with the following properties:*

(i) *$W \in \mathcal{A}_{1,w}$;*

(ii) *the identity*

$$W \circ K_{\mathfrak{F}}(x, y) = \int_X W(x, z) K_{\mathfrak{F}}(z, y) \, d\mu(z) = K_{\mathfrak{F}}(x, y) \tag{5.4.5}$$

holds true for all $x, y \in X$;

(iii) *For any $\delta > 0$ there exists an admissible covering $\mathcal{U} = \mathcal{U}(\delta)$ such that $\mathrm{osc}_{\mathcal{U}} W \in \mathcal{A}_{1,w}$ and*

$$\|\mathrm{osc}_{\mathcal{U}} W\|_{\mathcal{A}_{1,w}} \leqslant \delta. \tag{5.4.6}$$

Under this assumption the following important properties are true.

Lemma 5.4.3. *Let Assumption 5.4.2 be fulfilled. Then the following hold true.*

(i) *Assumption 5.2.2 is fulfilled, that is, $\mathrm{span}\{K_{\mathfrak{F}}(\cdot, x)\}_{x \in X}$ is dense in the reproducing kernel space $\mathcal{M}_{r,m}$ defined in (5.1.21).*

(ii) *For any $\delta > 0$ there exists an admissible covering $\mathcal{U} = \mathcal{U}(\delta)$ such that for all $f \in \mathcal{M}_{r,m}$ we have*

$$\|\mathrm{osc}_{\mathcal{U}} f\|_{L_{r,m}} \leqslant \delta \cdot \|f\|_{L_{r,m}}. \tag{5.4.7}$$

Here, the $\mathcal{U}$-oscillation of a function f on X is defined as

$$\mathrm{osc}_{\mathcal{U}} f(x) := \sup_{y \in Q_x} |f(x) - f(y)| \tag{5.4.8}$$

with $Q_x = \bigcup_{x \in U_i} U_i$.

Proof. It is enough to show part (ii), since (i) is an application of Lemma 5.2.7.

First we will prove that the integral operator associated with W suffices $W(f) = f$ for all $f \in \mathcal{M}_{r,m}$. This can be shown as follows. Using the reproducing formula (5.4.5) we have

$$W\left(K_{\mathfrak{F}}(\cdot, x)\right)(y) = \int_X W(y, z) K_{\mathfrak{F}}(z, x) \, d\mu(z) = K_{\mathfrak{F}}(y, x)$$

for all $x, y \in X$, so that $W(f) = f$ for all $f \in \mathrm{span}\{K_{\mathfrak{F}}(\cdot, x)\}_{x \in X}$. By part (i) and the continuity of the kernel operator $W \colon \mathcal{M}_{r,m} \to L_{r,m}(X, \mu)$ — which is a consequence of part (i) of Assumption 5.4.2 and Corollary 1.8.6 — the identity $W(f) = f$ extends to all $\mathcal{M}_{r,m}$.

Now let $f \in \mathcal{M}_{r,m}$, then for any $x \in X$, $y \in Q_x$ we have

$$
\begin{aligned}
|f(x) - f(y)| &= |W(f)(x) - W(f)(y)| \\
&\leqslant \int_X |W(x, z) - W(y, z)| \cdot |f(z)| \, d\mu(z) \\
&\leqslant \int_X |\mathrm{osc}_{\mathcal{U}} W(x, z)| \cdot |f(z)| \, d\mu(z) = |\mathrm{osc}_{\mathcal{U}} W|(|f|)(x),
\end{aligned}
$$

thus

$$\mathrm{osc}_{\mathcal{U}} f(x) \leqslant |\mathrm{osc}_{\mathcal{U}} W|(|f|)(x)$$

for all $x \in X$. Using part (i) of Assumption 5.4.2 and Corollary 1.8.6 this implies

$$\|\mathrm{osc}_{\mathcal{U}} f\|_{L_{r,m}} = \||\mathrm{osc}_{\mathcal{U}} W|(|f|)\|_{L_{r,m}} \leqslant \|\mathrm{osc}_{\mathcal{U}} W\|_{\mathcal{A}_{1,w}} \cdot \|f\|_{L_{r,m}} \leqslant \delta \cdot \|f\|_{L_{r,m}},$$

which proves the claim. $\qquad\square$

In the following lemma we show that a certain sampling operator is bounded on the reproducing kernel space $\mathcal{M}_{r,m}$. We will later us the fact that $\mathcal{M}_{r,m}$ is isomorphic to the coorbit space $\mathrm{Co}(L_{r,m})$ and transfer this result.

Lemma 5.4.4. *Let Assumption 5.4.2 be fulfilled for some $\delta > 0$ and $\mathcal{U}(\delta) = \mathcal{U} = (U_i)_{i \in I}$ and choose arbitrary points $x_i \in U_i$ for all $i \in I$. Then, the sampling operator*

$$\mathrm{Samp} \colon \mathcal{M}_{r,m} \to \ell_{r,m}(I), \quad f \mapsto (f(x_i))_{i \in I} \tag{5.4.9}$$

is well-defined and bounded.

Proof. Let us first show that Samp is well-defined. For any $i \in I$ and $f \in \mathcal{M}_{r,m}$ we have

$$|f(x_i)| = |K_{\mathfrak{F}}(f)(x_i)| \leqslant \int_X |K_{\mathfrak{F}}(x_i, y) f(y)| \, d\mu(y) \leqslant \|K_{\mathfrak{F}}(x_i, \cdot)\|_{L_{r',m^{-1}}} \cdot \|f\|_{L_{r,m}},$$

where $1/r + 1/r' = 1$, hence $1 < r' < \infty$. Since $m(x)^{-1} \leqslant m(x_i)^{-1} w(x, x_i)$, we have

$$\|K_{\mathfrak{F}}(x_i, \cdot)\|_{L_{r',m^{-1}}} \leqslant \mathop{\mathrm{ess\,sup}}_{x \in X} \|K_{\mathfrak{F}}(x, \cdot) \cdot w(x, \cdot)\|_{L_{r'}} \leqslant \|K_{\mathfrak{F}}\|_{\mathcal{A}_{r',w}},$$

implying $|f(x_i)| < \infty$. Therefore each entry of the sequence $(f(x_i))_{i \in I}$ in (5.4.9) is well-defined.

Next we show the boundedness of Samp. Again, let $f \in \mathcal{M}_{r,m}$, then for any $i \in I$ and $x \in U_i$ we have

$$|f(x_i)| \leqslant |f(x)| + |f(x_i) - f(x)| \leqslant |f(x)| + \mathrm{osc}_{\mathcal{U}} f(x) =: F(x).$$

Applying this estimate to the sampling operator and using the notation $I(x) = \{i \in I : x \in U_i\}$ gives us

$$
\begin{aligned}
\|\mathrm{Samp} f\|_{\ell_{r,m}}^r &= \sum_{i \in I} |f(x_i)|^r \mu(U_i) m(x_i)^r \\
&= \int_X \sum_{i \in I(x)} |f(x_i)|^r m(x_i)^r \chi_{U_i}(x) \, d\mu(x) \\
&\leqslant \int_X \sum_{i \in I} F(x)^r m(x)^r w(x, x_i)^r \chi_{U_i}(x) \, d\mu(x) \\
&\leqslant C_{w,\mathcal{U}}^r \cdot N \cdot \int_X F(x)^r m(x)^r \, d\mu(x) = C_{w,\mathcal{U}}^r \cdot N \cdot \|F\|_{L_{r,m}}^r.
\end{aligned}
$$

By part (iii) of Assumption 5.4.2 we have

$$
\|F\|_{L_{r,m}} = \||f| + \mathrm{osc}_{\mathcal{U}} f\|_{L_{r,m}} \leqslant \|f\|_{L_{r,m}} + \|\mathrm{osc}_{\mathcal{U}} f\|_{L_{r,m}} \leqslant (1 + \delta) \|f\|_{L_{r,m}},
$$

hence

$$
\|\mathrm{Samp} f\|_{\ell_{r,m}} \leqslant C_{w,\mathcal{U}} \cdot N^{1/r} \cdot (1 + \delta) \cdot \|f\|_{L_{r,m}}.
$$

This proves $\mathrm{Samp} f \in \ell_{r,m}(I)$ for all $f \in \mathcal{M}_{r,m}$ as well as the boundedness of the sampling operator. $\qquad\square$

We use this result to show the existence of Banach frames for the reproducing kernel space $\mathcal{M}_{r,m}$ for sufficiently small coverings of X.

Proposition 5.4.5. *Let Assumption 5.4.2 be satisfied for some $\delta > 0$, where*

$$
\|K_{\mathfrak{F}}\|_{L_{r,m} \to L_{r,m}} < \delta^{-1}. \tag{5.4.10}
$$

Furthermore let $(x_i)_{i \in I} \subset X$ such that $x_i \in U_i$ for all $i \in I$ and denote with $\Phi = (\varphi_i)_{i \in I}$ the partition of unity subordinate to $\mathcal{U} = \mathcal{U}(\delta)$. Then there is a bounded linear reconstruction operator $\mathrm{Reco} : \ell_{r,m}(I) \to \mathcal{M}_{r,m}$ which satisfies

$$
\mathrm{Reco} \circ \mathrm{Samp} = \mathrm{id}_{\mathcal{M}_{r,m}}
$$

for the sampling operator Samp from (5.4.9).

In other words, the family of point evaluations $(\delta_i)_{i \in I} \subset \mathcal{M}'_{r,m}$ forms a Banach frame for $\mathcal{M}_{r,m}$ with coefficient space $\ell_{r,m}(I)$.

Proof. Let Samp be the sampling operator form (5.4.9) and Synth_{Φ} the synthesis operator from (5.4.4), then we define

$$
B := \mathrm{Synth}_{\Phi} \circ \mathrm{Samp} \colon \mathcal{M}_{r,m} \to L_{r,m}(X, \mu), \quad f \mapsto \sum_{i \in I} f(x_i) \varphi_i.
$$

Furthermore we set

$$
A = K_{\mathfrak{F}} \circ B \colon \mathcal{M}_{r,m} \to \mathcal{M}_{r,m},
$$

which is well-defined since $K_{\mathfrak{F}}$ is a projection form $L_{r,m}(X, \mu)$ onto $\mathcal{M}_{r,m}$. Then, for any $f \in \mathcal{M}_{r,m}$ we have

$$
\|f - Af\|_{L_{r,m}} = \|K_{\mathfrak{F}}(f - Bf)\|_{L_{r,m}} \leqslant \|K_{\mathfrak{F}}\|_{L_{r,m} \to L_{r,m}} \cdot \|f - Bf\|_{L_{r,m}}.
$$

Because of $f = \sum_{i \in I} \varphi_i f$, for any $x \in X$ we have

$$|f(x) - Bf(x)| = \left| \sum_{i \in I} f(x)\varphi_i(x) - \sum_{i \in I} f(x_i)\varphi_i(x) \right|$$

$$\leqslant \sum_{i \in I(x)} \varphi_i(x) \cdot |f(x) - f(x_i)|$$

$$\leqslant \sum_{i \in I} \phi_i(x) \cdot \sup_{y \in Q_x} |f(x) - f(y)| = \mathrm{osc}_{\mathcal{U}} f(x),$$

where $I(x) = \{i \in I : x \in U_i\}$. By part (iii) of Assumption (5.4.2) and (5.4.10) this implies

$$\|f - Af\|_{L_{r,m}} \leqslant \|K_{\mathfrak{F}}\|_{L_{r,m} \to L_{r,m}} \cdot \|\mathrm{osc}_{\mathcal{U}} f\|_{L_{r,m}} \leqslant \|K_{\mathfrak{F}}\|_{L_{r,m} \to L_{r,m}} \cdot \delta \cdot \|f\|_{L_{r,m}} < \|f\|_{L_{r,m}},$$

hence

$$\|\mathrm{id}_{\mathcal{M}_{r,m}} - A\|_{L_{r,m} \to L_{r,m}} < 1.$$

Therefore a Neumann-series argument, see [4, Sect. 5.7], shows that the bounded linear operator $H := \sum_{n=0}^{\infty} (\mathrm{id}_{\mathcal{M}_{r,m}} - A)^n \colon \mathcal{M}_{r,m} \to \mathcal{M}_{r,m}$ satisfies

$$\mathrm{id}_{\mathcal{M}_{r,m}} = H \circ A = H \circ K_{\mathfrak{F}} \circ \mathrm{Synth}_{\Phi} \circ \mathrm{Samp}.$$

Thus, $\mathrm{Reco} := H \circ R \circ \mathrm{Synth}_{\Phi}$ is the desired bounded linear reconstruction operator. Note that the action of the operator A is independent of r and m, since the same holds for the defining operators $K_{\mathfrak{F}}$, Synth_{Φ} and Samp. Therefore also H is independent of r and m. $\qquad\square$

Remark 5.4.6. The proof above shows that the action of the reconstruction operator is *independent* of the choice of r, m.

In other words, if the condition (5.4.10) is satisfied for $L_{r_1,m_1}(X, \mu)$ and $L_{r_2,m_2}(X, \mu)$ and if $\mathrm{Reco}_1 \colon \ell_{r_1,m_1}(I) \to \mathcal{M}_{r_1,m_1}$ and $\mathrm{Reco}_2 \colon \ell_{r_2,m_2}(I) \to \mathcal{M}_{r_2,m_2}$ denote the respective reconstruction operators, then $\mathrm{Reco}_1 c = \mathrm{Reco}_2 c$ for all $c \in \ell_{r_1,m_1}(I) \cap \ell_{r_2,m_2}(I)$.

The same holds true by the correspondance principle for the reconstruction operator in the following theorem.

Finally, we use the correspondence principle to show that we have Banach frames for the coorbit space $\mathrm{Co}(L_{r,m})$ as well.

Theorem 5.4.7. *Let Assumption 5.4.2 be satisfied for some $\delta > 0$, where*

$$\|K_{\mathfrak{F}}\|_{L_{r,m} \to L_{r,m}} < \delta^{-1}. \tag{5.4.11}$$

Furthermore let $(x_i)_{i \in I} \subset X$ such that $x_i \in U_i$ for all $i \in I$. Then the family $(\psi_{x_i})_{i \in I} \subset \mathrm{Co}(L_{r,m})'$ forms a Banach frame for $\mathrm{Co}(L_{r,m})$ with coefficient space $\ell_{r,m}(I)$.

In particular, the sampling operator

$$\mathrm{Samp}_{\mathrm{Co}} \colon \mathrm{Co}(L_{r,m}) \to \ell_{r,m}(I), \quad f \mapsto (V_{e,\mathfrak{F}} f(x_i))_{i \in I} = \left(\langle f, \psi_{x_i} \rangle_{\mathcal{H}'_m \times \mathcal{H}_m} \right)_{i \in I} \tag{5.4.12}$$

is well-defined and bounded and there exists a bounded linear reconstruction operator denoted by $\mathrm{Reco}_{\mathrm{Co}} \colon \ell_{r,m}(I) \to \mathrm{Co}(L_{r,m})$ satisfying $\mathrm{Reco}_{\mathrm{Co}} \circ \mathrm{Samp}_{\mathrm{Co}} = \mathrm{id}_{\mathrm{Co}(L_{r,m})}$.

Finally, the action of the reconstruction operator $\mathrm{Reco}_{\mathrm{Co}}$ is independent of the choice of r, m, that is, if the assumptions of the current theorem are satisfied for $L_{r_1,m_1}(X, \mu)$ and for $L_{r_2,m_2}(X, \mu)$ and if $\mathrm{Reco}_{\mathrm{Co},1}$ and $\mathrm{Reco}_{\mathrm{Co},2}$ denote the corresponding reconstruction operators, then $\mathrm{Reco}_{\mathrm{Co},1} c = \mathrm{Reco}_{\mathrm{Co},2} c$ for all $c \in \ell_{r_1,m_1}(I) \cap \ell_{r_2,m_2}(I)$.

Proof. The correspondence principle, see Proposition 5.1.9, states that the extended voice transform $V_{e,\mathfrak{F}}\colon \mathrm{Co}(L_{r,m}) \to \mathcal{M}_{r,m}$ is an isometric isomorphism. Using the sampling operator Samp from (5.4.9) we have

$$(\mathrm{Samp} \circ V_{e,\mathfrak{F}})f = (V_{e,\mathfrak{F}}f(x_i))_{i\in I} = \big(\langle f, \psi_{x_i}\rangle_{\mathcal{H}'_m \times \mathcal{H}_m}\big)_{i\in I} = \mathrm{Samp}_{\mathrm{Co}}f$$

for all $f \in \mathrm{Co}(L_{r,m})$. Thus, by Proposition 5.4.5, $\mathrm{Samp}_{\mathrm{Co}}\colon \mathrm{Co}(L_{r,m}) \to \ell_{r,m}(I)$ is indeed well-defined and bounded.

Now, with the reconstruction operator $\mathrm{Reco}\colon \ell_{r,m} \to \mathrm{M}_{r,m}$ from Proposition 5.4.5, we define $\mathrm{Reco}_{\mathrm{Co}} = V_{e,\mathfrak{F}}^{-1} \circ \mathrm{Reco}\colon \ell_{r,m}(I) \to \mathrm{Co}(L_{r,m})$. Then,

$$\mathrm{Reco}_{\mathrm{Co}} \circ \mathrm{Samp}_{\mathrm{Co}} = V_{e,\mathfrak{F}}^{-1} \circ \mathrm{Reco} \circ \mathrm{Samp} \circ V_{e,\mathfrak{F}} = V_{e,\mathfrak{F}}^{-1} \circ V_{e,\mathfrak{F}} = \mathrm{id}_{\mathrm{Co}(L_{r,m})}$$

as desired. Finally, since the action of Reco is independent of the choice of r, m, so is the action of $\mathrm{Reco}_{\mathrm{Co}}$. $\qquad\square$

5.4.2   Atomic Decompositions

Under similar assumptions as in the previous subsection we present atomic decompositions for generalized coorbit spaces $\mathrm{Co}(L_{r,m})$, which is the content of Theorem 5.4.11. As before we fix $1 < r < \infty$ and let m be a w-moderate weight on X satisfying (5.4.1).

We propose the following assumption. A main difference in the assumption is that we now require $K_{\mathfrak{F}} = W \circ K_{\mathfrak{F}}$ for a second kernel W where in the previous subsection it was the other way round.

Assumption 5.4.8. *Suppose that the kernel $K_{\mathfrak{F}}$ from (5.1.3) satisfies Assumption 5.1.2, i.e., $K_{\mathfrak{F}} \in \mathcal{A}_{q,w}$ for all $1 < q < \infty$, and that the associated kernel operator is bounded on $L_{r,m}(X,\mu)$. Assume further there exists a kernel $W\colon X \times X \to \mathbb{C}$ and a moderate admissible covering $\mathcal{U} = (U_i)_{i\in I}$ of X satisfying $0 < C_{\mathcal{U}} \leqslant \mu(U_i) \leqslant D_{\mathcal{U}} < \infty$ with $W, \mathrm{osc}_{\mathcal{U}}W \in \mathcal{A}_{1,w}$ and*

$$K_{\mathfrak{F}} \circ W(x,y) = \int_X K_{\mathfrak{F}}(x,z)W(z,y)\,d\mu(z) = K_{\mathfrak{F}}(x,y) \tag{5.4.13}$$

for all $x, y \in X$.

The following lemma paves the way for the definition of suitable reconstruction and coefficient operators we need for an atomic decomposition.

Lemma 5.4.9. *Let Assumption 5.4.8 be fulfilled and denote with $\mathcal{U} = (U_i)_{i\in I}$ the corresponding moderate admissible covering. Choose points $x_i \in U_i$ for all $i \in I$ and let $\Phi = (\varphi_i)_{i\in I}$ be the bounded admissible partition of unity subordinate to $\mathcal{U}$. Then*

(i) the analysis operator

$$\mathrm{Ana}_\Phi\colon L_{r,m}(X,\mu) \to \ell_{r,m}(I), \quad f \mapsto (\langle f, \varphi_i\rangle_{L_2})_{i\in I} \tag{5.4.14}$$

is a well-defined and bounded linear map;

(ii) the synthesis operator

$$\mathrm{Synth}_W\colon \ell_{r,m}(I) \to L_{r,m}(X,\mu), \quad c = (c_i)_{i\in I} \mapsto \sum_{i\in I} c_i \cdot W(\cdot, x_i) \tag{5.4.15}$$

is a well-defined and bounded linear map, where the defining series is almost everywhere absolutely convergent.

Proof. We start by showing (i). By definition of Φ we have supp $\varphi_i \subset U_i$ as well as $0 \leqslant \varphi_i \leqslant 1$ for all $i \in I$. Fixing some $f \in L_{r,m}(X, \mu)$ and $i \in I$ we obtain by following the lines of the proof of Proposition 5.2.12 and using Jensen's inequality:

$$|\langle f, \varphi_i \rangle_{L_2}|^r \cdot m(x_i)^r \leqslant \left(\int_{U_i} |f(x)| \varphi_i(x) m(x_i) \frac{d\mu(x)}{\mu(U_i)} \right)^r \mu(U_i)^r$$

$$\leqslant \int_{U_i} |f(x)|^r \varphi_i(x) m(x)^r \frac{d\mu(x)}{\mu(U_i)} \cdot \sup_{x,y \in U_i} w(x,y)^r \cdot \mu(U_i)^r$$

$$\leqslant \int_X |f(x)|^r \varphi_i(x) m(x)^r \, d\mu(x) \cdot C_{w,\mathcal{U}}^r \cdot \mu(U_i)^{r-1}.$$

Summing over all $i \in I$ and applying the monotone convergence theorem, we thus get

$$\|\mathrm{Ana}_\Phi f\|_{\ell_{r,m}}^r = \sum_{i \in I} |\langle f, \phi_i \rangle_{L_2}|^r \cdot \mu(U_i) \cdot m(x_i)^r$$

$$\leqslant \sum_{i \in I} \int_X |f(x)|^r m(x)^r \varphi_i(x) \, d\mu(x) \cdot C_{w,\mathcal{U}}^r \cdot \mu(U_i)^r$$

$$= \|f\|_{L_{r,m}}^r \cdot C_{w,\mathcal{U}}^r \cdot D_{\mathcal{U}}^r < \infty,$$

proving the well-definedness and boundedness of the analysis operator.

It remains to prove (ii), for this let $c \in \ell_{r,m}(I)$ and $i \in I$ and consider

$$|c_i| \cdot |W(x, x_i)| = \mu(U_i)^{-1} \int_X |c_i| \cdot |w(x, x_i)| \cdot \chi_{U_i}(y) \, d\mu(y)$$

$$\leqslant \mu(U_i)^{-1} \int_X |c_i| \cdot \chi_{U_i}(y) \cdot \sup_{z \in U_i} |Q(x, z)| d\mu(y).$$

Now, for any $y, z \in U_i$ it holds that

$$|W(x, z)| \leqslant |W(x, y)| + |W(x, y) - W(x, z)| \leqslant |W(x, y)| + \mathrm{osc}_{\mathcal{U}} W(x, y),$$

implying

$$|c_i| \cdot |Q(x, x_i)| \leqslant \mu(U_i)^{-1} \int_X |c_i| \cdot \chi_{U_i}(y) \cdot [|W(x, y)| + \mathrm{osc}_{\mathcal{U}} W(x, y)] \, d\mu(y)$$

$$= [|W| + \mathrm{osc}_{\mathcal{U}} W] \left(|c_i| \cdot \mu(U_i)^{-1} \cdot \chi_{U_i} \right)(x).$$

Summing over $i \in I$ and using the monotone convergence theorem as well as Corollary 1.8.6 yields

$$\|\mathrm{Synth}_W c\|_{L_{r,m}} \leqslant \left\| \sum_{i \in I} |c_i| \cdot |W(\cdot, x_i)| \right\|_{L_{r,m}}$$

$$\leqslant \left\| \sum_{i \in I} [|W| + \mathrm{osc}_{\mathcal{U}} W] \left(|c_i| \cdot \mu(U_i)^{-1} \cdot \chi_{U_i} \right) \right\|_{L_{r,m}}$$

$$\leqslant \left\| [|W| + \mathrm{osc}_{\mathcal{U}} W] \left(\sum_{i \in I} |c_i| \cdot \mu(U_i)^{-1} \cdot \chi_{U_i} \right) \right\|_{L_{r,m}}$$

$$\leqslant \left(\|W\|_{\mathcal{A}_{1,w}} + \|\mathrm{osc}_{\mathcal{U}} W\|_{\mathcal{A}_{1,w}} \right) \cdot \left\| \sum_{i \in I} |c_i| \cdot \mu(U_i)^{-1} \cdot \chi_{U_i} \right\|_{L_{r,m}}.$$

Elementary calculations show

$$\left\| \sum_{i \in I} |c_i| \cdot \mu(U_i)^{-1} \cdot \chi_{U_i} \right\|_{L_{r,m}} \leqslant C \cdot \|c\|_{\ell_{r,m}}$$

for a constant $C = C(w, \mathcal{U}, r)$ independent of c. Thus the synthesis operator is well-defined and bounded the defining series is almost everywhere absolutely convergent. $\square$

Using this lemma we can now find an atomic decomposition for the reproducing kernel space $\mathcal{M}_{r,m}$.

Proposition 5.4.10. *Let Assumption 5.4.8 be satisfied and assume further*

$$\|K_{\mathfrak{F}}\|_{L_{r,m} \to L_{r,m}} \cdot \|\operatorname{osc}_{\mathcal{U}} W\|_{\mathcal{A}_{1,w}} < 1. \tag{5.4.16}$$

Moreover let $(x_i)_{i \in I}$ be a family such that $x_i \in U_i$ for all $i \in I$ and $\Phi = (\phi_i)_{i \in I}$ be a bounded admissible partition of unity subordinate to $\mathcal{U}$.

Then the family $(K_{\mathfrak{F}}(\cdot, x_i))_{i \in I}$ forms an atomic decomposition for $\mathcal{M}_{r,m}$ with associated sequence space $\ell_{r,m}(I)$. By virtue of Definition 1.9.2 this means:

(i) The synthesis operator

$$\operatorname{Synth}_K : \ell_{r,m}(I) \to \mathcal{M}_{r,m}, \quad (c_i)_{i \in I} \mapsto \sum_{i \in I} c_i K_{\mathfrak{F}}(\cdot, x_i) \tag{5.4.17}$$

is well-defined and bounded, with unconditional convergence of the defining series. This even holds without assuming (5.4.16).

(ii) There exists a bounded coefficient operator $\operatorname{Coef} : \mathcal{M}_{r,m} \to \ell_{r,m}(I)$ such that

$$\operatorname{Synth}_K \circ \operatorname{Coef} = \operatorname{id}_{\mathcal{M}_{r,m}}.$$

Proof. We first show part (i), i.e., the well-definedness and boundedness of the synthesis operator Synth_K. For this we recall the definition of $\operatorname{Synth}_W : \ell_{r,m}(I) \to L_{r,m}(X, \mu)$ in (5.4.15) and note that this operator is bounded. By Lemma 5.4.9 the defining series of Synth_W converges unconditionally in $L_{r,m}(X, \mu)$. Since bounded operators preserve unconditional convergence, Assumption 5.4.8 yields

$$K_{\mathfrak{F}}(\operatorname{Synth}_W c)(x) = K_{\mathfrak{F}}\left(\sum_{i \in I} c_i W(\cdot, x_i) \right)(x) = \sum_{i \in I} c_i K_{\mathfrak{F}}(W(\cdot, x_i))(x) = \sum_{i \in I} c_i \cdot K_{\mathfrak{F}}(x, x_i)$$

for all $x \in X$, with unconditional convergence of the series. This shows $\operatorname{Synth}_K = K_{\mathfrak{F}} \circ \operatorname{Synth}_W$ and since $K_{\mathfrak{F}}$ is a projection onto $\mathcal{M}_{r,m}$ this implies that the synthesis operator Synth_K is well-defined and bounded.

To prove (ii) we first show the following auxiliary result:

$$f = K_{\mathfrak{F}}(W(f)) \quad \text{for all } f \in \mathcal{M}_{r,m}. \tag{5.4.18}$$

By the assumptions on W we have

$$f(x) = K_{\mathfrak{F}} f(x) = \int_X K_{\mathfrak{F}}(x, y) f(y) \, d\mu(y)$$

$$= \int_X \left(\int_X K_{\mathfrak{F}}(x, z) W(z, y) \, d\mu(z) \right) f(y) \, d\mu(y) = (K_{\mathfrak{F}} \circ W) f(x) \tag{5.4.19}$$

for any $x \in X$. If we can thus justify the change of integration, we have shown (5.4.18). In other words, we intend to show the associativity of the multiplication of the kernel operators $K_{\mathfrak{F}}$ and W. Fix $x \in X$, $f \in \mathcal{M}_{r,m}$ and $s \in (r, \infty)$, then

$$\int_X \int_X |K_{\mathfrak{F}}(x,z)| \cdot |W(z,y)| \cdot |f(y)| \, d\mu(z) \, d\mu(y) = |K_{\mathfrak{F}}| \left(|W| \left(|f| \right) \right)(x).$$

Since $|W| \in \mathcal{A}_{1,w}$, we have $|W|(|f|) \in L_{r,m}(X, \mu)$ by Corollary 1.8.6. Furthermore, this implies $|K_{\mathfrak{F}}| \left(|W| \left(|f| \right) \right) \in L_{s,m}(X, \mu)$ by Corollary 1.8.9, as a function it is therefore bounded almost everywhere. By Fubini's theorem we can therefore interchange the integrals in (5.4.19) for almost all $x \in X$, which gives the desired reproducing formula (5.4.18).

Now let $\mathrm{Ana}_\Phi \colon L_{r,m}(X, \mu) \to \ell_{r,m}(I)$ and $\mathrm{Synth}_W \colon \ell_{r,m}(I) \to L_{r,m}(X, \mu)$ be defined as in Lemma 5.4.9, and set $A := \mathrm{Synth}_W \circ \mathrm{Ana}_\Phi \colon L_{r,m}(X, \mu) \to L_{r,m}(X, \mu)$. This is a well-defined and bounded operator and we intend to show

$$\|Wf - Af\|_{L_{r,m}} \leqslant \|W\|_{\mathcal{A}_{1,w}} \cdot \|f\|_{L_{r,m}}$$

for all $f \in L_{r,m}(X, \mu)$. We recall that the integral operator W acts as a bounded operator on $L_{r,m}(X, \mu)$ and therefore $Wf(x) < \infty$ for almost all $x \in X$. We fix such a $x \in X$ and some $f \in L_{r,m(X, \mu)}$ and obtain

$$|Wf(x) - Af(x)| = \left| \int_X W(x,y)f(y) \sum_{i \in I} \varphi_i(y) \, d\mu(y) - \sum_{i \in I} \langle f, \varphi_i \rangle_{L_2} \cdot W(x, x_i) \right|$$

$$= \left| \sum_{i \in I} \int_X W(x,y)f(y)\varphi_i(y) \, d\mu(y) - \sum_{i \in I} \int_X W(x, x_i)f(y)\varphi_i(y) \, d\mu(y) \right|$$

$$\leqslant \sum_{i \in I} \int_X |f(y)| \cdot \varphi_i(y) \cdot |W(x,y) - W(x, x_i)| \, d\mu(y).$$

By the properties of Φ we have $\mathrm{supp}\, \varphi_i \subseteq U_i$ for all $i \in I$, hence $y \in U_i$ for all $y \in X$ with $\varphi_i(y) \neq 0$. This means $|W(x,y) - W(x, x_i)| \leqslant \mathrm{osc}_\mathcal{U} W(x,y)$ for all $y \in \mathrm{supp}\, \varphi_i$. Combining this with the monotone convergence theorem, we get

$$|Wf(x) - Af(x)| \leqslant \sum_{i \in I} \int_X |f(y)| \cdot \phi_i(y) \cdot \mathrm{osc}_\mathcal{U} W(x,y) \, d\mu(y)$$

$$= \int_X |f(y)| \cdot \mathrm{osc}_\mathcal{U} W(x,y) \, d\mu(y) = \mathrm{osc}_\mathcal{U} W(f)(x).$$

Since this holds true for almost all $x \in X$, it implies

$$\|Wf - Af\|_{L_{r,m}} \leqslant \|\mathrm{osc}_\mathcal{U} W(f)\|_{L_{r,m}} \leqslant \|\mathrm{osc}_\mathcal{U} W\|_{\mathcal{A}_{1,w}} \cdot \|f\|_{L_{r,m}}.$$

Finally, we construct the coefficient operator $\mathrm{Coef} \colon \mathcal{M}_{r,m} \to \ell_{r,m}(I)$. For this define the operator

$$B := K_{\mathfrak{F}} \circ A|_{\mathcal{M}_{r,m}} \colon \mathcal{M}_{r,m} \to \mathcal{M}_{r,m},$$

which is well-defined since $K_{\mathfrak{F}}$ is a projection onto $\mathcal{M}_{r,m}$. Then, for any $f \in \mathcal{M}_{r,m}$ by the reproducing formula and the results above we have

$$\|f - Bf\|_{L_{r,m}} = \|K_{\mathfrak{F}}(Wf - Af)\|_{L_{r,m}} \leqslant \|K_{\mathfrak{F}}\|_{L_{r,m} \to L_{r,m}} \cdot \|W\|_{\mathcal{A}_{1,w}} \cdot \|f\|_{L_{r,m}} < \|f\|_{L_{r,m}}.$$

As before, a Neumann series argument yields that the operator $H := \sum_{n=0}^{\infty}(\mathrm{id}_{\mathcal{M}_{r,m}} - B)^n$ defines a bounded linear operator $H \colon \mathcal{M}_{r,m} \to \mathcal{M}_{r,m}$ such that

$$\mathrm{id}_{\mathcal{M}_{r,m}} = B \circ H = K_{\mathfrak{F}} \circ A|_{\mathcal{M}_{r,m}} \circ H = K_{\mathfrak{F}} \circ \mathrm{Synth}_W \circ \mathrm{Ana}_\Phi|_{\mathcal{M}_{r,m}} \circ H$$
$$= \mathrm{Synth}_K \circ \mathrm{Ana}_\Phi|_{\mathcal{M}_{r,m}} \circ H.$$

Thus, $\mathrm{Coef} := \mathrm{Ana}|_{\mathcal{M}_{r,m}} \circ H$ is the desired bounded linear coefficient operator.

Similarly to the proof of Proposition 5.4.10 it can be shown that the action of the operator Coef is independent of the choice of r, m.   □

Using the correspondence principle we can transfer the result above from the reproducing kernel space $\mathcal{M}_{r,m}$ to the generalized coorbit space $\mathrm{Co}(L_{r,m})$.

Theorem 5.4.11. *Under the same assumptions as in Proposition 5.4.10 the family of functions $(\psi_{x_i})_{i \in I} \subset \mathrm{Co}(L_{r,m})$ forms a family of atoms for $\mathrm{Co}(L_{r,m})$ with coefficient space $\ell_{r,m}(I)$.*

More precisely, the synthesis operator

$$\mathrm{Synth}_{\mathrm{Co}} \colon \ell_{r,m}(I) \to \mathrm{Co}(L_{r,m}), \quad c = (c_i)_{i \in I} \mapsto \sum_{i \in I} c_i \cdot \psi_{x_i}$$

is well-defined and bounded with unconditional convergence of the defining series, and there is a bounded linear coefficient operator $\mathrm{Coef}_{\mathrm{Co}} \colon \mathrm{Co}(L_{r,m}) \to \ell_{r,m}(I)$ which satisfies the identity $\mathrm{Synth}_{\mathrm{Co}} \circ \mathrm{Coef}_{\mathrm{Co}} = \mathrm{id}_{\mathrm{Co}(L_{r,m})}$.

Finally, the action of the coefficient operator $\mathrm{Coef}_{\mathrm{Co}}$ is independent of the choice of r, m. In other words, if the assumptions of the current theorem are satisfied for $L_{r_1, m_1}(X, \mu)$ and for $L_{r_2, m_2}(X, \mu)$ and if C_1, C_2 denote the corresponding coefficient operators, then $C_1 f = C_2 f$ for all $f \in \mathrm{Co}(L_{r_1, m_1}) \cap \mathrm{Co}(L_{r_2, m_2})$.

Proof. Since the extended voice transform $V_{e, \mathfrak{F}} \colon \mathrm{Co}(L_{r,m}) \to \mathcal{M}_{r,m}$ is an isometric isomorphism, see Proposition 5.1.9, it preserves unconditional convergence of series. By using the synthesis operator Synth_K from (5.4.17) we obtain

$$(V_{e, \mathfrak{F}}^{-1} \circ \mathrm{Synth}_K)(c) = V_{e, \mathfrak{F}}^{-1}\left(\sum_{i \in I} c_i \cdot V_{\mathfrak{F}} \psi_{x_i}\right) = \sum_{i \in I} c_i \cdot \psi_{x_i} = \mathrm{Synth}_{\mathrm{Co}}(c)$$

for all $c \in \ell_{r,m}(I)$, where we used $V_e \psi_{x_i} = K_{\mathfrak{F}}(\cdot, x_i)$. Thus the operator

$$\mathrm{Synth}_{\mathrm{Co}} = V_{e, \mathfrak{F}}^{-1} \circ \mathrm{Synth}_K \colon \ell_{r,m}(I) \to \mathrm{Co}(L_{r,m})$$

is well-defined and bounded, which proves (i).

To show (ii) recall the bounded operator $\mathrm{Coef} \colon \mathcal{M}_{r,m} \to \ell_{r,m}(I)$ from Proposition 5.4.10 and define

$$\mathrm{Coef}_{\mathrm{Co}} := \mathrm{Coef} \circ V_{e, \mathfrak{F}} \colon \mathrm{Co}(L_{r,m}) \to \ell_{r,m}(I).$$

Then, $\mathrm{Synth}_{\mathrm{Co}} \circ \mathrm{Coef}_{\mathrm{Co}} = V_{e, \mathfrak{F}}^{-1} \circ \mathrm{Synth}_K \circ \mathrm{Coef} \circ V_{e, \mathfrak{F}} = V_{e, \mathfrak{F}}^{-1} \circ V_{e, \mathfrak{F}} = \mathrm{id}_{\mathrm{Co}(L_{r,m})}$ as desired. Since the action of Coef is independent of the choice of r, m, the same applies to $\mathrm{Coef}_{\mathrm{Co}}$.   □

5.5 An Example: Inhomogeneous Shearlet Coorbit Spaces

This final section is dedicated to an example of an application of the theory above, namely the so-called inhomogeneous shearlet coorbit spaces. As mentioned before, the main idea is to merge Subsection 2.3.2 and Section 4.3 and define function spaces via coorbit theory that are based on the shearlet transform but possess an inhomogeneous structure similar to the inhomogeneous Besov spaces. We will show in Theorem 5.5.13 that the corresponding reproducing kernel is contained in all weighted Lebesgue spaces $L_{q,w}(X,\mu)$ for $1 < q < \infty$, so the theory developed in this chapter is applicable. Theorem 5.5.22 further shows that oscillations of the reproducing kernel with respect to certain coverings are integrable as well, which paves the way for a first discretization result of these spaces.

Let $d \geqslant 2$ be fixed in this section and $\mathcal{H} = L_2(\mathbb{R}^d)$. For $a \in \mathbb{R}^*$ and $s = (s_1,\ldots,s_{d-1}) \in \mathbb{R}^{d-1}$ we recall the *parabolic dilation matrix*, or *scaling matrix*, $A_a \in \mathbb{R}^{d\times d}$ given by

$$A_a := \begin{pmatrix} a & 0_{d-1}^T \\ 0_{d-1} & \mathrm{sgn}(a)|a|^{\frac{1}{d}}I_{d-1} \end{pmatrix} \tag{5.5.1}$$

and for $s = (s_1,\ldots,s_{d-1}) \in \mathbb{R}^{d-1}$ the *shear matrix* $S_s \in \mathbb{R}^{d\times d}$ is defined via

$$S_s := \begin{pmatrix} 1 & s^T \\ 0_{d-1} & I_{d-1} \end{pmatrix}. \tag{5.5.2}$$

The inverses of these matrices are given by

$$A_a^{-1} = A_{a^{-1}} = \begin{pmatrix} a^{-1} & 0_{d-1}^T \\ 0_{d-1} & \mathrm{sgn}(a)|a|^{-\frac{1}{d}}I_{d-1} \end{pmatrix} \quad \text{and} \quad S_s^{-1} = S_{-s} = \begin{pmatrix} 1 & -s^T \\ 0_{d-1} & I_{d-1} \end{pmatrix}, \tag{5.5.3}$$

see also (2.3.20). As shown in (2.3.19) the multiplication of dilation and shear matrices is not commutative and it is

$$S_s A_a S_{s'} A_{a'} = S_{s+|a|^{1-\frac{1}{d}}s'} A_{aa'} \quad \text{for all } a, a' \in \mathbb{R}^*, \ s, s' \in \mathbb{R}^{d-1}. \tag{5.5.4}$$

Using the dilation and shear matrices the set $\mathbb{S} = \mathbb{R}^* \times \mathbb{R}^{d-1} \times \mathbb{R}^d$ endowed with the operation

$$(a,s,t) \circ (a',s',t') := (aa', s + |a|^{1-\frac{1}{d}}s', t + S_s A_a t'), \quad (a,s,t), (a',s',t') \in \mathbb{S}$$

forms a non-abelian locally compact topological group, the so-called *full shearlet group*, see Lemma 2.3.5.

5.5.1 Inhomogeneous Shearlet Frame

Using the group $\mathbb{S}$ above we will define a frame for $L_2(\mathbb{R}^d)$ which we call the *inhomogeneous shearlet frame*. The main idea for the construction of this frame is to transfer the construction of the *inhomogeneous Besov frame* in Section 4.3 to the shearlet setting. Since the full shearlet group $\mathbb{S}$ defined above and the full affine group $\mathbb{A}$ in Subsection 2.3.1 are very similar, this transfer seems natural.

As a first step we choose a closed subset of $\mathbb{S}$ by restricting the dilation parameter to the set $[-1,1]^*$, thereby only covering the higher frequence content of a signal. To analyze the polynomial and lower frequency part of a signal a second function is introduced, which is not dilated. We therefore choose the parameter set X as follows.

Definition 5.5.1. Let

$$X := \left(\{\infty\} \times \mathbb{R}^{d-1} \times \mathbb{R}^d\right) \cup \left([-1,1]^* \times \mathbb{R}^{d-1} \times \mathbb{R}^d\right) \tag{5.5.5}$$

with ∞ representing an isolated point in $\mathbb{R}$ and $[-1,1]^* = [-1,1]\backslash\{0\}$ as usual. Furthermore let the measure μ on X be defined via

$$\int_X F(x)\,d\mu(x) := \int_{\mathbb{R}^d}\int_{\mathbb{R}^{d-1}} F(\infty,s,t)\,dt + \int_{\mathbb{R}^d}\int_{\mathbb{R}^{d-1}}\int_{-1}^1 F(a,s,t)\,\frac{da}{|a|^{d+1}}\,ds\,dt \tag{5.5.6}$$

for any complex valued function F on X which is measurable with respect to the Borel σ-algebra.

The measure μ above is indeed a Radon measure with supp $\mu = X$. By the properties of $\mathbb{S}$, X is locally compact and also a σ-finite and σ-compact Hausdorff space. Therefore all assumptions posed in Section 5.1 are fulfilled.

By comparing (5.5.6) with (2.3.21), it is easy to see that the measure μ on X is heavily inspired by the Haar measure of the shearlet group $\mathbb{S}$. Next, we define how an element $x \in X$ acts on a function on $\mathbb{R}^d$. For this we need *translation, shearing* and *dilation* operators acting on a function f on $\mathbb{R}^d$ by

$$T_t f(y) = f(y - t), \tag{5.5.7}$$

$$D_s f(y) = f(S_s^{-1} y), \tag{5.5.8}$$

$$D_a f(y) = |\det A_a|^{-1/2} f(A_a^{-1} y), \tag{5.5.9}$$

for all $y \in \mathbb{R}^d$. It is easy to verify that these three operators are isometries on $L_2(\mathbb{R}^d)$. Let us quickly recall the definition of an *admissible shearlet* $\Psi \in L_2(\mathbb{R}^d)$. By (2.3.28), Ψ is admissible if and only if the condition

$$0 < C_\Psi := \int_{\mathbb{R}^d} \frac{|\widehat{\Psi}(\omega)|^2}{|\omega_1|^d} \, d\omega < \infty \tag{5.5.10}$$

is fulfilled.

Definition 5.5.2. Let $\Phi, \Psi \in L_2(\mathbb{R}^d)$ with Ψ being an *admissible shearlet*. Then we define the family $\mathfrak{F} := \{\psi_x\}_{x \in X}$ via

$$\psi_{(\infty,s,t)} := T_t D_s \Phi = \Phi(S_s^{-1}(\cdot - t)) \quad \text{and} \tag{5.5.11}$$

$$\psi_{(a,s,t)} := T_t D_s D_a \Psi = |\det A_a|^{-1/2} \Psi(A_a^{-1} S_s^{-1}(\cdot - t)). \tag{5.5.12}$$

The main theorem of this subsection states that $\mathfrak{F}$ is indeed a continuous frame for $L_2(\mathbb{R}^d)$ in the sense of Definition 1.7.1, and under certain assumptions we even have a Parseval frame. Before we state the theorem we recall the following technical identities that we will use time and time again in this section. We omit the proof since it is an easy generalization of (2.3.24) and (2.3.25).

Lemma 5.5.3. *For all* $(a, s, t) \in X$ *with* $a \in [-1, 1]^*$ *or* $a = \infty$ *and* $f, \psi \in L_2(\mathbb{R}^d)$ *we have*

$$\langle f, \psi_{(a,s,t)} \rangle_{L_2} = (f * \psi_{(a,s,0)}^*)(t). \tag{5.5.13}$$

Furthermore, for any $a \in \mathbb{R}^*$, $s \in \mathbb{R}^{d-1}$ *and* $t \in \mathbb{R}^d$ *as well as* $\xi \in \mathbb{R}^d$ *we have*

$$\mathcal{F}(D_s f)(\xi) = \widehat{f}(S_s^T \xi) \quad \text{and} \tag{5.5.14}$$

$$\mathcal{F}(D_s D_a f)(\xi) = |\det A_a|^{1/2} \widehat{f}(A_a S_s^T \xi). \tag{5.5.15}$$

The following theorem is the main statement of this subsection and identifies conditions on Φ and Ψ for $\mathfrak{F}$ being a frame for $L_2(\mathbb{R}^d)$.

Theorem 5.5.4. *Let* $\Psi \in L_1(\mathbb{R}^d) \cap L_2(\mathbb{R}^d)$ *be an admissible shearlet and let* $\Phi \in L_1(\mathbb{R}^d) \cap L_2(\mathbb{R}^d)$ *such that there exist constants* $0 < A \leqslant B < \infty$ *with*

$$A \leqslant \int_{\mathbb{R}^{d-1}} \frac{|\widehat{\Phi}(y, \sigma)|^2}{|y|^{d-1}} \, d\sigma + \int_{\mathbb{R}^{d-1}} \int_{-|y|}^{|y|} \frac{|\widehat{\Psi}(\xi_1, \widetilde{\xi})|^2}{|\xi_1|^d} \, d\xi_1 \, d\widetilde{\xi} \leqslant B \tag{5.5.16}$$

for almost every $y \in \mathbb{R}$. Then the family $\mathfrak{F} = (\psi_x)_{x \in X}$ constitutes a continuous frame *of $L_2(\mathbb{R}^d)$, i.e.,*

$$A \cdot \|f\|_{L_2}^2 \leqslant \int_X |\langle f, \psi_x \rangle_{L_2}|^2 \, d\mu(x) \leqslant B \cdot \|f\|_{L_2}^2 \quad \text{for all } f \in L_2(\mathbb{R}^d). \tag{5.5.17}$$

If $A = B$ then $\mathfrak{F}$ is a tight frame *and if $A = B = 1$ then $\mathfrak{F}$ is a* Parseval frame *of $L_2(\mathbb{R}^d)$. We call $\mathfrak{F}$ an* inhomogeneous shearlet frame.

Proof. We first note that the mapping $x \to |\langle f, \psi_x \rangle_{L_2}|^2$ is measurable. By applying (5.5.6), Fubini's and Plancherel's theorem we obtain

$$\int_X |\langle f, \psi_x \rangle_{L_2}|^2 \, d\mu(x)$$
$$= \int_{\mathbb{R}^d} \int_{\mathbb{R}^{d-1}} |\langle f, \psi_{(\infty,s,t)} \rangle_{L_2}|^2 \, ds \, dt + \int_{\mathbb{R}^d} \int_{\mathbb{R}^{d-1}} \int_{-1}^1 |\langle f, \psi_{(a,s,t)} \rangle_{L_2}|^2 \, \frac{da}{|a|^{d+1}} \, ds \, dt$$
$$= \int_{\mathbb{R}^{d-1}} \int_{\mathbb{R}^d} |\langle f, \psi_{(\infty,s,t)} \rangle_{L_2}|^2 \, dt \, ds + \int_{\mathbb{R}^{d-1}} \int_{-1}^1 \int_{\mathbb{R}^d} |\langle f, \psi_{(a,s,t)} \rangle_{L_2}|^2 \, dt \, \frac{da}{|a|^{d+1}} \, ds$$
$$= \int_{\mathbb{R}^{d-1}} \|\langle f, \psi_{(\infty,s,\cdot)} \rangle_{L_2}\|_{L_2(\mathbb{R}^d)}^2 \, ds + \int_{\mathbb{R}^{d-1}} \int_{-1}^1 \|\langle f, \psi_{(a,s,\cdot)} \rangle_{L_2}\|_{L_2(\mathbb{R}^d)}^2 \, \frac{da}{|a|^{d+1}} \, ds$$
$$= \int_{\mathbb{R}^{d-1}} \|\mathcal{F}(\langle f, \psi_{(\infty,s,\cdot)} \rangle_{L_2})\|_{L_2(\mathbb{R}^d)}^2 \, ds + \int_{\mathbb{R}^{d-1}} \int_{-1}^1 \|\mathcal{F}(\langle f, \psi_{(a,s,\cdot)} \rangle_{L_2})\|_{L_2(\mathbb{R}^d)}^2 \, \frac{da}{|a|^{d+1}} \, ds$$
$$= \int_{\mathbb{R}^{d-1}} \int_{\mathbb{R}^d} |\mathcal{F}(\langle f, \psi_{(\infty,s,\cdot)} \rangle_{L_2})(\xi)|^2 \, d\xi \, ds + \int_{\mathbb{R}^{d-1}} \int_{-1}^1 \int_{\mathbb{R}^d} |\mathcal{F}(\langle f, \psi_{(a,s,\cdot)} \rangle_{L_2})(\xi)|^2 \, d\xi \, \frac{da}{|a|^{d+1}} \, ds.$$

Using Lemma 5.5.3, Fubini's theorem and the fact that $\mathcal{F}(f * g) = \hat{f} \cdot \hat{g}$ leads to

$$\int_X |\langle f, \psi_x \rangle_{L_2}|^2 \, d\mu(x)$$
$$= \int_{\mathbb{R}^{d-1}} \int_{\mathbb{R}^d} |\mathcal{F}(f * \psi_{(\infty,s,0)}^*)(\xi)|^2 \, d\xi \, ds + \int_{\mathbb{R}^{d-1}} \int_{-1}^1 \int_{\mathbb{R}^d} |\mathcal{F}(f * \psi_{(a,s,0)}^*)(\xi)|^2 \, d\xi \, \frac{da}{|a|^{d+1}} \, ds$$
$$= \int_{\mathbb{R}^{d-1}} \int_{\mathbb{R}^d} |\hat{f}(\xi)|^2 \cdot |\mathcal{F}(\psi_{(\infty,s,0)}^*)(\xi)|^2 \, d\xi \, ds + \int_{\mathbb{R}^{d-1}} \int_{-1}^1 \int_{\mathbb{R}^d} |\hat{f}(\xi)|^2 \cdot |\mathcal{F}(\psi_{(a,s,0)}^*)(\xi)|^2 \, d\xi \, \frac{da}{|a|^{d+1}} \, ds$$
$$= \int_{\mathbb{R}^d} |\hat{f}(\xi)|^2 \cdot \left(\int_{\mathbb{R}^{d-1}} |\mathcal{F}(\psi_{(\infty,s,0)})(\xi)|^2 \, ds + \int_{\mathbb{R}^{d-1}} \int_{-1}^1 |\mathcal{F}(\psi_{(a,s,0)})(\xi)|^2 \, \frac{da}{|a|^{d+1}} \, ds \right) d\xi.$$

Thus, if we can prove

$$A \leqslant \int_{\mathbb{R}^{d-1}} |\mathcal{F}(\psi_{(\infty,s,0)})(\xi)|^2 \, ds + \int_{\mathbb{R}^{d-1}} \int_{-1}^1 |\mathcal{F}(\psi_{(a,s,0)})(\xi)|^2 \, \frac{da}{|a|^{d+1}} \, ds \leqslant B \tag{5.5.18}$$

for almost every $\xi \in \mathbb{R}^d$, the assertion follows, since then

$$\int_X |\langle f, \psi_x \rangle_{L_2}|^2 d\mu(x) \leqslant B \cdot \int_{\mathbb{R}^d} |\hat{f}(\xi)|^2 \, d\xi = B \cdot \|\hat{f}\|_{L_2(\mathbb{R}^d)}^2 = B \cdot \|f\|_{L_2(\mathbb{R}^d)}^2$$

and the same inequality holds true for the lower bound in (5.5.17). Hence, it remains to show (5.5.18).

Assuming $\xi_1 \neq 0$ we use Lemma 5.5.3 to obtain

$$\int_{\mathbb{R}^{d-1}} |\mathcal{F}(\psi_{(\infty,s,0)})(\xi)|^2 \, \mathrm{d}s + \int_{\mathbb{R}^{d-1}} \int_{-1}^{1} |\mathcal{F}(\psi_{(a,s,0)})(\xi)|^2 \frac{\mathrm{d}a}{|a|^{d+1}} \, \mathrm{d}s$$

$$= \int_{\mathbb{R}^{d-1}} |\mathcal{F}(D_{S_s}\Phi)(\xi)|^2 \, \mathrm{d}s + \int_{\mathbb{R}^{d-1}} \int_{-1}^{1} |\mathcal{F}(D_{S_s} D_{A_a}\Psi)(\xi)|^2 \frac{\mathrm{d}a}{|a|^{d+1}} \, \mathrm{d}s$$

$$= \int_{\mathbb{R}^{d-1}} |\widehat{\Phi}(S_s^T \xi)|^2 \, \mathrm{d}s + \int_{\mathbb{R}^{d-1}} \int_{-1}^{1} |\det A_a| \cdot |\widehat{\Psi}(A_a S_s^T \xi)|^2 \frac{\mathrm{d}a}{|a|^{d+1}} \, \mathrm{d}s$$

$$= \int_{\mathbb{R}^{d-1}} |\widehat{\Phi}(\xi_1, \widetilde{\xi} + \xi_1 s)|^2 \, \mathrm{d}s + \int_{\mathbb{R}^{d-1}} \int_{-1}^{1} |\det A_a| \cdot |\widehat{\Psi}(a\xi_1, \mathrm{sgn}(a)|a|^{1/d}(\widetilde{\xi} + \xi_1 s))|^2 \frac{\mathrm{d}a}{|a|^{d+1}} \, \mathrm{d}s,$$

using the notation $\xi = (\xi_1, \widetilde{\xi})^T$, $\widetilde{\xi} \in \mathbb{R}^{d-1}$. Substituting $\sigma := \widetilde{\xi} + \xi_1 s$ and $\theta = (\theta_1, \widetilde{\theta}) := (a\xi_1, \mathrm{sgn}(a)|a|^{\frac{1}{d}}(\widetilde{\xi} + \xi_1 s))$, we end up with

$$\int_{\mathbb{R}^{d-1}} |\mathcal{F}(\psi_{(\infty,s,0)})(\xi)|^2 \, \mathrm{d}s + \int_{\mathbb{R}^{d-1}} \int_{-1}^{1} |\mathcal{F}(\psi_{(a,s,0)})(\xi)|^2 \frac{\mathrm{d}a}{|a|^{d+1}} \, \mathrm{d}s$$

$$= \int_{\mathbb{R}^{d-1}} |\xi_1|^{-(d-1)} \cdot |\widehat{\Phi}(\xi_1, \sigma)|^2 \, \mathrm{d}\sigma + \int_{\mathbb{R}^{d-1}} \int_{-|\xi_1|}^{-|\xi_1|} |\theta_1|^{-d} \cdot |\widehat{\Psi}(\theta_1, \widetilde{\theta})|^2 \, \mathrm{d}\theta_1 \, \mathrm{d}\widetilde{\theta}$$

$$= \int_{\mathbb{R}^{d-1}} \frac{|\widehat{\Phi}(\xi_1, \sigma)|^2}{|\xi_1|^{d-1}} \, \mathrm{d}\sigma + \int_{\mathbb{R}^{d-1}} \int_{-|\xi_1|}^{|\xi_1|} \frac{|\widehat{\Psi}(\theta_1, \widetilde{\theta})|^2}{|\theta_1|^d} \, \mathrm{d}\theta_1 \, \mathrm{d}\widetilde{\theta},$$

and (5.5.18) follows from assumption (5.5.16). $\qquad\square$

We add a remark concerning the practical construction of functions Φ and Ψ satisfying the assumption (5.5.16).

Remark 5.5.5. For a given shearlet Ψ it is still necessary to show that one can satisfy condition (5.5.16) for some function $\Phi \in L_1(\mathbb{R}^d) \cap L_2(\mathbb{R}^d)$. To this end we restrict ourselves to odd dimensions, i.e., $\frac{d-1}{2} \in \mathbb{N}_0$, since then the map $\xi \mapsto \xi^{\frac{d-1}{2}}$ is smooth. Now we define $\widehat{\Phi}: \mathbb{R}^d \to \mathbb{C}$ for almost all $\xi \in \mathbb{R}^d$ by

$$\widehat{\Phi}(\xi) := \xi_1^{\frac{d-1}{2}} \left[\int_{\mathbb{R}\setminus[-|\xi_1|,|\xi_1|]} \frac{|\widehat{\Psi}(\omega_1, \widetilde{\xi})|^2}{|\omega_1|^d} \, \mathrm{d}\omega_1 \right]^{1/2}. \tag{5.5.19}$$

If we assume Ψ to be admissible in the sense of (5.5.10) with admissibility constant $0 < C_\Psi < \infty$ it is straightforward to see that Φ fulfills (5.5.16) with $A = B = C_\Psi$. Moreover, $\Phi \in L_2(\mathbb{R}^d)$ is immediate and $\Phi \in L_1(\mathbb{R}^d)$ can be shown if $\widehat{\Psi} \in \mathcal{C}_0^\infty(\mathbb{R}^d)$ is chosen suitable, see Example 5.5.14 of Subsection 5.5.3.

It is therefore sufficient to fix an admissible shearlet Ψ and let Φ be defined via (5.5.19). In the following we will always assume this construction.

Having shown in Theorem 5.5.4 that when assuming (5.5.16) for $A = B = 1$ the family $\mathfrak{F} = (\psi_x)_{x \in X}$ is a Parseval frame, we can now define the *voice transform* associated with $\mathfrak{F}$ following the lines of Definition 5.1.1.

Definition 5.5.6. Let Φ and Ψ satisfy the assumptions from Theorem 5.5.4 and let $\mathfrak{F} = (\psi_x)_{x \in X}$ be a Parseval frame given by Definition 5.5.2. Then the *inhomogeneous shearlet transform* $\mathcal{SH}_{\mathfrak{F}}$ associated with $\mathfrak{F}$ is defined as

$$\mathcal{SH}_{\mathfrak{F}}: L_2(\mathbb{R}^d) \to L_2(X, \mu), \quad \mathcal{SH}_{\mathfrak{F}}f(x) := \langle f, \psi_x \rangle_{L_2}. \tag{5.5.20}$$

Correspondingly, the *reproducing kernel* $K_{\mathfrak{F}}$ is the map

$$K_{\mathfrak{F}}: X \times X \to \mathbb{C}, \quad K_{\mathfrak{F}}(x, y) := \mathcal{SH}_{\mathfrak{F}}\psi_y(x) = \langle \psi_y, \psi_x \rangle_{L_2}. \tag{5.5.21}$$

5.5.2 Conditions on the Reproducing Kernel

The next step towards the definition of coorbit spaces $\mathrm{Co}(L_{r,m})$, $1 \leqslant r < \infty$, associated with the inhomogeneous shearlet frame is to show that Assumption 5.1.2 is fulfilled for certain weights w on X. Recall that by Remark 5.1.10 this is sufficient for Proposition 5.1.9 to hold, which states that the coorbit spaces in Definition 5.1.8 are well-defined Banach spaces. Therefore we will show the kernel $K_{\mathfrak{F}}$ is contained in the kernel spaces $\mathcal{A}_{q,w}$ for all $1 < q < \infty$ and suitable weights w.

Before we show this we need a couple of preparing results. We start by showing the following preparing lemma.

Lemma 5.5.7. *Let* $a, a' \in [-1,1]^*$, $s, s' \in \mathbb{R}^{d-1}$, $t, t' \in \mathbb{R}^d$ *and*

$$\varphi_{(a,s,t)} := |\det A_a|^{-1/2}\Phi(A_a^{-1}S_s^{-1}(\cdot - t)).$$

Then, the following equations hold true:

$$|\langle \psi_{(\infty,s,t)}, \psi_{(\infty,s',t')}\rangle_{L_2}| = |(\mathcal{SH}_{\mathfrak{F}}\Phi)(\infty, s - s', S_{s'}^{-1}(t - t'))|, \tag{5.5.22}$$

$$|\langle \psi_{(\infty,s,t)}, \psi_{(a',s',t')}\rangle_{L_2}| = |\langle \Psi, \varphi_{(a'^{-1},|a'|^{\frac{1}{d}-1}(s-s'),A_{a'}^{-1}S_{s'}^{-1}(t-t'))}\rangle_{L_2}|, \tag{5.5.23}$$

$$|\langle \psi_{(a,s,t)}, \psi_{(\infty,s',t')}\rangle_{L_2}| = |(\mathcal{SH}_{\mathfrak{F}}\Phi)(a, s - s', S_{s'}^{-1}(t - t'))|, \tag{5.5.24}$$

$$|\langle \psi_{(a,s,t)}, \psi_{(a',s',t')}\rangle_{L_2}| = |(\mathcal{SH}_{\mathfrak{F}}\Psi)(aa'^{-1}, |a'|^{\frac{1}{d}-1}(s - s'), A_{a'}^{-1}S_{s'}^{-1}(t - t'))|. \tag{5.5.25}$$

Proof. We only state the proof for (5.5.25) in detail, (5.5.22)–(5.5.24) can be proven analogously.

We recall the translation operator T_t from (5.5.7) and introduce the operator D_A defined via $D_A f(x) := |\det A|^{-1/2} f(A^{-1}x)$ for any invertible matrix $A \in \mathbb{R}^{d \times d}$. Obviously, this generalizes the definition of D_a and D_s from (5.5.1) and (5.5.8), respectively. We now observe that these operators satisfy the following properties. For this we let $f, g \in L_2$, then

$$\langle T_t f, g\rangle_{L_2} = \int_{\mathbb{R}^d} f(x - t)\,\overline{g(x)}\,\mathrm{d}x = \int_{\mathbb{R}^d} f(y)\,\overline{g(y + t)}\,\mathrm{d}y = \langle f, T_{-t}g\rangle_{L_2}$$

for any $t \in \mathbb{R}^d$. Similarly,

$$\begin{aligned}
\langle D_A f, g\rangle_{L_2} &= |\det A|^{-1/2} \int_{\mathbb{R}^d} f(A^{-1}x)\,\overline{g(x)}\,\mathrm{d}x \\
&= |\det A^{-1}|^{-1/2} \int_{\mathbb{R}^d} f(y)\,\overline{g(Ay)}\,\mathrm{d}y \\
&= \langle f, D_{A^{-1}}g\rangle_{L_2}
\end{aligned}$$

for any $A \in \mathrm{GL}(\mathbb{R}^d)$. Moreover, for two matrices $A, B \in \mathrm{GL}(\mathbb{R}^d)$ and $t \in \mathbb{R}^d$ it is easy to see that

$$\begin{aligned}
D_A D_B f(x) &= |\det A|^{-1/2}|\det B|^{-\frac{1}{2}} f(B^{-1}A^{-1}x) \\
&= |\det AB|^{-1/2} f((AB)^{-1}x) \\
&= D_{AB}f(x)
\end{aligned}$$

and

$$\begin{aligned}
D_A T_t f(x) &= |\det A|^{-1/2} f(A^{-1}x - t) \\
&= |\det A|^{-1/2} f(A^{-1}(x - At)) \\
&= T_{At} D_A f(x).
\end{aligned}$$

Furthermore, for $a \in \mathbb{R}^*$ and $s \in \mathbb{R}^{d-1}$ the relation

$$A_a S_s = S_{|a|^{1-\frac{1}{d}}} A_a$$

holds. With this at hand and using $A_a^{-1} = A_{a^{-1}}$, $S_s^{-1} = S_{-s}$ it follows that

$$
\begin{aligned}
|\langle \psi_{(a,s,t)}, \psi_{(a',s',t')} \rangle_{L_2}| &= |\langle T_t D_{S_s} D_{A_a} \Psi, T_{t'} D_{S_{s'}} D_{A_{a'}} \Psi \rangle_{L_2}| \\
&= |\langle D_{A_{a'-1}} D_{S_{-s'}} T_{-t'} T_t D_{S_s} D_{A_a} \Psi, \Psi \rangle_{L_2}| \\
&= |\langle D_{A_{a'-1} S_{-s'}} T_{t-t'} D_{S_s A_a} \Psi, \Psi \rangle_{L_2}| \\
&= |\langle T_{A_{a'-1} S_{-s'}(t-t')} D_{A_{a'-1} S_{s-s'} A_a} \Psi, \Psi \rangle_{L_2}| \\
&= |\langle T_{A_{a'-1} S_{-s'}(t-t')} D_{S_{|a'|^{\frac{1}{d}-1}(s-s')}} D_{aa'^{-1}} \Psi, \Psi \rangle_{L_2}| \\
&= |(\mathcal{SH}\Psi)(aa'^{-1}, |a'|^{\frac{1}{d}-1}(s-s'), A_{a'}^{-1} S_{s'}^{-1}(t-t'))|,
\end{aligned}
$$

which proves (5.5.25). $\qquad\square$

Using the result above we can show the following lemma concerning the $\mathcal{A}_{q,w}$-norm of the kernel $K_{\mathfrak{F}}$ for certain weights w, see Definition 1.8.2 for the definition of the kernel space $\mathcal{A}_{q,w}$.

Lemma 5.5.8. *Let $K_{\mathfrak{F}}$ be the kernel associated to the inhomogeneous shearlet frame defined as in Definition 5.5.6 and assume the symmetric weight $w\colon X \times X \to \mathbb{R}_+$ to only depend on $a \in \{\infty\} \cup [-1,1]^*$, i.e., $w((a,s,t),(a',s',t')) = w(a,a') = w(a',a)$ for all $(a,s,t),(a',s',t') \in X$. Then for all $1 < q < \infty$ the following identity holds true:*

$$
\begin{aligned}
\|K_{\mathfrak{F}}\|_{\mathcal{A}_{q,w}}^q &= \operatorname*{ess\,sup}_{(\alpha,\sigma,\tau)\in X} \int_X |K_{\mathfrak{F}}((\alpha,\sigma,\tau),(a,s,t))|^q \cdot w((\alpha,\sigma,\tau),(a,s,t))^q \, d\mu(a,s,t) \\
&= \max\left\{ \int_{\mathbb{R}^d} \int_{\mathbb{R}^{d-1}} \left(w(\infty,\infty)^q \cdot |\langle \Phi, \psi_{(\infty,s',t')}\rangle_{L_2}|^q \right. \right. \\
&\qquad\qquad\qquad + \left. \int_{-1}^1 w(\infty,a)^q \cdot |\langle \Phi, \psi_{(a,s',t')}\rangle_{L_2}|^q \, \frac{da}{|a|^{d+1}} \right) ds'\, dt', \\
&\qquad \operatorname*{ess\,sup}_{\alpha\in[-1,1]^*} \int_{\mathbb{R}^d} \int_{\mathbb{R}^{d-1}} \left(w(\alpha,\infty)^q \cdot |\langle \Phi, \psi_{(\alpha,s',t')}\rangle_{L_2}|^q \right. \\
&\qquad\qquad\qquad + \left.\left. \int_{-|\alpha|^{-1}}^{|\alpha|^{-1}} w(\alpha,\alpha a')^q \cdot |\langle \Psi, \psi_{(a',s',t')}\rangle_{L_2}|^q \, \frac{da'}{|a'|^{d+1}} \right) ds'\, dt' \right\}.
\end{aligned}
$$

(5.5.26)

Proof. The first identity is a consequence of the symmetries $|K_{\mathfrak{F}}(x,y)| = |K_{\mathfrak{F}}(y,x)|$ and $w(x,y) = w(y,x)$ for all $x,y \in X$.

To prove the second equality fix $(\alpha,\sigma,\tau) \in X$. Using (5.5.22) if $\alpha = \infty$ and (5.5.24) if $\alpha \in [-1,1]^*$ we obtain

$$
\begin{aligned}
\int_{\mathbb{R}^d} \int_{\mathbb{R}^{d-1}} & w(\alpha,\infty)^q \cdot |\langle \psi_{(\infty,s,t)}, \psi_{(\alpha,\sigma,\tau)}\rangle_{L_2}|^q \, ds\, dt \\
&= \int_{\mathbb{R}^d} \int_{\mathbb{R}^{d-1}} w(\alpha,\infty)^q \cdot |\langle \Phi, \psi_{(\alpha,\sigma-s,S_s^{-1}(\tau-t))}\rangle_{L_2}|^q \, ds\, dt.
\end{aligned}
$$

Substituting $s' = \sigma - s$ and $t' = S_{\sigma-s'}^{-1}(\tau - t)$ then leads to

$$
\int_{\mathbb{R}^d} \int_{\mathbb{R}^{d-1}} w(\alpha, \infty)^q \cdot |\langle \psi_{(\infty,s,t)}, \psi_{(\alpha,\sigma,\tau)}\rangle_{L_2}|^q \, \mathrm{d}s \, \mathrm{d}t
$$

$$
= \int_{\mathbb{R}^d} \int_{\mathbb{R}^{d-1}} w(\alpha, \infty)^q \cdot \left|\langle \Phi, \psi_{(\alpha,s',S_{\sigma-s'}^{-1}(\tau-t))}\rangle_{L_2}\right|^q \, \mathrm{d}s' \, \mathrm{d}t \qquad (5.5.27)
$$

$$
= \int_{\mathbb{R}^d} \int_{\mathbb{R}^{d-1}} w(\alpha, \infty)^q \cdot |\langle \Phi, \psi_{(\alpha,s',t')}\rangle_{L_2}|^q \, \mathrm{d}s' \, \mathrm{d}t'.
$$

Analogously, we see that

$$
\int_{\mathbb{R}^d} \int_{\mathbb{R}^{d-1}} \int_{-1}^{1} w(\infty, a)^q \cdot |\langle \psi_{(a,s,t)}, \psi_{(\infty,\sigma,\tau)}\rangle|^q \, \frac{\mathrm{d}a}{|a|^{d+1}} \, \mathrm{d}s \, \mathrm{d}t
$$

$$
= \int_{\mathbb{R}^d} \int_{\mathbb{R}^{d-1}} \int_{-1}^{1} w(\infty, a)^q \cdot |\langle \Phi, \psi_{(a,s',t')}\rangle_{L_2}|^q \, \frac{\mathrm{d}a}{|a|^{d+1}} \, \mathrm{d}s' \, \mathrm{d}t' \qquad (5.5.28)
$$

for any $\sigma \in \mathbb{R}^{d-1}$ and $\tau \in \mathbb{R}^d$. Now fix $\alpha \in [-1,1]^*$, then (5.5.25) and the substitution $a' := a\alpha^{-1}$ yields

$$
\int_{\mathbb{R}^d} \int_{\mathbb{R}^{d-1}} \int_{-1}^{1} w(\alpha, a)^q \cdot |\langle \psi_{(a,s,t)}, \psi_{(\alpha,\sigma,\tau)}\rangle_{L_2}|^q \, \frac{\mathrm{d}a}{|a|^{d+1}} \, \mathrm{d}s \, \mathrm{d}t
$$

$$
= \int_{\mathbb{R}^d} \int_{\mathbb{R}^{d-1}} \int_{-1}^{1} w(\alpha, a)^q \cdot \left|\left\langle \Psi, \psi_{(a\alpha^{-1},|\alpha|^{\frac{1}{d}-1}(s-\sigma),A_\alpha^{-1}S_\sigma^{-1}(t-\tau))}\right\rangle_{L_2}\right|^q \, \frac{\mathrm{d}a}{|a|^{d+1}} \, \mathrm{d}s \, \mathrm{d}t
$$

$$
= \int_{\mathbb{R}^d} \int_{\mathbb{R}^{d-1}} \int_{-|\alpha|^{-1}}^{|\alpha|^{-1}} w(\alpha, \alpha a')^q \cdot \left|\left\langle \Psi, \psi_{(a',|\alpha|^{\frac{1}{d}-1}(s-\sigma),A_\alpha^{-1}S_\sigma^{-1}(t-\tau))}\right\rangle_{L_2}\right|^q \, \frac{1}{|\alpha|^d} \, \frac{\mathrm{d}a'}{|a'|^{d+1}} \, \mathrm{d}s \, \mathrm{d}t.
$$

Finally, we substitute $s' := |\alpha|^{\frac{1}{d}-1}(s-\sigma)$ and $t' := A_\alpha^{-1}S_\sigma^{-1}(t-\tau)$ and obtain

$$
\int_{\mathbb{R}^d} \int_{\mathbb{R}^{d-1}} \int_{-1}^{1} w(\alpha, a)^q \cdot |\langle \psi_{(a,s,t)}, \psi_{(\alpha,\sigma,\tau)}\rangle_{L_2}|^q \, \frac{\mathrm{d}a}{|a|^{d+1}} \, \mathrm{d}s \, \mathrm{d}t
$$

$$
= \int_{\mathbb{R}^d} \int_{\mathbb{R}^{d-1}} \int_{-|\alpha|^{-1}}^{|\alpha|^{-1}} w(\alpha, \alpha a')^q \cdot |\langle \Psi, \psi_{(a',s',A_\alpha^{-1}S_\sigma^{-1}(t-\tau))}\rangle_{L_2}|^q \cdot |\alpha|^{\frac{1}{d}-2} \, \frac{\mathrm{d}a'}{|a'|^{d+1}} \, \mathrm{d}s' \, \mathrm{d}t
$$

$$
= \int_{\mathbb{R}^d} \int_{\mathbb{R}^{d-1}} \int_{-|\alpha|^{-1}}^{|\alpha|^{-1}} w(\alpha, \alpha a')^q \cdot |\langle \Psi, \psi_{(a',s',t')}\rangle_{L_2}|^q \, \frac{\mathrm{d}a'}{|a'|^{d+1}} \, \mathrm{d}s' \, \mathrm{d}t'. \qquad (5.5.29)
$$

Using the three equalities (5.5.27), (5.5.28), and (5.5.29), we now have

$$
\|K_{\mathfrak{F}}\|_{\mathcal{A}_{q,w}} = \operatorname*{ess\,sup}_{(\alpha,\sigma,\tau)\in X} \int_{\mathbb{R}^d} \int_{\mathbb{R}^{d-1}} \left(w(\alpha, \infty)^q \cdot |\langle \psi_{(\infty,s,t)}, \psi_{(\alpha,\sigma,\tau)}\rangle_{L_2}|^q \right.
$$

$$
\left. + \int_{-1}^{1} w(\alpha, a)^q \cdot |\langle \psi_{(a,s,t)}, \psi_{(\alpha,\sigma,\tau)}\rangle_{L_2}|^q \, \frac{\mathrm{d}a}{|a|^{d+1}} \right) \mathrm{d}s \, \mathrm{d}t
$$

$$
= \max \left\{ \operatorname*{ess\,sup}_{(\sigma,\tau)\in\mathbb{R}^{d-1}\times\mathbb{R}^d} \int_{\mathbb{R}^d} \int_{\mathbb{R}^{d-1}} \left(w(\infty, \infty)^q \cdot |\langle \psi_{(\infty,s,t)}, \psi_{(\infty,\sigma,\tau)}\rangle_{L_2}|^q \right. \right.
$$

$$
\left. + \int_{-1}^{1} w(\infty, a)^q \cdot |\langle \psi_{(a,s,t)}, \psi_{(\infty,\sigma,\tau)}\rangle_{L_2}|^q \, \frac{\mathrm{d}a}{|a|^{d+1}} \right) \mathrm{d}s \, \mathrm{d}t,
$$

$$
\operatorname*{ess\,sup}_{(\alpha,\sigma,\tau)\in[-1,1]^*\times\mathbb{R}^{d-1}\times\mathbb{R}^d} \int_{\mathbb{R}^d} \int_{\mathbb{R}^{d-1}} \left(w(\alpha, \infty)^q \cdot |\langle \psi_{(\infty,s,t)}, \psi_{(\alpha,\sigma,\tau)}\rangle|^q \right.
$$

$$+ \int_{-1}^{1} w(\alpha, a)^q \cdot |\langle \psi_{(a,s,t)}, \psi_{(\alpha,\sigma,\tau)} \rangle_{L_2}|^q \, \frac{da}{|a|^{d+1}} \Bigg) ds \, dt \Bigg\}$$

$$= \max \Bigg\{ \int_{\mathbb{R}^d} \int_{\mathbb{R}^{d-1}} \Bigg(w(\infty, \infty)^q \cdot |\langle \Phi, \psi_{(\infty,s',t')} \rangle_{L_2}|^q$$

$$+ \int_{-1}^{1} w(\infty, a)^q \cdot |\langle \Phi, \psi_{(a,s',t')} \rangle_{L_2}|^q \, \frac{da}{|a|^{d+1}} \Bigg) ds' \, dt',$$

$$\operatorname*{ess\,sup}_{\alpha \in [-1,1]^*} \int_{\mathbb{R}^d} \int_{\mathbb{R}^{d-1}} \Bigg(w(\alpha, \infty)^q \cdot |\langle \Phi, \psi_{(\alpha,s',t')} \rangle_{L_2}|^q$$

$$+ \int_{-|\alpha|^{-1}}^{|\alpha|^{-1}} w(\alpha, \alpha a')^q \cdot |\langle \Psi, \psi_{(a',s',t')} \rangle_{L_2}|^q \, \frac{da'}{|a'|^{d+1}} \Bigg) ds' \, dt' \Bigg\},$$

which proves (5.5.26). $\qquad \square$

In Subsection 2.3.2 we have seen that band-limited admissible shearlets have nice properties, see Theorem 2.3.7. We have shown specifically, that the integrability of the reproducing kernel can be easily shown. Let us transfer these ideas to the present setting and suppose that Ψ is a Schwartz function such that

$$\operatorname{supp} \hat{\Psi} \subseteq ([-a_1, -a_0] \cup [a_0, a_1]) \times Q_b$$

for some $0 < a_0 < a_1$ and $Q_b = [-b_1, b_1] \times \ldots \times [-b_{d-1}, b_{d-1}]$ for $b \in \mathbb{R}_+^{d-1}$. Setting $\hat{\Phi}$ as in (5.5.19) we see that $\hat{\Phi}(\xi_1, \tilde{\xi}) = 0$ for all $\tilde{\xi} \notin Q_b$. In addition, fix some $\tilde{\xi} \notin Q_b$ and let $|\xi_1| > a_1$, then $\hat{\Psi}(\omega_1, \tilde{\xi}) = 0$ for all $|\omega_1| \geq |\xi_1|$, which implies $\hat{\Phi}(\xi) = 0$. So altogether we have

$$\operatorname{supp} \hat{\Phi} \subseteq [-a_1, a_1] \times Q_b.$$

Following the lines of Section 5.1 we define a weight $m_r \colon X \to \mathbb{R}_+$ for $r \in \mathbb{R}_{\geq 0}$ via

$$m_r(a, s, t) = m_r(a) := \begin{cases} 1, & a = \infty, \\ |a|^{-r}, & a \in [-1, 1]^*. \end{cases} \tag{5.5.30}$$

Furthermore, we define the weight $w_r \colon X \times X \to \mathbb{R}_+$ for the same $r \in \mathbb{R}_{\geq 0}$ as above only depending on a, that is, $w_r((a, s, t), (a', s', t')) = w_r(a, a')$ for all $(a, s, t), (a', s', t') \in X$ and for all $a, a' \in [-1, 1]^*$ we set

$$w_r(\infty, \infty) = 1, \tag{5.5.31}$$

$$w_r(a, \infty) = w_r(\infty, a) = |a|^{-r}, \tag{5.5.32}$$

$$w_r(a, a') = \max \left\{ \frac{|a|}{|a'|}, \frac{|a'|}{|a|} \right\}^r. \tag{5.5.33}$$

It is easy to see that w is admissible in the sense of (5.1.7)–(5.1.9) with $w_r(a, a) = C_w = 1$ for all $a \in [-1, 1]^*$ and that it fulfills the assumptions in Lemma 5.5.7. Moreover, m_r is a w_r-moderate weight satisfying $m_r \geq 1$, so that we are in the setting of Section 5.1.

The following lemma concers the support properties of Ψ and Φ in the frequency domain, similar to [28, Lemma 3.1], see also the proof of Theorem 2.3.7.

Lemma 5.5.9. *Let $0 < a_0 < a_1$ and $b \in \mathbb{R}_+^{d-1}$. Then with Ψ and Φ defined as above for any $a \in \mathbb{R}^*$ and $s \in \mathbb{R}^{d-1}$ we have the following assertions:*

(i) $\widehat{\Psi}(\cdot)\widehat{\Psi}(A_a S_s^T \cdot) \not\equiv 0$ *implies*

$$a \in \left[-\frac{a_1}{a_0}, -\frac{a_0}{a_1} \right] \cup \left[\frac{a_0}{a_1}, \frac{a_1}{a_0} \right] \quad and \quad s \in Q_{d_1} \tag{5.5.34}$$

with $d_1 := (a_0^{-1} + a_0^{-(1+\frac{1}{d})} a_1^{\frac{1}{d}}) b$;

(ii) assume $|a| \leqslant 1$*, then* $\widehat{\Phi}(\cdot)\widehat{\Psi}(A_a S_s^T \cdot) \not\equiv 0$ *implies*

$$a \in \left[-1, -\frac{a_0}{a_1} \right] \cup \left[\frac{a_0}{a_1}, 1 \right] \quad and \quad s \in Q_{d_2} \tag{5.5.35}$$

with $d_2 := (a_0^{-1} + a_0^{-(1+\frac{1}{d})} a_1^{\frac{1}{d}}) b$;

(iii) we have $\operatorname{supp} \widehat{\Phi}(\cdot)\widehat{\Phi}(S_s^T \cdot) \subseteq \Omega_s$ *with*

$$\Omega_s := \Big\{ \xi \in \mathbb{R}^d \ : \ |\xi_1| \leqslant a_1, \tag{5.5.36}$$

$$\max\{-b_i, -b_i - s_{i-1}\xi_1\} \leqslant \xi_i \leqslant \min\{b_i, b_i - s_{i-1}\xi_1\}, \ i = 2, \dots, d \Big\}.$$

Proof. The proof of part (i) can be found in [28, Lemma 3.1].

To prove part (ii) we assume there exists a $\xi \in \operatorname{supp} \widehat{\Phi} \cap \operatorname{supp} \widehat{\Psi}(A_a S_s^T \cdot)$ which means $\xi \in \operatorname{supp} \widehat{\Phi}$ and $A_a S_s^T \xi \in \operatorname{supp} \widehat{\Psi}$. This leads to

$$|\xi_1| \leqslant a_1, \tag{5.5.37}$$

$$-b_i \leqslant \xi_{i+1} \leqslant b_i, \tag{5.5.38}$$

$$a_0 \leqslant |a||\xi_1| \leqslant a_1, \tag{5.5.39}$$

$$-b_i |a|^{-\frac{1}{d}} - \xi_1 s_i \leqslant \xi_{i+1} \leqslant b_i |a|^{-\frac{1}{d}} - \xi_1 s_i, \tag{5.5.40}$$

for $i = 1, \dots, d - 1$. By (5.5.37) and (5.5.39) it follows that $|a| \geqslant \frac{a_0}{a_1}$ which implies $a \in [-1, -\frac{a_0}{a_1}] \cup [\frac{a_0}{a_1}, 1]$. Using (5.5.39) and $|a| \leqslant 1$, it follows that $a_0 \leqslant |a||\xi_1| \leqslant |\xi_1|$. Also, with (5.5.40) and (5.5.38) we obtain

$$-b_i |a|^{-\frac{1}{d}} - b_i \leqslant \xi_1 s_i \leqslant b_i |a|^{-\frac{1}{d}} + b_i,$$

which leads to

$$|s_i| \leqslant |\xi_1|^{-1}(b_i |a|^{-\frac{1}{d}} + b_i) \leqslant a_0^{-1} b_i \left(\frac{a_0}{a_1} \right)^{-\frac{1}{d}} + a_0^{-1} b_i$$

for $i = 1, \dots, d - 1$ which proves (ii).

To prove part (iii) we assume there exists $\xi \in \operatorname{supp} \widehat{\Phi} \cap \operatorname{supp} \widehat{\Phi}(S_s^T \cdot)$ which means $\xi \in \operatorname{supp} \widehat{\Phi}$ and $S_s^T \xi \in \operatorname{supp} \widehat{\Phi}$. This leads to

$$|\xi_1| \leqslant a_1, \tag{5.5.41}$$

$$-b_i \leqslant \xi_{i+1} \leqslant b_i, \tag{5.5.42}$$

$$-b_i \leqslant \xi_1 s_i + \xi_{i+1} \leqslant b_i \tag{5.5.43}$$

for all $i = 1, \dots, d - 1$, which means $\xi \in \Omega_s$. □

The following two auxiliary lemmas are of technical nature only and the proof of Lemma 5.5.10 is based on a draft by Steidl, Dahlke, Häuser and Teschke or can be found in [66, Appendix B.1] in a similar form.

Lemma 5.5.10. *For all $y, z \in \mathbb{R}$, $\lambda, \lambda' > 0$ and $k > 1$ the following integral estimate holds true*

$$\int_{\mathbb{R}} (1 + \lambda|x - y|)^{-k} (1 + \lambda'|x - z|)^{-k} \, dx \leqslant C \cdot \max\{\lambda, \lambda'\}^{-1} \left(1 + \min\{\lambda, \lambda'\} \cdot |y - z|\right)^{-k},$$

where the constant $C > 0$ only depends on k.

Proof. Let $y, z \in \mathbb{R}$ be arbitrary and assume without loss of generality that $\lambda \leqslant \lambda'$. Assume further that $|y - z| \leqslant \lambda^{-1}$, then

$$(1 + \lambda|x - y|)^{-k} \leqslant 1 \leqslant 2^k (1 + \lambda|y - z|)^{-k}$$

and thus

$$\begin{aligned}
\int_{\mathbb{R}} (1 + \lambda|x - y|)^{-k} (1 + \lambda'|x - z|)^{-k} \, dx &\lesssim (1 + \lambda|y - z|)^{-k} \int_{\mathbb{R}} (1 + \lambda'|x - z|)^{-k} \, dx \\
&= (1 + \lambda|y - z|)^{-k} \frac{1}{\lambda'} \int_{\mathbb{R}} (1 + |x|)^{-k} \, dx \\
&\lesssim \frac{1}{\lambda'} (1 + \lambda|y - z|)^{-k}. \tag{5.5.44}
\end{aligned}$$

On the other hand if $|y - z| > \lambda^{-1}$ let H_y and H_z be the two half-axes containing the points y and z respectively, such that $H_y \cap H_z = \{\frac{y+z}{2}\}$. Then, for every $x \in H_z$ it holds that $|x - y| \geqslant \frac{1}{2}|y - z|$ and thus

$$\begin{aligned}
\int_{H_z} (1 + \lambda|x - y|)^{-k} (1 + \lambda'|x - z|)^{-k} \, dx &\leqslant \left(1 + \frac{\lambda}{2}|y - z|\right)^{-k} \int_{H_z} (1 + \lambda'|x - z|)^{-k} \, dx \\
&\lesssim (1 + \lambda|y - z|)^{-k} \frac{1}{\lambda'} \int_{\mathbb{R}} (1 + |x|)^{-k} \, dx \\
&\lesssim \frac{1}{\lambda'} (1 + \lambda|y - z|)^{-k} \tag{5.5.45}
\end{aligned}$$

similarly for every $x \in H_y$ it holds $|x - z| \geqslant \frac{1}{2}|y - z|$ and since $|y - z| > \lambda^{-1}$ we first deduce

$$\begin{aligned}
(1 + \lambda'|x - z|)^{-k} &\leqslant \left(\frac{\lambda'}{2}|y - z|\right)^{-k} \\
&\lesssim \left(\frac{\lambda}{\lambda'}\right)^k (\lambda|y - z|)^{-k} \\
&\lesssim \left(\frac{\lambda}{\lambda'}\right)^k (1 + \lambda|y - z|)^{-k} \\
&\leqslant \frac{\lambda}{\lambda'} (1 + \lambda|y - z|)^{-k}
\end{aligned}$$

and hence we derive the estimate

$$\begin{aligned}
\int_{H_y} (1 + \lambda|x - y|)^{-k} (1 + \lambda'|x - z|)^{-k} \, dx &\leqslant \frac{\lambda}{\lambda'} (1 + \lambda|y - z|)^{-k} \int_{H_y} (1 + \lambda|x - y|)^{-k} \, dx \\
&\leqslant \frac{1}{\lambda'} (1 + \lambda|y - z|)^{-k} \int_{\mathbb{R}} (1 + |x|)^{-k} \, dx \\
&\lesssim \frac{1}{\lambda'} (1 + \lambda|y - z|)^{-k}. \tag{5.5.46}
\end{aligned}$$

Combining (5.5.45) and (5.5.46) thus yields

$$\int_{\mathbb{R}} (1 + \lambda|x - y|)^{-k} (1 + \lambda'|x - z|)^{-k} \, dx = \left(\int_{H_y} + \int_{H_z} \right) (1 + \lambda|x - y|)^{-k} (1 + \lambda'|x - z|)^{-k} \, dx$$

$$\lesssim \frac{1}{\lambda'} (1 + \lambda|y - z|)^{-k}$$

and together with (5.5.44) this completes the proof. $\square$

We extend this result in the following form to integrals with three factors.

Lemma 5.5.11. *For all $y, z \in \mathbb{R}$, $\lambda \neq 0$ and $k > 1$ the following integral estimate holds true:*

$$\int_{\mathbb{R}} (1 + |x|)^{-k} (1 + |x - y|)^{-k} (1 + |\lambda x - z|)^{-k} \, dx$$

$$\leqslant C \cdot (1 + |y|)^{-k} \max\{1, |\lambda|\}^{-1} \cdot \left[\left(1 + \min\{1, |\lambda|\} \cdot \left| y - \frac{z}{\lambda} \right| \right)^{-k} + \left(1 + \min\{1, |\lambda|\} \cdot \left| \frac{z}{\lambda} \right| \right)^{-k} \right],$$

where the constant $C > 0$ only depends on k.

Proof. We use the ideas of the proof of [69, Lemma 11.1.1], as well as Lemma 5.5.10 and define the set $N_y := \{ x \in \mathbb{R} : |x - y| \leqslant \frac{|y|}{2} \}$. For all $x \in N_y$ it follows that $|x| \geqslant \frac{|y|}{2}$ and thus

$$(1 + |x|)^{-k} \leqslant \left(1 + \frac{|y|}{2} \right)^{-k} \leqslant 2^k (1 + |y|)^{-k}.$$

On the other hand if $x \in N_y^c$ one has $(1 + |x - y|)^{-k} \leqslant (1 + \frac{|y|}{2})^{-k}$. Hence, with Lemma 5.5.10 we can derive

$$\int_{\mathbb{R}} (1 + |x|)^{-k} (1 + |x - y|)^{-k} (1 + |\lambda x - z|)^{-k} \, dx$$

$$= \left(\int_{N_y} + \int_{N_y^c} \right) (1 + |x|)^{-k} (1 + |x - y|)^{-k} (1 + |\lambda x - z|)^{-k} \, dx$$

$$\lesssim (1 + |y|)^{-k} \int_{\mathbb{R}} (1 + |x - y|)^{-k} \left(1 + |\lambda| \cdot \left| x - \frac{z}{\lambda} \right| \right)^{-k} \, dx$$

$$+ (1 + |y|)^{-k} \int_{\mathbb{R}} (1 + |x|)^{-k} \left(1 + |\lambda| \cdot \left| x - \frac{z}{\lambda} \right| \right)^{-k} \, dx$$

$$\lesssim (1 + |y|)^{-k} \max\{1, |\lambda|\}^{-1} \left(1 + \min\{1, |\lambda|\} \cdot \left| y - \frac{z}{\lambda} \right| \right)^{-k}$$

$$+ (1 + |y|)^{-k} \max\{1, |\lambda|\}^{-1} \left(1 + \min\{1, |\lambda|\} \cdot \left| \frac{z}{\lambda} \right| \right)^{-k},$$

which concludes the proof. $\square$

Using the result above the next lemma holds true. Recall that the shearing matrix $S_s \in \mathbb{R}^{d \times d}$ for $s \in \mathbb{R}^{d-1}$ defined in (5.5.2) is given by

$$S_s := \begin{pmatrix} 1 & s^T \\ 0_{d-1} & I_{d-1} \end{pmatrix} \quad \text{with} \quad S_s^{-1} = S_{-s} = \begin{pmatrix} 1 & -s^T \\ 0_{d-1} & I_{d-1} \end{pmatrix}.$$

Lemma 5.5.12. *Fix $0 < q_0 < 1$ and set*

$$I_s(t) := \int_{\mathbb{R}^d} \prod_{i=1}^{d} \left[(1 + |x_i + t_i|)^{-k} (1 + |(S_{-s}x)_i|)^{-k} \right] \, dx \tag{5.5.47}$$

for $s \in \mathbb{R}^{d-1}$ and $t \in \mathbb{R}^d$, where $(S_{-s}x)_i$ denotes the i-th entry of the vector $S_{-s}x \in \mathbb{R}^d$. Then,

$$\int_{\mathbb{R}^d} I_s(t)^{q_0}\, dt \leqslant C \cdot (1 + \|s\|_2)^{1-q_0} \int_{\mathbb{R}^d} \prod_{i=1}^{d} (1 + |t_i|)^{-kq_0}\, dt \qquad (5.5.48)$$

for almost all $s \in \mathbb{R}^{d-1}$ where the constant $C > 0$ is independent of q_0 and s.

Proof. The proof is structured as follows. We first show an auxiliary estimate from above for $d = 3$, which we will then generalize to arbitrary dimensions. This estimate is then used to prove (5.5.48). To illustrate our method we differentiate between the following four cases for $s \in \mathbb{R}^2$ with $s_1, s_2 \neq 0$.

Case 1: $|s_1|, |s_2| \leqslant 1$. We apply Lemma 5.5.10 with $\lambda = \lambda' = 1$, $y = -t_1$, $z = s_1 x_2 + s_2 x_3$ and Lemma 5.5.11 with $\lambda = s_1$, $y = -t_2$, $z = -t_1 - s_2 x_3$ and obtain

$$I_s(t) = \int_{\mathbb{R}^3} (1 + |x_1 + t_1|)^{-k}(1 + |x_1 - s_1 x_2 - s_2 x_3|)^{-k}$$
$$\times (1 + |x_2 + t_2|)^{-k}(1 + |x_2|)^{-k}(1 + |x_3 + t_3|)^{-k}(1 + |x_3|)^{-k}\, d(x_1, x_2, x_3)$$
$$\lesssim \int_{\mathbb{R}^2} (1 + |t_1 + s_1 x_2 + s_2 x_3|)^{-k}(1 + |x_2 + t_2|)^{-k}$$
$$\times (1 + |x_2|)^{-k}(1 + |x_3 + t_3|)^{-k}(1 + |x_3|)^{-k}\, d(x_2, x_3)$$
$$\lesssim (1 + |t_2|)^{-k} \int_{\mathbb{R}} (1 + |-s_1 t_2 + t_1 + s_2 x_3|)^{-k}(1 + |x_3 + t_3|)^{-k}(1 + |x_3|)^{-k}\, dx_3$$
$$+ (1 + |t_2|)^{-k} \int_{\mathbb{R}} (1 + |t_1 + s_2 x_3|)^{-k}(1 + |x_3 + t_3|)^{-k}(1 + |x_3|)^{-k}\, dx_3.$$

By applying 5.5.11 twice with $\lambda = s_2$, $y = -t_3$, $z = s_1 t_2 - t_1$ and $\lambda = s_2$, $y = -t_3$, $z = -t_1$ respectively we get

$$I_s(t) \lesssim (1 + |t_2|)^{-k}(1 + |t_3|)^{-k}\big[(1 + |s_2 t_3 + s_1 t_2 - t_1|)^{-k}$$
$$+ (1 + |s_1 t_2 - t_1|)^{-k} + (1 + |s_2 t_3 - t_1|)^{-k} + (1 + |t_1|)^{-k}\big]. \qquad (5.5.49)$$

Case 2: $|s_1| \leqslant 1, |s_2| > 1$. We use the same first two estimates from above and then apply Lemma 5.5.11 twice with $\lambda = s_2$, $y = -t_3$, $z = s_1 t_2 - t_1$ and $\lambda = s_2$, $y = -t_3$, $z = -t_1/s_2$ respectively, yielding

$$I_s(t) \lesssim (1 + |t_2|)^{-k} \int_{\mathbb{R}} (1 + |-s_1 t_2 + t_1 + s_2 x_3|)^{-k}(1 + |x_3 + t_3|)^{-k}(1 + |x_3|)^{-k}\, dx_3$$
$$+ (1 + |t_2|)^{-k} \int_{\mathbb{R}} (1 + |t_1 + s_2 x_3|)^{-k}(1 + |x_3 + t_3|)^{-k}(1 + |x_3|)^{-k}\, dx_3$$
$$\lesssim |s_2|^{-1}(1 + |t_2|)^{-k}(1 + |t_3|)^{-k}\big[(1 + |-t_3 + s_1 s_2^{-1} t_2 - s_2^{-1} t_1|)^{-k} \qquad (5.5.50)$$
$$+ (1 + |s_1 s_2^{-1} t_2 - s_2^{-1} t_1|)^{-k} + (1 + |t_3 - s_2^{-1} t_1|)^{-k} + (1 + |s_2^{-1} t_1|)^{-k}\big].$$

Case 3: $|s_1| > 1, |s_2| \leqslant |s_1|$. Similarly to case 1 we apply Lemma 5.5.10 with $\lambda = \lambda' = 1$, $y = -t_1$, $z = s_1 x_2 + s_2 x_3$ and Lemma 5.5.11 with $\lambda = s_1$, $y = -t_2$, $z = -t_1 - s_2 x_3$ and obtain

$$I_s(t) \lesssim \int_{\mathbb{R}^2} (1 + |t_1 + s_1 x_2 + s_2 x_3|)^{-k}(1 + |x_2 + t_2|)^{-k}$$
$$\times (1 + |x_2|)^{-k}(1 + |x_3 + t_3|)^{-k}(1 + |x_3|)^{-k}\, d(x_2, x_3)$$
$$\lesssim |s_1|^{-1}(1 + |t_2|)^{-k} \int_{\mathbb{R}} (1 + |-t_2 + s_1^{-1} t_1 + s_1^{-1} s_2 x_3|)^{-k}(1 + |x_3 + t_3|)^{-k}(1 + |x_3|)^{-k}\, dx_3$$
$$+ |s_1|^{-1}(1 + |t_2|)^{-k} \int_{\mathbb{R}} (1 + |s_1^{-1} t_1 + s_1^{-1} s_2 x_3|)^{-k}(1 + |x_3 + t_3|)^{-k}(1 + |x_3|)^{-k}\, dx_3.$$

Applying Lemma 5.5.11 twice with $\lambda = s_1^{-1}s_2$, $y = -t_3$, $z = t_2 - s_1^{-1}t_1$ and $\lambda = s_1^{-1}s_2$, $y = -t_3$, $z = -s_1^{-1}t_1$ respectively gives us

$$I_s(t) \lesssim |s_1|^{-1}(1 + |t_2|)^{-k}(1 + |t_3|)^{-k}\big[(1 + |s_1^{-1}s_2t_3 + t_2 - s_1^{-1}t_1|)^{-k}$$
$$+ (1 + |t_2 - s_1^{-1}t_1|)^{-k} + (1 + |s_1^{-1}s_2t_3 - s_1^{-1}t_1|)^{-k} + (1 + |s_1^{-1}t_1|)^{-k}\big]. \quad (5.5.51)$$

Case 4: $|s_1| > 1, |s_2| > |s_1|$. Finally, we use the same first two estimates as above and apply Lemma 5.5.11 twice with $\lambda = s_1^{-1}s_2$, $y = -t_3$, $z = t_2 - s_1^{-1}t_1$ and $\lambda = s_1^{-1}s_2$, $y = -t_3$, $z = -s_1^{-1}t_3$ to derive

$$I_s(t) \lesssim |s_1|^{-1}(1 + |t_2|)^{-k}\int_{\mathbb{R}} (1 + |-t_2 + s_1^{-1}t_1 + s_1^{-1}s_2x_3|)^{-k}(1 + |x_3 + t_3|)^{-k}(1 + |x_3|)^{-k}\,\mathrm{d}x_3$$
$$+ |s_1|^{-1}(1 + |t_2|)^{-k}\int_{\mathbb{R}} (1 + |s_1^{-1}t_1 + s_1^{-1}s_2x_3|)^{-k}(1 + |x_3 + t_3|)^{-k}(1 + |x_3|)^{-k}\,\mathrm{d}x_3.$$
$$\lesssim |s_2|^{-1}(1 + |t_2|)^{-k}(1 + |t_3|)^{-k}\big[(1 + |t_3s_1s_2^{-1}t_2 - s_2^{-1}t_1|)^{-k} \quad (5.5.52)$$
$$+ (1 + |s_1s_2^{-1}t_2 - s_2^{-1}t_1|)^{-k} + (1 + |t_3 - s_2^{-1}t_1|)^{-k} + (1 + |s_2^{-1}t_1|)^{-k}\big].$$

The four cases (5.5.49), (5.5.50), (5.5.51), (5.5.52) yield the estimate

$$I_s(t) \lesssim \min(s)\left[\prod_{j=2}^{3}(1 + |t_j|)^{-k}\right]\sum_{i=1}^{4}(1 + |v_i(s) \cdot t|)^{-k} \quad (5.5.53)$$

for all $s \in \mathbb{R}^2$ satisfying $s_1 \neq 0$, $s_2 \neq 0$ and $t \in \mathbb{R}^3$, where the vectors $v_i(s) \in \mathbb{R}^3$, $i = 1, \ldots, 4$, depend on $s \in \mathbb{R}^2$ and satisfy the condition $(v_i(s))_1 = \min(s) := \max\{1, |s_1|, |s_2|\}^{-1}$. In particular it follows from the four cases above that

$$\min(s) = \begin{cases} 1, & |s_1|, |s_2| \leq 1, \\ |s_2|^{-1}, & |s_1| \leq 1, |s_2| > 1, \\ |s_1|^{-1}, & |s_1| > 1, |s_2| \leq |s_1|, \\ |s_2|^{-1}, & |s_1| > 1, |s_2| > |s_1| \end{cases} = \max\{1, |s_1|, |s_2|\}^{-1}.$$

The other two entries of $v_i(s)$ are less or equal to one in absolute value.

We now intend to show, that this result holds for arbitrary dimension. For this we fix $d \geq 3$ as well as $s \in \mathbb{R}^{d-1}$ with $s_i \neq 0$ for all $i = 1, \ldots, d-1$ and assume that there exist vectors $v_i(s)$, $1 \leq i \leq 2^{d-1}$, with $(v_i(s))_1 = \min(s) = \max\{1, |s_1|, \ldots, |s_{d-1}|\}^{-1} = \min\{1, |s_1|^{-1}, \ldots, |s_{d-1}|^{-1}\}$ and all other entries are less or equal to one in absolut value such that the estimate

$$I_s(t) \lesssim \min(s)\left[\prod_{j=2}^{d}(1 + |t_j|)^{-k}\right]\sum_{i=1}^{2^{d-1}}(1 + |v_i(s) \cdot t|)^{-k} \quad (5.5.54)$$

holds true for fixed d. As shown in (5.5.53), this readily is the case for $d = 3$. We now intend to show that the estimate (5.5.54) also holds for $d + 1$. Then, (5.5.54) will hold for arbitrary dimension by induction over the dimension d. To this end we fix $s \in \mathbb{R}^d$ with $s_i \neq 0$ for all $i = 1, \ldots, d$ and define $\widetilde{x} := (x_1, \ldots, x_d)$, $\widetilde{s} := (s_1, \ldots, s_{d-1})$, $\widetilde{t} := (t_1, \ldots, t_d)$ and $u = u(x_{d+1}) :=$

$(t_1 + s_d x_{d+1}, t_2, \ldots, t_d)$. Then we deduce from (5.5.54) the estimate

$$
I_s(t) = \int_{\mathbb{R}^{d+1}} \prod_{i=1}^{d+1} \left[(1 + |x_i + t_i|)^{-k} (1 + |(S_{-s} x)_i|)^{-k} \right] \, dx
$$

$$
= \int_{\mathbb{R}} (1 + |x_{d+1} + t_{d+1}|)^{-k} (1 + |x_{d+1}|)^{-k}
$$

$$
\times \left(\int_{\mathbb{R}^d} \prod_{i=1}^{d} \left[(1 + |x_i + u_i|)^{-k} (1 + |(S_{-\tilde{s}} \tilde{x})_i|)^{-k} \right] d\tilde{x} \right) dx_{d+1}
$$

$$
= \int_{\mathbb{R}} I_{\tilde{s}}(u(x_{d+1})) \cdot (1 + |x_{d+1} + t_{d+1}|)^{-k} (1 + |x_{d+1}|)^{-k} \, dx_{d+1}
$$

$$
\lesssim \min(\tilde{s}) \cdot \left[\prod_{j=2}^{d} (1 + |t_j|)^{-k} \right] \tag{5.5.55}
$$

$$
\times \sum_{i=1}^{2^{d-1}} \int_{\mathbb{R}} (1 + |v_i(\tilde{s}) \cdot u(x_{d+1})|)^{-k} (1 + |x_{d+1} + t_{d+1}|)^{-k} (1 + |x_{d+1}|)^{-k} \, dx_{d+1}.
$$

We fix $1 \leqslant i \leqslant 2^{d-1}$, look at one summand above and apply Lemma 5.5.11 with $\lambda = \min(\tilde{s}) s_d$, $y = -t_{d+1}$, $z = v_i(\tilde{s}) \cdot \tilde{t}$, yielding

$$
\min(\tilde{s}) \cdot \int_{\mathbb{R}} (1 + |v_i(\tilde{s}) \cdot u(x_{d+1})|)^{-k} (1 + |x_{d+1} + t_{d+1}|)^{-k} (1 + |x_{d+1}|)^{-k} \, dx_{d+1}
$$

$$
= \min(\tilde{s}) \cdot \int_{\mathbb{R}} (1 + |v_i(\tilde{s}) \cdot \tilde{t} + \min(\tilde{s}) s_d x_{d+1}|)^{-k} (1 + |x_{d+1} + t_{d+1}|)^{-k} (1 + |x_{d+1}|)^{-k} \, dx_{d+1}
$$

$$
\lesssim \min(\tilde{s}) \cdot (1 + |t_{d+1}|)^{-k} \cdot \max\{1, |\min(\tilde{s}) s_d|\}^{-1}
$$

$$
\times \left[\left(1 + \left| \min\{1, |\min(\tilde{s}) s_d|\} \, t_{d+1} - \frac{\min\{1, |\min(\tilde{s}) s_d|\}}{|\min(\tilde{s}) s_d|} (v_i(\tilde{s}) \cdot \tilde{t}) \right| \right)^{-k} \right.
$$

$$
\left. + \left(1 + \frac{\min\{1, |\min(\tilde{s}) s_d|\}}{|\min(\tilde{s}) s_d|} |v_i(\tilde{s}) \cdot \tilde{t}| \right)^{-k} \right]
$$

$$
= \max\{\min(\tilde{s})^{-1}, |s_d|\}^{-1} \cdot \left[(1 + |t_{d+1}|)^{-k} (1 + |v_i^1(s) \cdot t|)^{-k} + (1 + |t_{d+1}|)^{-k} (1 + |v_i^2(s) \cdot t|)^{-k} \right]
$$

for some vectors $v_i^1(s), v_i^2(s) \in \mathbb{R}^{d+1}$ depending on s, where

$$
(v_i^1(s))_1 = (v_i^2(s))_1 = (v_i(\tilde{s}))_1 \cdot \left(\frac{\min\{1, |\min(\tilde{s}) s_d|\}}{|\min(\tilde{s}) s_d|} \right)
$$

$$
= \min\{|s_d|^{-1}, \min(\tilde{s})\} = \min(s).
$$

All other entries of $v_i^1(s)$ and $v_i^2(s)$ are less or equal to one in absolut value. Since $\max\{\min(\tilde{s})^{-1}, |s_d|\}^{-1} = \max\{1, |s_1|, \ldots, |s_d|\}^{-1} = \min(s)$ we derive together with (5.5.55) the estimate (5.5.54) for $d + 1$. Hence, (5.5.54) holds true for arbitrary dimension.

With this at hand we return to an arbitrary dimension $d \geqslant 3$ and estimate

$$
\min(s)^{-1} = \max\{1, |s_1|, \ldots, |s_{d-1}|\} \leqslant 1 + \max\{|s_1|, \ldots, |s_{d-1}|\} \lesssim 1 + \|s\|_2.
$$

Therefore the following estimate holds true for any $0 < q_0 < 1$ and almost every $s \in \mathbb{R}^d$:

$$\int_{\mathbb{R}^d} I_s(t)^{q_0} \, dt \lesssim \min(s)^{q_0} \int_{\mathbb{R}^d} \left(\left[\prod_{j=2}^{d} (1 + |t_j|)^{-k} \right] \sum_{i=1}^{2^{d-1}} 1 + |v_i(s) \cdot t|)^{-k} \right)^{q_0} dt$$

$$\leqslant \min(s)^{q_0} \sum_{i=1}^{2^{d-1}} \int_{\mathbb{R}^d} (1 + |v_i(s) \cdot t|)^{-kq_0} \prod_{j=2}^{d} (1 + |t_j|)^{-kq_0} \, dt.$$

We now substitute $w_1 = v_i(s) \cdot t$ and $w_j = t_j$ for $j = 2, \ldots, d$ and every $1 \leqslant i \leqslant 2^{d-1}$, so that $w = A_i(s)t$, where the matrices $A_i(s)$ have determinant $(v_i(s))_1 = \min(s)$. Therefore we obtain

$$\int_{\mathbb{R}^d} I_s(t)^{q_0} \, dt \lesssim \min(s)^{q_0-1} \int_{\mathbb{R}^d} \prod_{j=1}^{d} (1 + |w_j|)^{-kq_0} \, dw$$

$$\lesssim (1 + \|s\|_2)^{1-q_0} \int_{\mathbb{R}^d} \prod_{j=1}^{d} (1 + |t_j|)^{-kq_0} \, dt,$$

which proves (5.5.48). Note that all constants involved in this proof are independent of both q_0 and s. □

With this at hand we are finally ready to proof the main result of this subsection.

Theorem 5.5.13. *Let $d \geqslant 3$ be odd and $\Psi \in L_1(\mathbb{R}^d) \cap L_2(\mathbb{R}^d)$ an admissible shearlet with $C_\Psi = 1$ such that*

$$\operatorname{supp} \widehat{\Psi} \subseteq ([-a_1, -a_0] \cup [a_0, a_1]) \times Q_b$$

for some $0 < a_0 < a_1$ and $b \in \mathbb{R}_+^{d-1}$. Let $\widehat{\Phi}$ be chosen as in (5.5.19) so that condition (5.5.16) is fulfilled and assume further $\widehat{\Phi} \in C_0^\infty(\mathbb{R}^d)$, i.e., $\Phi \in \mathcal{S}(\mathbb{R}^d) \subset L_1(\mathbb{R}^d) \cap L_2(\mathbb{R}^d)$. Then, for every $1 < q < \infty$ and $r \geqslant 0$ the kernel $K_{\mathfrak{F}}$ fulfills

$$K_{\mathfrak{F}} \in \mathcal{A}_{q,w_r}.$$

Proof. Fix $1 < q < \infty$ and recall that the weight w_r is defined via (5.5.31)–(5.5.33). Then, by invoking Lemma 5.5.8 the $\mathcal{A}_{q,w_r}$-norm of the kernel $K_{\mathfrak{F}}$ can be written as

$$\|K_{\mathfrak{F}}\|_{\mathcal{A}_{q,w_r}}^q = \max \Bigg\{ \int_{\mathbb{R}^d} \int_{\mathbb{R}^{d-1}} \left(|\langle \Phi, \psi_{(\infty,s,t)} \rangle_{L_2}|^q + \int_{-1}^{1} |a|^{-rq} \cdot |\langle \Phi, \psi_{(a,s,t)} \rangle_{L_2}|^q \frac{da}{|a|^{d+1}} \right) ds \, dt,$$

$$\operatorname*{ess\,sup}_{\alpha \in [-1,1]^*} \int_{\mathbb{R}^d} \int_{\mathbb{R}^{d-1}} \left(|\alpha|^{-rq} \cdot |\langle \Phi, \psi_{(\alpha,s,t)} \rangle_{L_2}|^q \right. \tag{5.5.56}$$

$$\left. + \int_{-|\alpha|^{-1}}^{|\alpha|^{-1}} \max\{|a|, |a|^{-1}\}^{rq} \cdot |\langle \Psi, \psi_{(a,s,t)} \rangle_{L_2}|^q \frac{da}{|a|^{d+1}} \right) ds \, dt \Bigg\}.$$

Let us examine all four summands in (5.5.56) individually and show that all of them are finite. First, since $\mathcal{F}(f^*) = \overline{\mathcal{F}(f)}$ and $\Phi * \psi_{(a,s,0)}^* \in L_1(\mathbb{R}^d)$, we obtain

$$\langle \Phi, \psi_{(a,s,t)} \rangle_{L_2} = (\Phi * \psi_{(a,s,0)}^*)(t)$$

$$= \mathcal{F}^{-1}(\mathcal{F}(\Phi * \psi_{(a,s,0)}^*))(t)$$

$$= \mathcal{F}^{-1}(\widehat{\Phi}\overline{\mathcal{F}(\psi_{(a,s,0)})}))(t)$$

which implies

$$\int_{\mathbb{R}^d} |\langle \Phi, \psi_{(a,s,t)} \rangle_{L_2}|^q \, dt = \left\| \mathcal{F}^{-1} (\hat{\Phi} \overline{\mathcal{F}(\psi_{(a,s,0)})}) \right\|_{L_q}^q.$$

Applying Lemma 5.5.9 (ii) we see that $\hat{\Phi} \mathcal{F}(\psi_{(a,s,0)}) \equiv 0$ for all $s \notin Q_{d_2}$ where

$$d_2 := (a_0^{-1} + a_0^{-(1+\frac{1}{d})} a_1^{\frac{1}{d}}) b,$$

or $a \notin [-1, -\frac{a_0}{a_1}] \cup [\frac{a_0}{a_1}, 1]$. Hence,

$$\left\| \mathcal{F}^{-1}(\hat{\Phi} \overline{\mathcal{F}(\psi_{(a,s,0)})}) \right\|_{L_q}^q = 0$$

for all $s \notin Q_{d_2}$ or $a \notin [-1, -\frac{a_0}{a_1}] \cup [\frac{a_0}{a_1}, 1]$. Looking at the third summand in (5.5.56) we can use Lemma 5.5.9 (ii) and Young's inequality, see Proposition 1.3.8, and estimate

$$\operatorname*{ess\,sup}_{\alpha \in [-1,1]^*} \int_{\mathbb{R}^d} \int_{\mathbb{R}^{d-1}} |\alpha|^{-rq} \cdot |\langle \Phi, \psi_{(\alpha,s,t)} \rangle_{L_2}|^q \, ds \, dt$$

$$= \operatorname*{ess\,sup}_{\alpha \in [-1,1]^*} |\alpha|^{-rq} \cdot \int_{\mathbb{R}^{d-1}} \left\| \mathcal{F}^{-1}(\hat{\Phi} \overline{\mathcal{F}(\psi_{(\alpha,s,0)})}) \right\|_{L_q}^q \, ds$$

$$= \operatorname*{ess\,sup}_{\frac{a_0}{a_1} \leqslant |\alpha| \leqslant 1} |\alpha|^{-rq} \int_{Q_{d_2}} \left\| \Phi * \psi_{(\alpha,s,0)}^* \right\|_{L_q}^q \, ds$$

$$\leqslant \|\Phi\|_{L_q}^q \cdot |Q_{d_2}| \cdot \operatorname*{ess\,sup}_{\substack{\frac{a_0}{a_1} \leqslant |\alpha| \leqslant 1 \\ s \in Q_{d_2}}} |\alpha|^{-rq} \cdot \|\psi_{(\alpha,s,0)}\|_{L_1}^q,$$

where $|Q_{d_2}|$ is the size of Q_{d_2}. Using the definition of $\psi_{(\alpha,s,0)}$ we see that

$$\|\psi_{(\alpha,s,0)}\|_{L_1} = \int_{\mathbb{R}^d} |\det A_\alpha|^{-1/2} |\Psi(A_\alpha^{-1} S_s^{-1} x)| \, dx$$

$$= |\det A_\alpha|^{1/2} \int_{\mathbb{R}^d} |\Psi(x)| \, dx = \alpha^{1 - \frac{1}{2d}} \cdot \|\Psi\|_{L_1},$$

therefore

$$\operatorname*{ess\,sup}_{\alpha \in [-1,1]^*} \int_{\mathbb{R}^d} \int_{\mathbb{R}^{d-1}} |\alpha|^{-rq} \cdot |\langle \Phi, \psi_{(\alpha,s,t)} \rangle_{L_2}|^q \, ds \, dt$$

$$\leqslant \|\Phi\|_{L_q}^q \cdot \|\Psi\|_{L_1}^q \cdot |Q_{d_2}| \cdot \sup_{\frac{a_0}{a_1} \leqslant \alpha \leqslant 1} \alpha^{q(1 - \frac{1}{2d} - r)} < \infty, \qquad (5.5.57)$$

because $\Phi \in \mathcal{S}(\mathbb{R}^d) \subset L_q(\mathbb{R}^d)$ and $\Psi \in \mathcal{S}(\mathbb{R}^d) \subset L_1(\mathbb{R}^d)$.

The fourth summand in (5.5.56) can be treated analogously. We recall from Lemma 5.5.9 (i) that

$$\left\| \mathcal{F}^{-1}(\hat{\Psi} \overline{\mathcal{F}(\psi_{(a,s,0)})}) \right\|_{L_q}^q = 0$$

for all $a \notin [-a_1/a_0, -a_0/a_1] \cup [a_0/a_1, a_1/a_0]$ or $s \notin Q_{d_1}$ with $d_1 := (a_0^{-1} + a_0^{-(1+\frac{1}{d})} a_1^{\frac{1}{d}})b$. Then, Lemma 5.5.9 (i) and Young's inequality imply

$$
\operatorname*{ess\,sup}_{\alpha \in [-1,1]^*} \int_{\mathbb{R}^d} \int_{\mathbb{R}^{d-1}} \int_{-|\alpha|^{-1}}^{|\alpha|^{-1}} \max\{|a|, |a|^{-1}\}^{rq} \cdot |\langle \Psi, \psi_{(a,s,t)} \rangle_{L_2}|^q \, \frac{da}{|a|^{d+1}} \, ds \, dt
$$

$$
\leqslant \int_{\mathbb{R}} \max\{|a|, |a|^{-1}\}^{rq} \int_{\mathbb{R}^{d-1}} \|\mathcal{F}^{-1}(\widehat{\Psi} \overline{\mathcal{F}(\psi_{(a,s,0)})})\|_{L_q}^q \, ds \, \frac{da}{|a|^{d+1}}
$$

$$
= \left(\int_{-\frac{a_1}{a_0}}^{-\frac{a_0}{a_1}} + \int_{\frac{a_0}{a_1}}^{\frac{a_1}{a_0}} \right) \max\{|a|, |a|^{-1}\}^{rq} \int_{Q_{d_1}} \|\Psi * \psi_{(a,s,0)}^*\|_{L_q}^q \, ds \, \frac{da}{|a|^{d+1}}
$$

$$
\leqslant \|\Psi\|_{L_1}^q \cdot \|\Psi\|_{L_q}^q \cdot |Q_{d_1}| \cdot 2 \int_{\frac{a_0}{a_1}}^{\frac{a_1}{a_0}} \max\{a, a^{-1}\}^{rq} a^{q(1-\frac{1}{2d})-d-1} \, da < \infty. \tag{5.5.58}
$$

Similarly, the second summand in (5.5.56) can be estimated using Lemma 5.5.9 (ii) and Young's inequality and we get

$$
\int_{\mathbb{R}^d} \int_{\mathbb{R}^{d-1}} \int_{-1}^{1} |a|^{-rq} \cdot |\langle \Phi, \psi_{(a,s,t)} \rangle_{L_2}|^q \, \frac{da}{|a|^{d+1}} \, ds \, dt
$$

$$
= \int_{-1}^{1} |a|^{-rq} \int_{\mathbb{R}^{d-1}} \|\mathcal{F}^{-1}(\widehat{\Phi} \overline{\mathcal{F}(\psi_{(a,s,0)})})\|_{L_q}^q \, ds \, \frac{da}{|a|^{d+1}}
$$

$$
= \left(\int_{-1}^{-\frac{a_0}{a_1}} + \int_{\frac{a_0}{a_1}}^{1} \right) |a|^{-rq} \int_{Q_{d_2}} \|\Phi * \psi_{(a,s,0)}^*\|_{L_q}^q \, ds \, \frac{da}{|a|^{d+1}}
$$

$$
\leqslant \|\Phi\|_{L_q}^q \cdot \|\Psi\|_{L_1}^q \cdot |Q_{d_2}| \cdot 2 \int_{\frac{a_0}{a_1}}^{1} a^{q(1-\frac{1}{2d}-r)-d-1} \, da < \infty \tag{5.5.59}
$$

It remains to show the that the first summand in (5.5.56) is finite. Unfortunately the same approach as above is not suitable for this summand, instead we choose $q_0 > 0$ and $q_1 > 0$ such that $q_0 + q_1 = q > 1$. We will specify the choice at the end of the proof. We calculate

$$
\int_{\mathbb{R}^d} \int_{\mathbb{R}^{d-1}} |\langle \Phi, \psi_{(\infty,s,t)} \rangle_{L_2}|^q \, ds \, dt
$$

$$
= \int_{\mathbb{R}^{d-1}} \int_{\mathbb{R}^d} |(\Phi * \psi_{(\infty,s,0)}^*)(-t)|^{q_0+q_1} \, dt \, ds
$$

$$
= \int_{\mathbb{R}^{d-1}} \int_{\mathbb{R}^d} |(\Phi * \psi_{(\infty,s,0)}^*)(-t)|^{q_0} \cdot |\mathcal{F}^{-1}(\widehat{\Phi} \overline{\mathcal{F}(\psi_{(\infty,s,0)})})(-t)|^{q_1} \, dt \, ds
$$

$$
\lesssim \int_{\mathbb{R}^{d-1}} \int_{\mathbb{R}^d} \left(\int_{\mathbb{R}^d} |\Phi(x+t) \psi_{(\infty,s,0)}(x)| \, dx \right)^{q_0} dt \left(\int_{\mathbb{R}^d} |\widehat{\Phi}(\omega) \mathcal{F}\psi_{(\infty,s,0)}(\omega)| \, \omega \right)^{q_1} ds
$$

$$
=: \int_{\mathbb{R}^{d-1}} I_0(s) I_1(s) \, ds. \tag{5.5.60}
$$

Note that the second integral is not concerned with translations by t. In the folloing we treat both integrals independently.

$I_0(s)$: Since $\Phi \in \mathcal{S}(\mathbb{R}^d)$, for every $k \in \mathbb{N}$ it follows that $|\Phi(x)| \leqslant C \cdot (1 + \|x\|_2)^{-k}$ for all $x \in \mathbb{R}^d$ with the constant $C > 0$ depending on k, d and Φ. Then,

$$
I_0(s) \leqslant \int_{\mathbb{R}^d} \left(\int_{\mathbb{R}^d} (1 + \|x+t\|_2)^{-k} (1 + \|S_{-s}x\|_2)^{-k} \, dx \right) dt
$$

$$
\lesssim \int_{\mathbb{R}^d} \left(\int_{\mathbb{R}^d} \prod_{i=1}^{d} \left[(1 + |x_i + t_i|)^{-k} (1 + |(S_{-s}x)_i|)^{-k} \right] dx \right) dt = \int_{\mathbb{R}^d} I_s(t)^{q_0} \, dt
$$

for all $s \in \mathbb{R}^{d-1}$ fixed and where $(S_{-s}x)_i$ denotes the i-th entry of the vector $S_{-s}x \in \mathbb{R}^d$. Recall that $I_s(t)$ was originally defined in (5.5.47). By invoking Lemma 5.5.12 we observe

$$I_0(s) \lesssim \int_{\mathbb{R}^d} I_s(t)^{q_0}\, dt \lesssim (1 + \|s\|_2)^{1-q_0} \int_{\mathbb{R}^d} \prod_{i=1}^{d} (1 + |t_i|)^{-kq_0}\, dt, \qquad (5.5.61)$$

where the constant is independent of q_0 and s.

$I_1(s)$: We shall now deal with the second factor in (5.5.59) for some $q_1 > 0$. By part (iii) of Lemma 5.5.9 and the definition of $\widehat{\Phi}$ we obtain

$$\begin{aligned}
I_1(s)^{1/q_1} &= \int_{\mathbb{R}^d} |\widehat{\Phi}(\omega)\widehat{\Phi}(S_s^T \omega)|\, d\omega \\
&\leq \int_{\Omega_s} |\omega_1|^{d-1} \left(\int_{\mathbb{R}} \frac{|\widehat{\Psi}(\xi_1, \widetilde{\omega})|^2}{|\xi_1|^d}\, d\xi_1 \right)^{1/2} \left(\int_{\mathbb{R}} \frac{|\widehat{\Psi}(\xi_1, \widetilde{S_s^T \omega})|^2}{|\xi_1|^d}\, d\xi_1 \right)^{1/2}\, d\omega \qquad (5.5.62)
\end{aligned}$$

with

$$\Omega_s = \left\{ \xi \in \mathbb{R}^d : |\xi_1| \leq a_1,\ \max\{-b_i, -b_i - s_{i-1}\xi_1\} \leq \xi_i \leq \min\{b_i, b_i - s_{i-1}\xi_1\},\ i = 2, \ldots, d \right\}.$$

Since $\widehat{\Psi}$ is compactly supported away from zero and continuous, i.e., $\|\widehat{\Psi}\|_{L_\infty} < \infty$, we have

$$\left(\int_{\mathbb{R}} \frac{|\widehat{\Psi}(\xi_1, \widetilde{\omega})|^2}{|\xi_1|^d} d\xi_1 \right)^{1/2} \left(\int_{\mathbb{R}} \frac{|\widehat{\Psi}(\xi_1, \widetilde{S_s^T \omega})|^2}{|\xi_1|^d} d\xi_1 \right)^{1/2} \leq 2\|\widehat{\Psi}\|_{L_\infty}^2 \cdot \int_{a_0}^{a_1} \frac{d\xi_1}{\xi_1^d} < \infty. \qquad (5.5.63)$$

Together with (5.5.62) this means

$$I_1(s)^{1/q_1} \lesssim \int_{\Omega_s} |\omega_1|^{d-1}\, d\omega. \qquad (5.5.64)$$

In the following we assume $s > 0$ componentwise, all other cases can be treated analogously by symmetry arguments. Then, for any $\omega \in \Omega_s$ it follows that $|\omega_1| \leq 2b_i s_{i-1}^{-1}$ for all $i = 2, \ldots, d$, hence, $|\omega_1| \lesssim (\max_{i=1,\ldots,d-1} s_i)^{-1} = \|s\|_\infty^{-1}$. Moreover, since $\omega \in \operatorname{supp}\widehat{\Phi}$, we derive $|\omega_1| \leq a_1$ and $|\omega_{i+1}| \leq b_i$ for all $i = 1, \ldots, d-1$. We can now estimate (5.5.64) in the following manner:

$$I_1(s)^{1/q_1} \lesssim \int_{|\omega_1| \leq \min\{a_1, \|s\|_\infty^{-1}\}} |\omega_1|^{d-1}\, d\omega_1.$$

Assume first that $\|s\|_\infty^{-1} \geq a_1$, then we have

$$I_1(s)^{1/q_1} \lesssim \int_{|\omega_1| \leq a_1} |\omega_1|^{d-1}\, d\omega_1 \lesssim a_1^d \lesssim (1 + \|s\|_2)^{-d}.$$

On the other hand if $\|s\|_\infty^{-1} < a_1$ it follows that

$$I_1(s)^{1/q_1} \lesssim \int_{|\omega_1| \leq \|s\|_\infty^{-1}} |\omega_1|^{d-1}\, d\omega_1 \lesssim \|s\|_\infty^{-d} \lesssim (1 + \|s\|_2)^{-d}.$$

In both cases we obtain

$$I_1(s) \lesssim (1 + \|s\|)^{-dq_1}, \qquad (5.5.65)$$

where the constant is independent of s.

The estimates (5.5.61) and (5.5.65) plugged into (5.5.59) yield

$$\int_{\mathbb{R}^d}\int_{\mathbb{R}^{d-1}}|\langle\Phi,\psi_{(\infty,s,t)}\rangle_{L_2}|^q\,\mathrm{d}s\,\mathrm{d}t\lesssim\int_{\mathbb{R}^{d-1}}I_0(s)I_1(s)\,\mathrm{d}s \qquad (5.5.66)$$

$$\lesssim\int_{\mathbb{R}^d}\prod_{i=1}^{d}(1+|t_i|)^{-kq_0}\,\mathrm{d}t\int_{\mathbb{R}^{d-1}}(1+\|s\|_2)^{1-q_0-dq_1}\,\mathrm{d}s.$$

For any choice of q_0 we can find a $k\in\mathbb{N}$, such that the first integral in (5.5.66) converges. The second integral in (5.5.66) is known to converge if and only if $q_0+dq_1>d$. This can be obtained by setting $q_0=\frac{q-1}{d}$ and $q_1=\frac{d-1}{d}q+\frac{1}{d}$. If $q>1$ this choice satisfies

$$q_0+q_1=\frac{q-1}{d}+\frac{d-1}{d}q+\frac{1}{d}=\frac{1}{d}(q-1+q(d-1)+1)=q$$

and

$$q_0+dq_1=\frac{q-1}{d}+(d-1)q+1=1+q\left(d-1+\frac{1}{d}\right)-\frac{1}{d}$$

$$=d-\left(d-1+\frac{1}{d}\right)+q\left(d-1+\frac{1}{d}\right)=d+(q-1)\left(d-1+\frac{1}{d}\right)>d$$

and we finally conclude

$$\int_{\mathbb{R}^d}\int_{\mathbb{R}^{d-1}}|\langle\Phi,\psi_{(\infty,s,t)}\rangle_{L_2}|^q\,\mathrm{d}s\,\mathrm{d}t<\infty.$$

Altogether with (5.5.57), (5.5.58) and (5.5.59) we have now shown that all four summands in (5.5.56) are finite and this concludes the proof. $\qquad\square$

Before we close this subsection and apply the result of Theorem 5.5.13, we show that there exist functions that satisfy the proposed assumptions. To be more precise, we show that there exists a compactly supported function $\widehat{\Psi}\in\mathcal{S}(\mathbb{R}^d)$, such the function $\widehat{\Phi}$ defined via (5.5.19) satisfies $\Psi\in\mathcal{S}(\mathbb{R}^d)$.

Example 5.5.14. We fix any odd dimension $d\geqslant 3$. Then, for $\xi=(\xi_1,\widetilde{\xi})\in\mathbb{R}^d$ let $\widehat{\Psi}(\xi):=\widehat{\psi_1}(\xi_1)\widehat{\psi_2}(\widetilde{\xi})$ with

$$\widehat{\psi_1}(\xi_1):=\begin{cases}|\xi_1|^{\frac{d}{2}}e^{\frac{1}{(\xi_1-1)(\xi_1-3)}}, & 1<\xi_1<3\\ |\xi_1|^{\frac{d}{2}}e^{\frac{1}{(\xi_1+1)(\xi_1+3)}}, & -3<\xi_1<-1\\ 0, & \text{otherwise}\end{cases}$$

and $\widehat{\psi_2}\in\mathcal{C}_0^\infty(\mathbb{R}^{d-1})$ satisfying $\widehat{\psi_2}\geqslant 0$ and $\|\widehat{\psi_2}\|_{L_2}=\|\gamma\|_{L_2}^{-1}$ with $\gamma(\xi_1):=\widehat{\psi_1}(\xi_1)\cdot|\xi_1|^{-d/2}$. The function Ψ is an admissible shearlet, i.e., it satisfies (5.5.10):

$$C_\Psi=\int_{\mathbb{R}^d}\frac{|\widehat{\Psi}(\omega)|^2}{|\omega_1|^d}\,\mathrm{d}\omega=\int_\mathbb{R}\frac{|\widehat{\psi_1}(\omega_1)|^2}{|\omega_1|^d}\,\mathrm{d}\omega_1\cdot\|\widehat{\psi_2}\|_{L_2}^2=\|\gamma\|_{L_2}^2\cdot\|\widehat{\psi_2}\|_{L_2}^2=1.$$

According to Remark 5.5.5 we set

$$\widehat{\Phi}(\xi):=\xi_1^{\frac{d-1}{2}}\left(\int_{\mathbb{R}\setminus[-|\xi_1|,|\xi_1|]}\frac{|\widehat{\Psi}(\omega_1,\widetilde{\xi})|^2}{|\omega_1|^d}\,\mathrm{d}\omega_1\right)^{1/2}$$

$$=\xi_1^{\frac{d-1}{2}}\cdot|\widehat{\psi_2}(\widetilde{\xi})|\cdot\left(2\int_{\max\{|\xi_1|,1\}}^3 e^{\frac{2}{(\omega_1-1)(\omega_1-3)}}\,\mathrm{d}\omega_1\right)^{1/2}=:\sqrt{2}\cdot\xi_1^{\frac{d-1}{2}}\cdot|\widehat{\psi_2}(\widetilde{\xi})|\cdot\widehat{\varphi_1}(\xi_1)$$

with $\hat{\Phi}(\xi) = 0$ for $|\xi_1| > 3$. Now we show that this function satisfies the required assumptions. The fact that $\widehat{\psi_1} \in C_0^\infty(\mathbb{R})$ and therefore $\hat{\Psi} \in C_0^\infty(\mathbb{R}^d)$ is immediately obvious. With the given construction, together with (5.5.19), we see that the necessary condition from Theorem 5.5.4 is satisfied, i.e., the functions Φ and Ψ generate a Parseval frame. Furthermore, if we assume $\hat{\Phi} \in C_0^\infty(\mathbb{R}^d) \subset \mathcal{S}(\mathbb{R}^d)$ then $\Phi \in \mathcal{S}(\mathbb{R}^d) \subset L_1(\mathbb{R}^d) \cap L_2(\mathbb{R}^d)$ and all necessary conditions on Φ are satisfied.

So we need to show that $\hat{\Phi} \in C_0^\infty(\mathbb{R}^d)$, which means that we will show that $\widehat{\varphi_1}$ is infinitely continuously differentiable since $\xi_1^{\frac{d-1}{2}}$, for d odd, is a monomial. To show this we need to prove that

$$\lim_{x \nearrow 3} \frac{\mathrm{d}^n}{\mathrm{d}x^n}(\widehat{\varphi_1}(x)) = 0$$

and

$$\lim_{x \searrow 1} \frac{\mathrm{d}^n}{\mathrm{d}x^n}(\widehat{\varphi_1}(x)) = 0$$

for all $n \in \mathbb{N}$. Since both statements are proven in an analogous manner, we will only show the proof of the first statement and for the remainder of this example we assume $2 < x < 3$.

Because we have $\widehat{\varphi_1}(x) = (f \circ g)(x)$ with $f(x) = \sqrt{x}$ and

$$g(x) = \int_x^3 e^{\frac{2}{(\omega-1)(\omega-3)}} \, \mathrm{d}\omega,$$

we can use Faà di Bruno's formula to get a closed expression for the n-th derivative. Recall that for two functions f and g the identity

$$\frac{\mathrm{d}^n}{\mathrm{d}x^n}((f \circ g)(x)) = \sum_{k=1}^n \frac{\mathrm{d}^k f}{\mathrm{d}x^k}(g(x))B_{n,k}\left(\frac{\mathrm{d}g}{\mathrm{d}x}(x), \frac{\mathrm{d}^2g}{\mathrm{d}x^2}(x), \dots, \frac{\mathrm{d}^{(n-k+1)}g}{\mathrm{d}x^{(n-k+1)}}(x)\right) \qquad (5.5.67)$$

holds true with $B_{n,k}$ being the Bell polynomials, i.e.,

$$B_{n,k}(x_1, x_2, \dots, x_{(n-k+1)}) = \sum \frac{n!}{j_1! \cdots j_{(n-k+1)}!}\left(\frac{x_1}{1!}\right)^{j_1} \times \cdots \times \left(\frac{x_{(n-k+1)}}{(n-k+1)!}\right)^{j_{(n-k+1)}}.$$

The sum in the above expression is taken over all multiindices $(j_1, \dots, j_{(n-k+1)})$ with $j_1 + \cdots + j_{(n-k+1)} = k$ and $j_1 + 2j_2 + \cdots + (n-k+1)j_{(n-k+1)} = n$.

The derivatives of the square root satisfy

$$\frac{\mathrm{d}^k f}{\mathrm{d}x^k}(x) = c_k x^{-k+\frac{1}{2}}$$

with c_k being some constant depending on k and since because of $1 < x < 3$ we have

$$\frac{\mathrm{d}g}{\mathrm{d}x}(x) = -e^{\frac{2}{(x-1)(x-3)}} \qquad (5.5.68)$$

this means that for all $k \in \mathbb{N}$ the derivatives of g satisfy

$$\frac{\mathrm{d}^k g}{\mathrm{d}x^k}(x) = Q_k(x) \cdot e^{\frac{2}{(x-1)(x-3)}}$$

with Q_k being some rational function without singularities in the interval $(1,3)$. At this point we restrict ourselves to values $x \in (3 - \varepsilon, 3)$ for $\varepsilon > 0$ sufficiently small. Thus, using (5.5.67) we now have

$$\frac{\mathrm{d}^n \widehat{\varphi_1}}{\mathrm{d}x^n}(x) = \sum_{k=1}^n c_k \big(g(x)\big)^{-k+\frac{1}{2}} \sum_{(j_1, \ldots, j_{(n-k+1)})} c_{n,k,j} \Big(Q_1(x) \cdot e^{\frac{2}{(x-1)(x-3)}}\Big)^{j_1} \times \ldots$$

$$\times \Big(Q_{(n-k+1)}(x) \cdot e^{\frac{2}{(x-1)(x-3)}}\Big)^{j_{(n-k+1)}}$$

$$= \sum_{k=1}^n \operatorname{sgn}(R_{k,n}) \cdot |R_{k,n}(x)| \cdot \big(g(x)\big)^{-k+\frac{1}{2}} \cdot \Big(e^{\frac{2}{(x-1)(x-3)}}\Big)^k$$

$$= \sum_{k=1}^n \operatorname{sgn}(R_{k,n}) \cdot \left(\frac{\widetilde{R}_{k,n}(x) \cdot \big(e^{\frac{2}{(x-1)(x-3)}}\big)^{1+\frac{1}{2k-1}}}{g(x)}\right)^{k-\frac{1}{2}}$$

where $R_{k,n}$ is a rational function for every $k = 1, \ldots, n$ possibly changing from line to line, $\widetilde{R}_{k,n}(x) := |R_{k,n}(x)|^{(k-\frac{1}{2})^{-1}}$ and $\operatorname{sgn}(R_{k,n})$ denotes the sign of $R_{k,n}$ on the interval $(3 - \varepsilon, 3)$. By choosing ε suitably small, the sign of $R_{k,n}$ is constant, whereat $\operatorname{sgn}(R_{k,n}) = 0$ if $R_{k,n} \equiv 0$. Since

$$\lim_{x \nearrow 3} \widetilde{R}_{k,n}(x) \cdot \big(e^{\frac{2}{(x-1)(x-3)}}\big)^{1+\frac{1}{2k-1}} = 0 \quad \text{and} \quad \lim_{x \nearrow 3} g(x) = 0$$

we use l'Hospital's rule to determine the limit of the fraction. For the derivative of the numerator we obtain

$$\frac{\mathrm{d}}{\mathrm{d}x}\Big(\widetilde{R}_{k,n}(x) \cdot \big(e^{\frac{2}{(x-1)(x-3)}}\big)^{1+\frac{1}{2k-1}}\Big)$$

$$= \frac{\mathrm{d}}{\mathrm{d}x}\widetilde{R}_{k,n}(x) \cdot \big(e^{\frac{2}{(x-1)(x-3)}}\big)^{1+\frac{1}{2k-1}} + \widetilde{R}_{k,n}(x) \cdot \frac{\mathrm{d}}{\mathrm{d}x}\big(e^{\frac{2}{(x-1)(x-3)}}\big)^{1+\frac{1}{2k-1}}$$

$$= Q(x) \cdot \big(e^{\frac{2}{(x-1)(x-3)}}\big)^{1+\frac{1}{2k-1}}$$

where Q is of the form $Q(x) = Q_2(x) \cdot (Q_1(x))^{\frac{-2k+3}{2k-1}} + Q_3(x) \cdot (Q_1(x))^{\frac{2}{2k-1}}$ with Q_1, Q_2, Q_3 being rational functions. This, together with (5.5.68), yields

$$\lim_{x \nearrow 3} \frac{\frac{\mathrm{d}}{\mathrm{d}x}\Big(\widetilde{R}_{k,n}(x) \cdot \big(e^{\frac{2}{(x-1)(x-3)}}\big)^{1+\frac{1}{2k-1}}\Big)}{\frac{\mathrm{d}}{\mathrm{d}x}\big(g(x)\big)} = \lim_{x \nearrow 3} Q(x) \cdot e^{\frac{2}{(2k-1)((x-1)(x-3))}} = 0.$$

Thus, with l'Hospital's rule we get

$$\lim_{x \nearrow 3} \frac{\mathrm{d}^n \widehat{\varphi_1}}{\mathrm{d}x^n}(x) = \lim_{x \nearrow 3} \sum_{k=1}^n \left(\frac{\widetilde{R}_{k,n}(x) \cdot \big(e^{\frac{2}{(x-1)(x-3)}}\big)^{1+\frac{1}{2k-1}}}{g(x)}\right)^{k-\frac{1}{2}}$$

$$= \sum_{k=1}^n \left(\lim_{x \nearrow 3} \frac{\widetilde{R}_{k,n}(x) \cdot \big(e^{\frac{2}{(x-1)(x-3)}}\big)^{1+\frac{1}{2k-1}}}{g(x)}\right)^{k-\frac{1}{2}}$$

$$= \sum_{k=1}^n \left(\lim_{x \nearrow 3} \frac{\frac{\mathrm{d}}{\mathrm{d}x}\Big(\widetilde{R}_{k,n}(x) \cdot \big(e^{\frac{2}{(x-1)(x-3)}}\big)^{1+\frac{1}{2k-1}}\Big)}{\frac{\mathrm{d}}{\mathrm{d}x}\big(g(x)\big)}\right)^{k-\frac{1}{2}} = 0.$$

With very similar arguments we can also show $\lim_{x \searrow 1} \frac{\mathrm{d}^n}{\mathrm{d}x^n}(\widehat{\varphi_1}(x)) = 0$. This proves $\widehat{\varphi_1} \in \mathcal{C}_0^\infty(\mathbb{R})$ and therefore $\widehat{\Phi} \in \mathcal{C}_0^\infty(\mathbb{R}^d)$.

5.5.3 Inhomogeneous Shearlet Coorbit Spaces

Now we are finally in the position to define coorbit spaces $\mathrm{Co}(L_{p,m_r})$ associated to the inhomogeneous shearlet frame $\mathfrak{F}$. Let first briefly recall the results of the previous subsection. Again, we look at odd dimensions $d \geqslant 3$ and set $\mathfrak{F} = (\psi_x)_{x \in X}$ as in Definition 5.5.2 with $\Psi \in \mathcal{S}(\mathbb{R}^d)$ satisfying the support property $\mathrm{supp}\,\widehat{\Psi} \subseteq ([-a_1, -a_0] \cup [a_0, a_1]) \times Q_b$ for some $0 < a_0 < a_1$ and $b \in \mathbb{R}_+^{d-1}$. With Φ as in Remark 5.5.5 the family $\mathfrak{F}$ is a Parseval frame due to Theorem 5.5.4 and by Theorem 5.5.13 the associated kernel satisfies $K_{\mathfrak{F}} \in \mathcal{A}_{q,w_r}$ for all $1 < q < \infty$, where the weight w_r is given by (5.5.31)–(5.5.33). With the corresponding w-moderate weight m_r, which is defined via

$$m_r(a, s, t) = \begin{cases} 1, & a = \infty, \\ |a|^{-r}, & a \in [-1, 1]^* \end{cases}$$

for $r \in \mathbb{R}_{\geqslant 0}$, the coorbit spaces $\mathrm{Co}(L_{p,m_r})$ are defined according to Definition 5.1.8.

Definition 5.5.15. For any $1 \leqslant p < \infty$ and $r \in \mathbb{R}_{\geqslant 0}$ the *inhomogeneous shearlet coorbit spaces* with respect to the Lebesgue space $L_{p,m_r}(X, \mu)$ are defined as

$$\mathcal{SC}_{\mathfrak{F},p}^r := \mathrm{Co}(L_{p,m_r}) = \left\{ f \in \mathcal{H}_{m_r}' \,:\, \mathcal{SH}_{\mathfrak{F}} f \in L_{p,m_r}(X, \mu) \right\}$$

endowed with the natural norm

$$\|f\|_{\mathcal{SC}_{\mathfrak{F},p}^r} := \|\mathcal{SH}_{\mathfrak{F}} f\|_{L_{p,m_r}}.$$

By the results of Section 5.1 these spaces are Banach spaces as the following Propositions states.

Proposition 5.5.16. *The spaces $\mathcal{SC}_{\mathfrak{F},p}^r$ with $1 \leqslant p < \infty$ and $r \in \mathbb{R}_{\geqslant 0}$ are well-defined Banach spaces and $\mathcal{SC}_{\mathfrak{F},p}^r$ is isometrically isomorphic to $\mathcal{M}_{p,m_r} = \{ F \in L_{p,m_r}(X, \mu) : K_{\mathfrak{F}} F = F \}$ via the map $\mathcal{SH}_{\mathfrak{F}}$.*

Proof. Due to Theorem 5.5.13 the condition $K_{\mathfrak{F}}(L_{p,m_r}(X, \mu)) \subset L_{r,m_r^{-1}}(X, \mu)$ is fulfilled, therefore Proposition 5.1.9 shows the assumption. $\qquad\square$

Before closing this subsection let us recall the following basic embeddings.

Corollary 5.5.17. *Let $1 \leqslant p < q < \infty$ and $0 \leqslant r < s < \infty$, then the following embeddings hold true:*

(i) $\mathcal{SH}_{\mathfrak{F},p}^r \hookrightarrow \mathcal{SH}_{\mathfrak{F},q}^r$,

(ii) $\mathcal{SH}_{\mathfrak{F},p}^s \hookrightarrow \mathcal{SH}_{\mathfrak{F},p}^r$.

Proof. Part (i) is an application of Corollary 5.1.11 (i) and part (ii) is a direct consequence of the pointwise estimate $m_r \leqslant m_2$. $\qquad\square$

5.5.4 Conditions on the Oscillation

In this subsection we intend to lay the foundation for a first discretization of the inhomogeneous shearlet coorbit spaces $\mathcal{SC}_{\mathfrak{F},p}^r$. To be more precise, we will apply Theorem 5.2.17 and for this we need to show that the necessary conditions are fulfilled. Recalling the conditions we thus have to show that Assumption 5.1.2 is fulfilled and the $\mathcal{U}$-oscillation of the kernel $K_{\mathfrak{F}}$ fulfills $\mathrm{osc}_{\mathcal{U}} K_{\mathfrak{F}} \in \mathcal{A}_{q,w_r}$ for some $q < r$ and a suitable covering $\mathcal{U}$ of X. Assumption 5.1.2 requires $K_{\mathfrak{F}} \in \mathcal{A}_{q,w_r}$ for all $1 < q < \infty$, which is readily satisfied by Theorem 5.5.13 for a specific family of functions $\mathfrak{F} = (\psi_x)_{x \in X}$. In the following we briefly recall these frames and show in the main theorem of this subsection that the second assumption is also fulfilled in this setting.

Let $d \geqslant 3$ be odd and $\Psi \in \mathcal{S}(\mathbb{R}^d) \subset L_1(\mathbb{R}^d) \cap L_2(\mathbb{R}^d)$ be an admissible shearlet with $C_\Psi = 1$ satisfying the support condition

$$\hat{\Psi} \subseteq ([-a_1, -a_0] \cup [a_0, a_1]) \times Q_b$$

for some $0 < a_0 < a_1$ and $b \in \mathbb{R}_+^{d-1}$. Furthermore, let $\Phi \in \mathcal{S}(\mathbb{R}^d)$ be chosen as in (5.5.19) such that condition (5.5.16) is fulfilled with $A = B = 1$. By Theorem 5.5.4 this means that the family $\mathfrak{F} = (\psi_x)_{x \in X}$ defined via

$$\psi_{(\infty,s,t)} := T_t D_s \Phi = \Phi(S_s^{-1}(\cdot - t)) \quad \text{and}$$
$$\psi_{(a,s,t)} := T_t D_s D_a \Psi = |\det A_a|^{-1/2} \Psi(A_a^{-1} S_s^{-1}(\cdot - t))$$

constitutes a Parseval frame for $L_2(\mathbb{R}^d)$. The kernel $K_{\mathfrak{F}}$ associated with $\mathfrak{F}$ is given by

$$K_{\mathfrak{F}}(x,y) = \langle \psi_y, \psi_x \rangle_{L_2}.$$

Recall that if the weight w_r is defined via (5.5.31)–(5.5.33), then Theorem 5.5.13 shows that the kernel $K_{\mathfrak{F}}$ is contained in all $\mathcal{A}_{q,w_r}$ for $1 < q < \infty$.

The next step is to find a suitable admissible covering $\mathcal{U}$ of X. For this we define the following sets, which are heavily inspired by the discretization results for homogeneous shearlet coorbit spaces in Subsection 2.3.2.

Definition 5.5.18. Let $\alpha > 1$ and $\beta, \tau > 0$, then we define the family of subsets $\mathcal{U} := (U_\lambda)_{\lambda \in \Lambda}$, where

$$\Lambda := \left(\{\infty\} \times \mathbb{Z}^{d-1} \times \mathbb{Z}^d\right) \cup \left(\{-1,1\} \times \mathbb{N}_0 \times \mathbb{Z}^{d-1} \times \mathbb{Z}^d\right) \tag{5.5.69}$$

and the subsets $U_\lambda \subset X$ are given by

$$U_{(\infty,k,m)} := \{\infty\} \times \beta \left(k + \left[-\frac{1}{2}, \frac{1}{2}\right)^{d-1}\right) \times \gamma S_{\beta k} \left(m + \left[-\frac{1}{2}, \frac{1}{2}\right)^d\right), \tag{5.5.70}$$

$$U_{(\varepsilon,j,k,m)} := \varepsilon \left(\alpha^{-j-1}, \alpha^{-j}\right] \times \alpha^{j(\frac{1}{d}-1)} \beta \left(k + \left[-\frac{1}{2}, \frac{1}{2}\right)^{d-1}\right)$$
$$\times \gamma S_{\alpha^{j(\frac{1}{d}-1)} \beta k} A_{\alpha^{-j}} \left(m + \left[-\frac{1}{2}, \frac{1}{2}\right)^d\right), \tag{5.5.71}$$

for $\varepsilon \in \{-1,1\}$, $j \in \mathbb{N}_0$, $k \in \mathbb{Z}^{d-1}$ and $m \in \mathbb{Z}^d$.

We collect some elementary properties of $\mathcal{U}$ in the following lemma.

Lemma 5.5.19. *For fixed $\alpha > 1$ and $\beta, \tau > 0$ the family $\mathcal{U} = (U_\lambda)_{\lambda \in \Lambda}$ has the following properties:*

(i) $\mathcal{U}$ is pairwise disjoint, i.e., $U_\lambda \cap U_{\lambda'} \neq \emptyset$ if and only if $\lambda = \lambda'$;

(ii) it holds that $\bigcup_{\lambda \in \Lambda} = X$;

(iii) for all $\varepsilon \in \{-1,1\}$, $j \in \mathbb{N}_0$, $k \in \mathbb{Z}^{d-1}$ and $m \in \mathbb{Z}^d$ we have

$$\mu(U_{(\infty,k,m)}) = \beta^{d-1} \tau^d \quad \text{and} \quad \mu(U_{(\varepsilon,j,k,m)}) = \frac{1}{d}(\alpha^d - 1)\beta^{d-1}\gamma^d.$$

Hence, $\mathcal{U}$ is a moderate admissible covering of X in the sense of Definition 1.5.5.

Proof. The first statement is obviously true.

To prove part (ii), let $(a, s, t) \in (0,1] \times \mathbb{R}^{d-1} \times \mathbb{R}^d$ be arbitrary. Since $\{(\alpha^{-j-1}, \alpha^{-j}]\}_{j \in \mathbb{N}_0}$ is a partition of $(0,1]$, there is a $j \in \mathbb{N}_0$ such that $a \in (\alpha^{-j-1}, \alpha^{-j}]$. Similarly, there is a unique $k \in \mathbb{Z}^{d-1}$ with $\alpha^{j(1-\frac{1}{d})}\beta^{-1}s - k \in [-1/2, 1/2)^{d-1}$. Because the shearing and dilation matrices are invertible, there is some $m \in \mathbb{Z}^d$ with $\tau^{-1}A_{\alpha^j}S_{-\alpha^j(\frac{1}{d}-1)\beta k} - m \in [-1/2, 1/2)^d$. By this construction it holds $x \in U_{(\varepsilon, j, k, m)}$. The other cases can be proven analogously.

To prove (iii) we write 1_d for a d-dimensional vector with ones only and calculate for $j \in \mathbb{N}_0$, $k \in \mathbb{Z}^{d-1}$ and $m \in \mathbb{Z}^d$:

$$
\mu(Q_{(1,j,k,m)}) = \int_{\alpha^{-(j+1)}}^{\alpha^{-j}} \frac{da}{|a|^{d+1}} \cdot \int_{\alpha^{j(\frac{1}{d}-1)}\beta(k-\frac{1}{2}1_{d-1})}^{\alpha^{j(\frac{1}{d}-1)}\beta(k+\frac{1}{2}1_{d-1})} ds \cdot \int_{\tau S_{\alpha^j(\frac{1}{d}-1)\beta k}A_{\alpha^{-j}}(m-\frac{1}{2}1_d)}^{\tau S_{\alpha^j(\frac{1}{d}-1)\beta k}A_{\alpha^{-j}}(m+\frac{1}{2}1_d)} dt
$$

$$
= \frac{1}{d}\alpha^{jd}(\alpha^d - 1) \cdot \alpha^{j(\frac{1}{d}-1)(d-1)}\beta^{d-1}\int_{-\frac{1}{2}1_{d-1}}^{\frac{1}{2}1_{d-1}} ds \cdot \tau^d |\det S_{\alpha^j(\frac{1}{d}-1)\beta k}\det A_{\alpha^{-j}}|\int_{-\frac{1}{2}1_d}^{\frac{1}{2}1_d} dt
$$

$$
= \frac{1}{d}(\alpha^d - 1)\beta^{d-1}\tau^d \cdot \alpha^{jd + j(2-\frac{1}{d}-d) + j(\frac{1}{d}-2)}
$$

$$
= \frac{1}{d}(\alpha^d - 1)\beta^{d-1}\tau^d,
$$

and analogously if $\varepsilon = -1$. In the same fashion we obtain

$$
\mu(Q_{(\infty,k,m)}) = \int_{\beta(k-\frac{1}{2}1_{d-1})}^{\beta(k+\frac{1}{2}1_{d-1})} ds \cdot \int_{\tau S_{\beta k}(m-\frac{1}{2}1_d)}^{\tau S_{\beta k}(m+\frac{1}{2}1_d)} dt = \beta^{d-1}\tau^d
$$

for any $k \in \mathbb{Z}^{d-1}$ and $m \in \mathbb{Z}^d$. □

As a direct consequence of Lemma 5.5.19 we note that (5.2.11) is fulfilled with

$$
C_{\mathcal{U}} = \frac{1}{d}(\alpha^d - 1)\beta^{d-1}\tau^d \quad \text{and} \quad D_{\mathcal{U}} = \beta^{d-1}\tau^d,
$$

at least when assuming $\alpha^d \leqslant d + 1$. Note further that by the disjointness of $\mathcal{U}$ we have $N_{\mathcal{U}} = 1$, where $N_{\mathcal{U}}$ is the bound in (5.2.10).

We add an additional technical result concerning the weight w_r, which we will need later on.

Corollary 5.5.20. *For all $\lambda \in \Lambda$ and $r \in \mathbb{R}_{\geqslant 0}$ we have*

$$
\sup_{x,y \in U_\lambda} w_r(x,y) \leqslant |\alpha|^r \tag{5.5.72}
$$

Proof. First, let $\lambda = (\infty, k, m)$ for some $k \in \mathbb{Z}^{d-1}$ and $m \in \mathbb{Z}^m$. Since w only depends on a this implies $w(x,y) = 1$ for all $x, y \in U_\lambda$.

Now let $\lambda = (\varepsilon, j, k, m)$ for some $\varepsilon \in \{-1, 1\}$, $j \in \mathbb{N}_0$, $k \in \mathbb{Z}^{d-1}$ and $m \in \mathbb{Z}^m$, then

$$
\sup_{x,y \in U_\lambda} w(x,y) = \sup_{\alpha^{-j-1} < a, a' \leqslant \alpha^{-j}} \max\left\{\frac{|a|}{|a'|}, \frac{|a'|}{|a|}\right\}^r \leqslant |\alpha|^r,
$$

which proves the assertion. □

We add the following very technical lemma.

Lemma 5.5.21. *Let $v, v' \in [-\frac{1}{2}, \frac{1}{2})^{d-1}$, $w, w' \in [-\frac{1}{2}, \frac{1}{2})^d$ as well as $j \in \mathbb{N}_0$, $k \in \mathbb{Z}^d$ and $m \in \mathbb{Z}^d$. For arbitrary $\alpha > 1$ and $\beta, \tau > 0$ let further $u, u' \in (\alpha^{-1}, 1]$ and set*

$$a' = \varepsilon \alpha^{-j} u, \qquad\qquad a'' = \varepsilon \alpha^{-j} u',$$
$$s' = \alpha^{j(\frac{1}{d}-1)}\beta(k+v), \qquad\qquad s'' = \alpha^{j(\frac{1}{d}-1)}\beta(k+v'),$$
$$t' = \tau S_{\alpha^{j(\frac{1}{d}-1)}\beta k} A_{\alpha^{-j}}(m+w), \qquad t'' = \tau S_{\alpha^{j(\frac{1}{d}-1)}\beta k} A_{\alpha^{-j}}(m+w').$$

Then, for any $\delta > 0$ we can chose α, β and τ sufficiently small, such that for all $s \in \mathbb{R}^{d-1}$ and $t \in \mathbb{R}^d$ the following inequalities are fulfilled:

$$|a'a''^{-1} - 1| \leq \delta, \tag{5.5.73}$$
$$\|A_{a''}^{-1} S_{s''}^{-1}(t' - t'')\|_2 \leq \delta \tag{5.5.74}$$
$$\||a'|^{\frac{1}{d}-1}(s - s') - |a''|^{\frac{1}{d}-1}(s - s'')\|_2 \leq \delta \cdot (\||a'|^{\frac{1}{d}-1}(s-s')\|_2 + 1), \tag{5.5.75}$$
$$\|A_{a'}^{-1} S_{s'}^{-1}(t - t') - A_{a''}^{-1} S_{s''}^{-1}(t - t'')\|_2 \leq \delta \cdot (\|A_{a'}^{-1} S_{s'}^{-1}(t-t')\|_2 + 1), \tag{5.5.76}$$
$$\|I_d - A_{a'a''^{-1}}^{-1} S_{|a''|^{\frac{1}{d}-1}(s'-s'')}^{-1}\|_{2 \to 2} \leq \delta. \tag{5.5.77}$$

Similarly, for $s' = \beta(k+v)$, $s'' = \beta(k+v')$ and $t' = S_{\beta k}(m+w)$, $t'' = S_{\beta k}(m+w')$ it holds that

$$\|s' - s''\|_2 \leq \delta, \tag{5.5.78}$$
$$\|I_d - S_{s'-s''}\|_{2 \to 2} \leq \delta, \tag{5.5.79}$$
$$\|S_{s''}^{-1}(t' - t'')\|_2 \leq \delta. \tag{5.5.80}$$

Proof. To show equation (5.5.73) we calculate

$$|a'a''^{-1} - 1| = u'^{-1}|u - u'| \leq 1 - \alpha^{-1}.$$

Next, we have

$$\|A_{a''}^{-1} S_{s''}^{-1}(t' - t'')\|_2 = \tau \|A_{u'} S_{\beta v'}(w - w')\|_2 \leq \tau \|A_{u'} S_{\beta v'}\|_{2 \to 2},$$

which can be made arbitrarily small and proves (5.5.74).

We now show (5.5.75). There exist $v_0, v_1 \in [-1, 1)^{d-1}$ and $k_0 \in \mathbb{Z}^{d-1}$ such that

$$s - s' = \alpha^{j(\frac{1}{d}-1)}\beta(k_0 + v_0) \quad \text{and} \quad s - s'' = \alpha^{j(\frac{1}{d}-1)}\beta(k_0 + v_1).$$

This implies

$$\begin{aligned}
\||a'|^{\frac{1}{d}-1}(s - s') - |a''|^{\frac{1}{d}-1}(s - s'')\|_2 &= \beta \|u(k_0 + v_0) - u'(k_0 + v_1)\|_2 \\
&\leq \beta |u - u'| \cdot \|k_0\|_2 + \beta |u - u'| \cdot \|v_0\|_2 + \beta u' \|v_0 - v_1\|_2 \\
&\lesssim \beta(1 - \alpha^{-1})\|k_0\|_2 + \beta(1 - \alpha^{-1}) + \beta.
\end{aligned}$$

The assertion now is a consequence of

$$\beta \|k_0\|_2 \leq u^{-1}\|u\beta(k_0 + v_0)\| + \beta\|v_0\|_2 \lesssim \alpha \||a'|^{\frac{1}{d}-1}(s - s')\|_2 + \beta.$$

To prove (5.5.76) we first note that by (2.3.19) we have

$$A_{a'}^{-1} S_{s'}^{-1} t' = \tau A_{\alpha^j u^{-1}} S_{-\alpha^{j(\frac{1}{d}-1)}\beta(k+v)} S_{\alpha^{j(\frac{1}{d}-1)}\beta k} A_{\alpha^{-j}}(m+w) = \tau A_u S_{\beta v}(m+w).$$

There exist $w_0, w_1 \in [-1,1)^d$ and $m_0 \in \mathbb{Z}^d$ such that

$$t - t' = S_{\alpha^j(\frac{1}{d}-1)\beta k} A_{\alpha^{-j}}(m_0 + w_0) \quad \text{and} \quad t - t'' = S_{\alpha^j(\frac{1}{d}-1)\beta k} A_{\alpha^{-j}}(m_0 + w_1),$$

which implies

$$\|A_{a'}^{-1} S_{s'}^{-1}(t - t') - A_{a''}^{-1} S_{s''}^{-1}(t - t'')\|_2$$
$$= \tau \|A_u S_{\beta v}(m_0 + w_0) - A_{u'} S_{\beta v'}(m_0 + w_1)\|_2$$
$$\leqslant \tau \|I_d - A_{u'} S_{\beta(v'-v)} A_u^{-1}\|_{2\to 2} \|A_u S_{\beta v} m_0\|_2 + \|A_u S_{\beta v} w_0 - A_{u'} S_{\beta v'} w_1\|_2$$
$$\leqslant \tau \delta \|A_u S_{\beta v} m_0\|_2 + \tau \|A_u S_{\beta v} - A_{u'} S_{\beta v'}\|_{2\to 2} \|w_0\|_2 + \tau \|A_{u'} S_{\beta v'}\|_{2\to 2} \|w_0 - w_1\|_2.$$

Since all entries are arbitrarily small, the norm of $A_u S_{\beta v} - A_{u'} S_{\beta v}$ is also small. On the other hand the matrix $A_{u'} S_{\beta v'}$ is nearly the identity and the norm therefore bounded. The assumption now follows from

$$\tau \|A_u S_{\beta v} m_0\|_2 \leqslant \|\tau A_u S_{\beta v}(m_0 + w_0)\|_2 + \tau \|A_u S_{\beta v}\|_{2\to 2}\|w_0\| \lesssim \|A_{a'}^{-1} S_{s'}^{-1}(t - t')\|_2 + \tau.$$

To show (5.5.77) we write

$$I_d - A_{a'a''-1}^{-1} S_{|a''|^{\frac{1}{d}-1}(s'-s'')}^{-1} = \begin{pmatrix} 1 - u^{-1}u' & \beta u'(v - v') \\ 0_{d-1} & I_{d-1}(1 - (u^{-1}u')^{\frac{1}{d}-1}) \end{pmatrix}.$$

By similar arguments as before all entries above are small for suitable α, β, which proves the assertion.

Finally, equations (5.5.78)–(5.5.80) are obviously true. $\qquad \square$

Recall the $\mathcal{U}$-oscillation of the kernel $K_{\mathfrak{F}}$ given by

$$\mathrm{osc}_{\mathcal{U}} K_{\mathfrak{F}}(x, y) = \sup_{z \in Q_y} |K_{\mathfrak{F}}(x, y) - K_{\mathfrak{F}}(x, z)| = \sup_{z \in Q_y} |\langle \psi_y - \psi_z, \psi_x \rangle_{L_2}|, \quad x, y \in X,$$

where $Q_y \in \mathcal{U}$ is the unique set U_λ, $\lambda \in \Lambda$, such that $y \in U_\lambda$. For Theorem 5.2.17 to hold true it remains to show that $\mathrm{osc}_{\mathfrak{F}} K_{\mathfrak{F}}$ is contained in some kernel spaces $\mathcal{A}_{q,w_r}$ for $q < p$ and $r \in \mathbb{R}_{\geqslant 0}$. This is shown in the next theorem.

Theorem 5.5.22. *Let $\mathfrak{F}$ and $\mathcal{U}$ be chosen as above, then for every $1 + \frac{1}{d-1} < p < \infty$ and $r \geqslant 0$ the $\mathcal{U}$-oscillation of $K_{\mathfrak{F}}$ fulfills*

$$\mathrm{osc}_{\mathcal{U}} K_{\mathfrak{F}} \in \mathcal{A}_{p,w_r}.$$

Proof. We introduce the notation $X_0 = [-1,1]^* \times \mathbb{R}^{d-1} \times \mathbb{R}^d$ and $X_1 = \{\infty\} \times \mathbb{R}^{d-1} \times \mathbb{R}^d$ such that $X = X_0 \cap X_1$, as well as $x = (a, s, t)$, $y = (a', s', t')$ and $z = (a'', s'', t'')$ for $x, y, z \in X$. Then,

$$\|\mathrm{osc}_{\mathcal{U}} K_{\mathfrak{F}}\|_{\mathcal{A}_{p,w_r}}^p = \max \left\{ \mathrm{ess\,sup}_{x \in X} \int_X \mathrm{osc}_{\mathcal{U}} K_{\mathfrak{F}}(x, y)^p w_r(x, y)^p \, \mathrm{d}\mu(y), \right.$$
$$\left. \mathrm{ess\,sup}_{y \in X} \int_X \mathrm{osc}_{\mathcal{U}} K_{\mathfrak{F}}(x, y)^p w_r(x, y)^p \, \mathrm{d}\mu(x) \right\}. \qquad (5.5.81)$$

Since both essential supremums can be handled analogously it suffices to look at the first. We write

$$\operatorname*{ess\,sup}_{x\in X}\int_X \operatorname{osc}_{\mathcal{U}} K_{\mathfrak{F}}(x,y)^p w_r(x,y)^p \mathrm{d}\mu(y)$$

$$\leqslant \max\Bigg\{\operatorname*{ess\,sup}_{x\in X_0}\int_{\mathbb{R}^d}\int_{\mathbb{R}^{d-1}}\int_{-1}^{1}\sup_{z\in Q_y}|\langle\psi_{(a,s,t)},\psi_{(a',s',t')}-\psi_{(a'',s'',t'')}\rangle_{L_2}|^p$$

$$\times\max\left\{\frac{|a|}{|a'|},\frac{|a'|}{|a|}\right\}^{pr}\frac{\mathrm{d}a'}{|a'|^{d+1}}\,\mathrm{d}s'\,\mathrm{d}t' \tag{5.5.82}$$

$$+\operatorname*{ess\,sup}_{x\in X_0}\int_{\mathbb{R}^d}\int_{\mathbb{R}^{d-1}}\sup_{z\in Q_y}|\langle\psi_{(a,s,t)},\psi_{(\infty,s',t')}-\psi_{(\infty,s'',t'')}\rangle_{L_2}|^p|a|^{-pr}\,\mathrm{d}s'\,\mathrm{d}t',$$

$$\operatorname*{ess\,sup}_{x\in X_1}\int_{\mathbb{R}^d}\int_{\mathbb{R}^{d-1}}\int_{-1}^{1}\sup_{z\in Q_y}|\langle\psi_{(\infty,s,t)},\psi_{(a',s',t')}-\psi_{(a'',s'',t'')}\rangle_{L_2}|^p|a'|^{-pr}\frac{\mathrm{d}a'}{|a'|^{d+1}}\,\mathrm{d}s'\,\mathrm{d}t'$$

$$+\operatorname*{ess\,sup}_{x\in X_1}\int_{\mathbb{R}^d}\int_{\mathbb{R}^{d-1}}\sup_{z\in Q_y}|\langle\psi_{(\infty,s,t)},\psi_{(\infty,s',t')}-\psi_{(\infty,s'',t'')}\rangle_{L_2}|^p\,\mathrm{d}s'\,\mathrm{d}t'\Bigg\}.$$

Before we look at the individual integrals, we first recall the following. By the definition of the covering $\mathcal{U}$, for any $y\in X_0$ there exist unique $\varepsilon\in\{-1,1\}$, $j\in\mathbb{N}_0$, $k\in\mathbb{Z}^{d-1}$ and $m\in\mathbb{Z}^d$, as well as unique $u,u'\in(\alpha^{-1},1]$, $v,v'\in[-\frac{1}{2},\frac{1}{2})^{d-1}$ and $w,w'\in[-\frac{1}{2},\frac{1}{2})^d$, such that $y=(a',s',t')$ and every $z=(a'',s'',t'')\in Q_y$ can be written as

$$a'=\varepsilon\alpha^{-j}u, \qquad\qquad a''=\varepsilon\alpha^{-j}u',$$
$$s'=\alpha^{j(\frac{1}{d}-1)}\beta(k+v), \qquad s''=\alpha^{j(\frac{1}{d}-1)}\beta(k+v'),$$
$$t'=\tau S_{\alpha^{j(\frac{1}{d}-1)}\beta k}A_{\alpha^{-j}}(m+w), \qquad t''=\tau S_{\alpha^{j(\frac{1}{d}-1)}\beta k}A_{\alpha^{-j}}(m+w').$$

Similarly this can be done for any $y\in X_1$ and Q_y. This means, we are in the position to apply Lemma 5.5.21 for some fixed $\delta>0$.

We will first find an estimate the fourth integral in (5.5.82), i.e., we look at the case $x=(\infty,s,t)$ and $y=(\infty,s',t')$. Due to the support properties of $\widehat{\Phi}$ this estimate is different than the estimates of the other integrals. Using the triangle inequality and Lemma 5.5.3 for any $x\in X_1$ we obtain

$$\int_{\mathbb{R}^d}\int_{\mathbb{R}^{d-1}}\sup_{z\in Q_y}|\langle\psi_{(\infty,s,t)},\psi_{(\infty,s',t')}-\psi_{(\infty,s'',t'')}\rangle_{L_2}|^p\,\mathrm{d}s'\,\mathrm{d}t'$$

$$=\int_{\mathbb{R}^d}\int_{\mathbb{R}^{d-1}}\sup_{z\in Q_y}\big|\big\langle\Phi,\Phi(S_{s-s'}^{-1}(\cdot-S_{s'}^{-1}(t-t')))-\Phi(S_{s-s''}^{-1}(\cdot-S_{s''}^{-1}(t-t'')))\big\rangle_{L_2}\big|^p\,\mathrm{d}s'\,\mathrm{d}t'$$

$$=\int_{\mathbb{R}^d}\int_{\mathbb{R}^{d-1}}\sup_{z\in Q_y}\big|\mathcal{F}^{-1}\big(\widehat{\Phi}\overline{\widehat{\Phi}}(S_{s-s'}^T\cdot)\big)(S_{s'}^{-1}(t-t'))-\mathcal{F}^{-1}\big(\widehat{\Phi}\overline{\widehat{\Phi}}(S_{s-s''}^T\cdot)\big)(S_{s''}^{-1}(t-t''))\big|^p\,\mathrm{d}s'\,\mathrm{d}t'$$

$$\leqslant\int_{\mathbb{R}^d}\int_{\mathbb{R}^{d-1}}\sup_{z\in Q_y}\big|\mathcal{F}^{-1}\big(\widehat{\Phi}\overline{\widehat{\Phi}}(S_{s-s'}^T\cdot)\big)(S_{s'}^{-1}(t-t'))-\mathcal{F}^{-1}\big(\widehat{\Phi}\overline{\widehat{\Phi}}(S_{s-s'}^T\cdot)\big)(S_{s''}^{-1}(t-t''))\big|^p\,\mathrm{d}s'\,\mathrm{d}t'$$

$$+\int_{\mathbb{R}^d}\int_{\mathbb{R}^{d-1}}\sup_{z\in Q_y}\big|\mathcal{F}^{-1}\big(\widehat{\Phi}\overline{\widehat{\Phi}}(S_{s-s'}^T\cdot)\big)(S_{s''}^{-1}(t-t''))-\mathcal{F}^{-1}\big(\widehat{\Phi}\overline{\widehat{\Phi}}(S_{s-s''}^T\cdot)\big)(S_{s''}^{-1}(t-t''))\big|^p\,\mathrm{d}s'\,\mathrm{d}t'$$

$$=:I_1+I_2. \tag{5.5.83}$$

We look at the integrals I_1 and I_2 separately and start with the first summand I_1. For fixed $s,s'\in\mathbb{R}^{d-1}$ write $G(x)=G_{s-s'}(x):=\mathcal{F}^{-1}(\widehat{\Phi}\overline{\widehat{\Phi}}(S_{s-s'}^T\cdot))(x)$. This implies $G\in\mathcal{S}(\mathbb{R}^d)$, since

$\hat{\Phi} \in \mathcal{C}_0^\infty \subset \mathcal{S}(\mathbb{R}^d)$ and $\mathcal{F}$ is an automorphism on $\mathcal{S}(\mathbb{R}^d)$. Then, we obtain using the fundamental theorem of calculus:

$$\left| G(S_{s'}^{-1}(t - t')) - G(S_{s''}^{-1}(t - t'')) \right|$$

$$\leq \int_0^1 \left| \nabla G(\theta S_{s'}^{-1}(t - t') + (1 - \theta) S_{s''}^{-1}(t - t'')) \times (S_{s'}^{-1}(t - t') - S_{s''}^{-1}(t - t'')) \right| d\theta$$

$$\leq \int_0^1 \left\| \nabla G(\theta S_{s'}^{-1}(t - t') + (1 - \theta) S_{s''}^{-1}(t - t'')) \right\|_2 d\theta \cdot \left\| S_{s'}^{-1}(t - t') - S_{s''}^{-1}(t - t'') \right\|_2. \qquad (5.5.84)$$

Denote with $G_i := \frac{\partial G}{\partial x_i}$ the partial derivatives of G for any fixed $i = 1, \dots, d$. Interchanging the roles of the Fourier transfrom and differentiation leads to

$$G_i = \frac{\partial}{\partial x_i} \mathcal{F}^{-1}\left(\hat{\Phi} \overline{\hat{\Phi}(S_{s-s'}^T \cdot)} \right) = \mathcal{F}^{-1}\left(\widehat{\frac{\partial \Phi}{\partial x_i}} \overline{\hat{\Phi}(S_{s-s'}^T \cdot)} \right).$$

Next, write $p = p_0 + p_1$ for some $p_0, p_1 > 0$ we will specify later. Then, similar to the proof of Theorem 5.5.13, for any $x \in \mathbb{R}^d$ we calculate

$$|G_i(x)|^p = \left| \left(\frac{\partial \Phi}{\partial x_i} * \psi_{(\infty, s-s', 0)}^* \right)(x) \right|^{p_0} \cdot \left| \mathcal{F}^{-1}\left(\widehat{\frac{\partial \Phi}{\partial x_i}} \overline{\hat{\Phi}(S_{s-s'}^T \cdot)} \right)(x) \right|^{p_1}$$

$$\leq \left(\int_{\mathbb{R}^d} \left| \frac{\partial \Phi}{\partial x_i}(t + x) \Phi(S_{s-s'}^{-1} t) \right| dt \right)^{p_0} \cdot \left(\int_{\mathbb{R}^d} \left| \widehat{\frac{\partial \Phi}{\partial x_i}}(\omega) \hat{\Phi}(S_{s-s'}^T \omega) \right| d\omega \right)^{p_1}$$

$$=: B_1(s - s', x)^{p_0} \cdot B_2(s - s')^{p_1} \qquad (5.5.85)$$

To estimate $B_1(\sigma, x)$ for any $\sigma \in \mathbb{R}^{d-1}$ and $x \in \mathbb{R}^d$ we use the fact that Φ is a Schwartz function, which implies $|\Phi(x)|, |\frac{\partial \Phi}{\partial x_i}(x)| \lesssim \prod_{i=1}^d (1 + |x_i|)^{-M}$ for any $M > 0$, $x \in \mathbb{R}^d$. Then, with (5.5.54), for any $\sigma \in \mathbb{R}^{d-1}$ and $x \in \mathbb{R}^d$ we obtain

$$B_1(\sigma, x) \lesssim \int_{\mathbb{R}^d} \prod_{i=1}^d (1 + |t_i + x_i|)^{-M} (1 + |(S_{-\sigma}t)_i)|)^{-M} dt$$

$$\lesssim \min(\sigma) \left[\prod_{j=2}^d (1 + |x_j|)^{-M} \right] \sum_{i=1}^{2^{d-1}} (1 + |v_i(\sigma) \cdot x|)^{-M}$$

$$=: \min(\sigma) \sum_{i=1}^{2^{d-1}} \prod_{j=1}^d (1 + |(V_i(\sigma)x)_j|)^{-M},$$

where $V_i(\sigma) \in \mathbb{R}^{d \times d}$ is a matrix and $(V_i(\sigma)x)_j$ the j-th entry of the vector $V_i(\sigma)x$. We can estimate this from above by

$$B_1(\sigma, x) \lesssim \min(\sigma) \sum_{i=1}^{2^{d-1}} \left(1 + \|V_i(\sigma)x\|_2 \right)^{-M}. \qquad (5.5.86)$$

Let us briefly recall some basic properties of $V_i(\sigma)$ that are a consequence of the proof of Lemma 5.5.12. First, the matrix is of the form

$$V_i(\sigma) = \begin{pmatrix} v_i(\sigma)_1 & v_i(\sigma)_2 & \dots & v_i(\sigma)_d \\ 0_{d-1} & & I_{d-1} & \end{pmatrix}, \qquad (5.5.87)$$

where all entries $v_i(\sigma)_1, \ldots, v_i(\sigma)_d$ are smaller or equal to one in absolut value. Furthermore, we have $v_i(\sigma)_1 = \min(s) \neq 0$. This implies

$$\|V_i(\sigma)\|_{2\to2} \lesssim \|V_i(\sigma)\|_1 = 2d - 1,$$

where the constant only depents on the dimension d. The inverse $V_i(\sigma)^{-1}$ exists and is given by

$$V_i(\sigma)^{-1} = \begin{pmatrix} v_i(\sigma)_1^{-1} & -v_i(\sigma)_2 v_i(\sigma)_1^{-1} & \cdots & -v_i(\sigma)_d v_i(\sigma)_1^{-1} \\ 0_{d-1} & & I_{d-1} & \end{pmatrix}. \tag{5.5.88}$$

We can easily check that

$$V_i(\sigma)(I_d - S_s)V_i(\sigma)^{-1} = v_i(\sigma)_1 \cdot (I_d - S_s)$$

for all $s \in \mathbb{R}^{d-1}$. By setting $\sigma = s - s'$ and $x = \theta S_{s'}^{-1}(t - t') + (1 - \theta)S_{s''}^{-1}(t - t'')$ for some fixed $\theta \in [0, 1]$. By the properties of $V_i(\sigma)$ and using the triangle inequality and (5.5.79), (5.5.80) we derive

$$
\begin{aligned}
&\left\|V_i(\sigma)S_{s'}^{-1}(t - t')\right\|_2 \\
&\leqslant \left\|(1 - \theta)V_i(\sigma)(S_{s'}^{-1}(t - t') - S_{s''}^{-1}(t - t''))\right\|_2 + \left\|V_i(\sigma)(\theta(S_{s'}^{-1}(t - t')) + (1 - \theta)(S_{s''}^{-1}(t - t'')))\right\|_2 \\
&\leqslant \left\|V_i(\sigma)(I_d - S_{s'-s''})S_{s'}^{-1}(t - t') - V_i(\sigma)S_{s''}^{-1}(t' - t'')\right\|_2 + \left\|V_i(\sigma)x\right\|_2 \\
&\leqslant \left\|V_i(\sigma)(I_d - S_{s'-s''})V_i(\sigma)^{-1}\right\|_{2\to2}\left\|V_i(\sigma)S_{s'}^{-1}(t - t')\right\|_2 + \left\|V_i(\sigma)\right\|_{2\to2}\left\|S_{s''}^{-1}(t' - t'')\right\|_2 + \left\|V_i(\sigma)x\right\|_2 \\
&\lesssim \delta \cdot \left\|V_i(\sigma)S_{s'}^{-1}(t - t')\right\|_2 + \delta \cdot (2d - 1) + \left\|V_i(\sigma)x\right\|_2, \tag{5.5.89}
\end{aligned}
$$

implying

$$1 + \left\|V_i(\sigma)x\right\|_2 \gtrsim 1 + \left\|V_i(\sigma)S_{s'}^{-1}(t - t')\right\|_2 \tag{5.5.90}$$

for δ sufficiently small. Plugging (5.5.90) into (5.5.86) with the same choice of σ and x as above yields

$$B_1(\sigma, x) \lesssim \min(\sigma) \sum_{i=1}^{2^d-1} \left(1 + \left\|V_i(\sigma)S_{s'}^{-1}(t - t')\right\|_2\right)^{-M}. \tag{5.5.91}$$

Note that at this point the equation (5.5.91) is independent of z.

We now turn to $B_2(\sigma)$ in (5.5.85), which we can rewrite and estimate with similar arguments as in (5.5.64) and get

$$B_2(\sigma) = \int_{\mathbb{R}^d} \left|\omega_i \cdot \widehat{\Phi}(\omega)\widehat{\Phi}(S_\sigma^T\omega)\right| \mathrm{d}\omega \lesssim \int_{\Omega_\sigma} |\omega_1|^{d-1}|\omega_i| \, \mathrm{d}\omega, \tag{5.5.92}$$

where

$$\Omega_\sigma = \left\{\xi \in \mathbb{R}^d \,:\, |\xi_1| \leqslant a_1, \ \max\{-b_i, -b_i - \sigma_{i-1}\xi_1\} \leqslant \xi_i \leqslant \min\{b_i, b_i - \sigma_{i-1}\xi_1\}, \ i = 2, \ldots, d\right\}.$$

For any $\omega \in \Omega_\sigma$ it holds that $|\omega_j| \leqslant b_j$ for all $j \geqslant 2$ as well as $|\omega_1| \leqslant a_1$. Using the same arguments that lead to (5.5.65) we conclude

$$B_2(\sigma) \lesssim \int_{\Omega_\sigma} |\omega_1|^{d-1} \, \mathrm{d}\omega \lesssim (1 + \|\sigma\|_2)^{-d}. \tag{5.5.93}$$

Finally, plugging (5.5.91), (5.5.93) and (5.5.85) into (5.5.84), and using (5.5.79) gives us

$$|G(S_{s'}^{-1}(t-t')) - G(S_{s''}^{-1}(t-t''))|^p$$

$$\lesssim \sum_{i=1}^{d} \int_0^1 |G_i(\theta S_{s'}^{-1}(t-t') + (1-\theta)S_{s''}^{-1}(t-t''))|^p \, d\theta \cdot \|S_{s'}^{-1}(t-t') - S_{s''}^{-1}(t-t'')\|_2^p$$

$$\lesssim \min(s-s')^{p_0} \sum_{i=1}^{2^{d-1}} \left(1 + \|V_i(s-s')S_{s'}^{-1}(t-t')\|_2\right)^{-Mp_0} \cdot \left(1 + \|s-s'\|_2\right)^{-dp_1}$$

$$\times \delta^p \cdot \left(1 + \|S_{s'}^{-1}(t-t')\|_2\right)^p \tag{5.5.94}$$

Since (5.5.94) is independent of the choice of $z = (\infty, s'', t'')$, we can estimate I_1 in (5.5.83) as follows. Recall that $\det V_i(s-s') = \min(s-s')$ and we can calculate $\|V_i(s-s')^{-1}\tau\|_2 \lesssim \|V_i(s-s')^{-1}\|_2 \cdot \|\tau\|_2 \lesssim (1 + \|s-s'\|_2) \cdot (1 + \|\tau\|_2)$ since $\min(\sigma)^{-1} \lesssim (1 + \|\sigma\|_2)$. Then, with $\tau = V_i(s-s')S_s'(t-t')$ and $\sigma = s - s'$, for $p_0 < 1$ we obtain

$$I_1 = \operatorname*{ess\,sup}_{x \in X_1} \int_{\mathbb{R}^d} \int_{\mathbb{R}^{d-1}} \sup_{z \in Q_y} \left|G_{s-s'}(S_{s'}^{-1}(t-t')) - G_{s-s''}(S_{s'}^{-1}(t-t''))\right|^p ds' \, dt'$$

$$\lesssim \delta^p \cdot \operatorname*{ess\,sup}_{x \in X_1} \int_{\mathbb{R}^{d-1}} \left(1 + \|s-s'\|_2\right)^{-dp_1} \min(s-s')^{p_0}$$

$$\times \sum_{i=1}^{2^{d-1}} \int_{\mathbb{R}^d} \left(1 + \|S_{s'}^{-1}(t-t')\|_2\right)^p \left(1 + \|V_i(s-s')S_{s'}^{-1}(t-t')\|_2\right)^{-Mp_0} dt' \, ds'$$

$$= \delta^p \cdot \operatorname*{ess\,sup}_{x \in X_1} \int_{\mathbb{R}^{d-1}} \left(1 + \|s-s'\|_2\right)^{-dp_1} \min(s-s')^{p_0-1}$$

$$\times \sum_{i=1}^{2^{d-1}} \int_{\mathbb{R}^d} \left(1 + \|V_i(s-s')^{-1}\tau\|_2\right)^p \left(1 + \|\tau\|_2\right)^{-Mp_0} d\tau \, ds'$$

$$\lesssim \delta^p \cdot \int_{\mathbb{R}^{d-1}} \left(1 + \|\sigma\|_2\right)^{-dp_1+1-p_0+p} d\sigma \cdot \int_{\mathbb{R}^d} \left(1 + \|\tau\|_2\right)^{p-Mp_0} d\tau. \tag{5.5.95}$$

The integrals in (5.5.95) are finite, if and only if

$$p - Mp_0 < -d \quad \text{and} \quad -dp_1 + 1 - p_0 + p = 1 - (d-1)p_1 < -(d-1).$$

The second inequality is equivalent to $p_1 > 1 + \frac{1}{d-1}$, whereas p_0 can be chosen arbitrarily small for sufficiently large M. Specifically, for any $p > 1 + \frac{1}{d-1}$ we choose $p_1 = 1 + \frac{1}{d-1} + \varepsilon$, where $0 < \delta < p - 1 - \frac{1}{d-1}$ and $p_0 = p - 1 - \frac{1}{d-1} - \varepsilon > 0$. Then the integrals in (5.5.95) are finite for large $M > \frac{p+d}{p_0}$ and it is $p_0 + p_1 = p$. We finally conclude

$$I_1 \leqslant C \cdot \varepsilon^p. \tag{5.5.96}$$

The next step is to estimate I_2 in (5.5.83) from above, and for this we first define for a fixed $x \in \mathbb{R}^d$ the function $F(\sigma) = F_x(\sigma) := \mathcal{F}^{-1}(\widehat{\Phi}\widehat{\Phi}(S_\sigma^T \cdot))(x) = \langle \Phi, \Phi(S_\sigma^{-1} \cdot -x)\rangle_{L_2}$ with $\sigma \in \mathbb{R}^{d-1}$. Since $\Phi \in \mathcal{S}(\mathbb{R}^d)$ and the function $\sigma \mapsto S_\sigma^{-1}y$ is smooth for any $y \in \mathbb{R}^d$, we have $F \in \mathcal{C}^\infty(\mathbb{R}^{d-1})$, see also [39, Theorem IV.5.7]. Then, by the fundamental theorem of calculus, we obtain

$$|F(s-s') - F(s-s'')| \leqslant \int_0^1 |\nabla F(\theta(s-s') + (1-\theta)(s-s'')) \times (s''-s')| \, d\theta$$

$$\leqslant \int_0^1 \|\nabla F(\theta(s-s') + (1-\theta)(s-s''))\|_2 \, d\theta \cdot \|s'-s''\|_2. \tag{5.5.97}$$

For any fixed $1 \leqslant i \leqslant d-1$ the partial derivative $\frac{\partial F}{\partial \sigma_i}(\sigma)$ for $\sigma \in \mathbb{R}^{d-1}$ is given by

$$
\begin{aligned}
\frac{\partial F}{\partial \sigma_i}(\sigma) &= \frac{\partial}{\partial \sigma_i} \big\langle \Phi, \Phi(S_\sigma^{-1} \cdot -x) \big\rangle_{L_2} \\
&= \int_{\mathbb{R}^d} \Phi(t) \overline{\frac{\partial}{\partial \sigma_i} \Phi(S_\sigma^{-1} t - x)} \, \mathrm{d}t \\
&= -\int_{\mathbb{R}^d} \Phi(t) \overline{\frac{\partial \Phi}{\partial x_1}(S_\sigma^{-1} t - x)} \cdot t_{i+1} \, \mathrm{d}t \\
&= -\Big\langle \Phi \cdot t_{i+1}, \frac{\partial \Phi}{\partial x_1}(S_\sigma^{-1} \cdot -x) \Big\rangle_{L_2}.
\end{aligned}
\tag{5.5.98}
$$

By a similar approach to (5.5.85) we write

$$
\begin{aligned}
\Big| \frac{\partial F}{\partial \sigma_i}(\sigma) \Big|^p &= \Big| \big\langle \Phi \cdot t_{i+1}, \frac{\partial \Phi}{\partial x_1}(S_\sigma^{-1} \cdot -x) \big\rangle_{L_2} \Big|^{p_0} \cdot \Big| \mathcal{F}^{-1}\Big(\widehat{\Phi \cdot t_{i+1}} \, \overline{\frac{\partial \Phi}{\partial x_1}(S_\sigma^T \cdot)} \Big)(x) \Big|^{p_1} \\
&\leqslant \Big(\int_{\mathbb{R}^d} \Big| \Phi(t) t_{i+1} \frac{\partial \Phi}{\partial x_1}(S_\sigma^{-1} t - x) \Big| \mathrm{d}t \Big)^{p_0} \cdot \Big(\int_{\mathbb{R}^d} \Big| \frac{\partial \widehat{\Phi}}{\partial \omega_{i+1}}(\omega) \widehat{\Phi}(S_\sigma^T \omega) \omega_1 \Big| \mathrm{d}\omega \Big)^{p_1} \\
&=: D_1(\sigma, x)^{p_0} \cdot D_2(\sigma)^{p_1},
\end{aligned}
\tag{5.5.99}
$$

for some $p_0, p_1 > 0$ satisfying $p_0 + p_1 = p$, wich will be specified later. Recall that we are interested in estimates for $\sigma = \theta(s-s') + (1-\theta)(s-s'')$ and $x = S_{s''}^{-1}(t-t'')$.

Let us start by estimating $D_1(\sigma, x)$ from above. Since Φ is a Schwartz function we have $|\Phi(t) t_{i+1}|, |\frac{\partial \Phi}{\partial x_1}(t)| \lesssim \prod_{j=1}^d (1+|t_j|)^{-M}$ for any $M > 0$. We can therefore estimate $D_1(\sigma, x)$ the same way as $B_1(\sigma, x)$ in (5.5.86) to obtain

$$
\begin{aligned}
D_1(\sigma, x) &\lesssim \min(\sigma) \sum_{k=1}^{2^{d-1}} \big(1 + \|V_k(\sigma) x\|_2\big)^{-M} \\
&\lesssim \min(\sigma) \sum_{k=1}^{2^{d-1}} \big(1 + \|V_k(s-s') V_k(\sigma)^{-1}\|_1^{-1} \cdot \|V_k(s-s') x\|_2\big)^{-M}.
\end{aligned}
\tag{5.5.100}
$$

Recall the definition of $V_k(\sigma)$ in (5.5.87) and its inverse $V_k(\sigma)^{-1}$ in (5.5.88), and let $\widetilde{v}_k(\sigma) = (v_k(\sigma)_2, \ldots, v_k(\sigma)_d)^T$, then

$$
\begin{aligned}
\|V_k(s-s') V_k(\sigma)^{-1}\|_1 &= \left\| \begin{pmatrix} \min(s-s') & \widetilde{v}_k(s-s')^T \\ 0_{d-1} & I_{d-1} \end{pmatrix} \begin{pmatrix} \min(\sigma)^{-1} & -\min(\sigma)^{-1}\widetilde{v}_k(\sigma)^T \\ 0_{d-1} & I_{d-1} \end{pmatrix} \right\|_1 \\
&= \left\| \begin{pmatrix} \min(s-s')\min(\sigma)^{-1} & \widetilde{v}_k(s-s')^T - \min(s-s')\min(\sigma)^{-1}\widetilde{v}_k(\sigma)^T \\ 0_{d-1} & I_{d-1} \end{pmatrix} \right\|_1 \\
&\leqslant \frac{\min(s-s')}{\min(\sigma)} + \|v_k(s-s')\|_1 + \frac{\min(s-s')}{\min(\sigma)} \|v_k(\sigma)\|_1 + d - 1 \\
&\leqslant 2(d-1) + d\frac{\min(s-s')}{\min(\sigma)}.
\end{aligned}
\tag{5.5.101}
$$

Using $\sigma - (s-s') = (1-\theta)(s'-s'')$ as well as (5.5.78) we have

$$
\begin{aligned}
\min(\sigma)^{-1} &= \max\{1, \|\sigma\|_\infty\} \\
&\leqslant \max\{1, \|s-s'\|_\infty + (1-\theta)\|s'-s''\|_\infty\} \\
&\leqslant \max\{1, \|s-s'\|_\infty + \delta\} \\
&\leqslant \min(s-s')^{-1} + \delta,
\end{aligned}
\tag{5.5.102}
$$

thus

$$\frac{\min(s-s')}{\min(\sigma)} \leqslant \min(s-s')\left[\min(s-s')^{-1} + \delta\right] \leqslant 1 + \delta. \tag{5.5.103}$$

And together with (5.5.101) this implies

$$\left\|V_k(s-s')V_k(\sigma)^{-1}\right\|_1 \leqslant 2(d-1) + d(1+\delta) \leqslant 4d - 2$$

and therefore

$$1 + \left\|V_k(s-s')V_k(\sigma)^{-1}\right\|_1^{-1} \cdot \|x\|_2 \geqslant (4d-2)^{-1}(1 + \|x\|_2)$$

for δ sufficiently small. By interchanging the rules of σ and $s - s'$ in (5.5.102) and (5.5.103) we also have $\min(\sigma) \leqslant (1 + \varepsilon)\min(s - s')$. Hence (5.5.100) can be estimated by

$$D_1(\sigma, x) \lesssim \min(s-s') \sum_{k=1}^{2^{d-1}} \left(1 + \|V_k(s-s')x\|_2\right)^{-M}. \tag{5.5.104}$$

Similarly to (5.5.89) we obtain by using the properties of V_k as well as (5.5.79) and (5.5.80):

$$
\begin{aligned}
&\left\|V_k(s-s')S_{s'}^{-1}(t-t')\right\|_2 \\
&\leqslant \left\|V_k(s-s')(S_{s'}^{-1}(t-t') - S_{s''}^{-1}(t-t''))\right\|_2 + \left\|V_k(s-s')x\right\|_2 \\
&\leqslant \left\|V_k(s-s')(I_d - S_{s''-s'}^{-1})S_{s'}^{-1}(t-t')\right\|_2 + \left\|V_k(s-s')S_{s''}^{-1}(t'-t'')\right\|_2 + \left\|V_k(s-s')x\right\|_2 \\
&\leqslant \left\|V_k(s-s')(I_d - S_{s''-s'}^{-1})V_k(s-s')^{-1}\right\|_{2\to2} \cdot \left\|V_k(s-s')S_{s'}^{-1}(t-t')\right\|_2 \\
&\qquad\qquad + \left\|V_k(s-s')\right\|_{2\to2} \cdot \left\|S_{s''}^{-1}(t'-t'')\right\|_2 + \left\|V_k(s-s')x\right\|_2 \\
&\lesssim \delta \cdot \left\|V_k(s-s')S_{s'}^{-1}(t-t')\right\|_2 + \delta \cdot (2d-1) + \left\|V_k(s-s')x\right\|_2, \tag{5.5.105}
\end{aligned}
$$

where $x = S_{s''}^{-1}(t - t'')$. This implies

$$1 + \left\|V_k(s-s')x\right\|_2 \geqslant \frac{1}{2} \cdot \left(1 + \left\|V_k(s-s')S_{s'}^{-1}(t-t')\right\|_2\right) \tag{5.5.106}$$

for sufficiently small δ. Hence, by virtue of (5.5.104) and (5.5.106) we have

$$D_1(\sigma, x) \lesssim \min(s-s') \sum_{k=1}^{2^{d-1}} \left(1 + \left\|V_k(s-s')S_{s'}(t-t')\right\|_2\right)^{-M}. \tag{5.5.107}$$

To estimate $D_2(\sigma)$ from above we use similar ideas as for $B_2(\sigma)$. Since $\operatorname{supp}\widehat{\Phi} \cap \operatorname{supp}\widehat{\Phi}(S_\sigma^T \cdot) \subseteq \Omega_\sigma$, with Ω_σ as in Lemma 5.5.9, it also holds that $\operatorname{supp}\frac{\partial\widehat{\Phi}}{\partial\omega_{i+1}} \cap \operatorname{supp}\widehat{\Phi}(S_\sigma^T \cdot) \subset \Omega_\sigma$. Additionally, by the construction of $\widehat{\Phi}$, see (5.5.19), and the properties of $\widehat{\Psi}$ we observe

$$\left|\frac{\partial\widehat{\Phi}}{\partial\omega_{i+1}}(\omega)\right| = |\omega_1|^{\frac{d-1}{2}} \left|\frac{\partial}{\partial\omega_{i+1}} \int_{\mathbb{R}\setminus[-|\omega_1|,|\omega_1|]} \frac{|\widehat{\Psi}(\xi_1,\tilde\omega)|^2}{|\xi_1|^d}\,d\xi_1\right|^2 \lesssim |\omega_1|^{\frac{d-1}{2}}$$

for all $1 \leqslant i \leqslant d - 1$. The same estimate holds for $|\widehat{\Phi}(S_\sigma^T \omega)|$ by (5.5.62) and (5.5.63). Therefore we conclude as in (5.5.93) that

$$D_2(\sigma) = \int_{\mathbb{R}^d} \left|\frac{\partial\widehat{\Phi}}{\partial\omega_{i+1}}(\omega)\widehat{\Phi}(S_\sigma^T \omega)\omega_1\right| d\omega \lesssim \int_{\Omega_\sigma} |\omega_1|^d\,d\omega_1 \lesssim (1 + \|\sigma\|_2)^{-(d+1)}. \tag{5.5.108}$$

Recalling $\sigma = \theta(s - s') + (1 - \theta)(s - s'')$ we intend to furtther estimate the expression above and by applying (5.5.78) we obtain

$$\|s - s'\|_2 \leqslant \|s - s' - (\theta(s - s') + (1 - \theta)(s - s''))\|_2 + \|\theta(s - s') + (1 - \theta)(s - s'')\|_2$$
$$= (1 - \theta) \cdot \|(s' - s'')\|_2 + \|\theta(s - s') + (1 - \theta)(s - s'')\|_2$$
$$\leqslant \delta + \|\theta(s - s') + (1 - \theta)(s - s'')\|_2,$$

hence

$$1 + \|\sigma\|_2 = 1 + \|\theta(s - s') + (1 - \theta)(s - s'')\|_2 \geqslant \frac{1}{2}(1 + \|s - s'\|_2)$$

for sufficiently small $\delta > 0$ and for any fixed $\theta \in [0, 1]$. By (5.5.108) this implies

$$D_2(\sigma) \lesssim (1 + \|\theta(s - s') + (1 - \theta)(s - s'')\|_2)^{-(d+1)} \lesssim (1 + \|s - s'\|_2)^{-(d+1)}. \tag{5.5.109}$$

Note that both (5.5.107) and (5.5.109) are independent of z. Now we can estimate I_2 from above for $p_0 < 1$ by inserting (5.5.107), (5.5.109) and (5.5.99) into (5.5.97) and we obtain

$$I_2 = \operatorname*{ess\,sup}_{x \in X_1} \int_{\mathbb{R}^d} \int_{\mathbb{R}^{d-1}} \sup_{z \in Q_y} |F(s - s') - F(s - s'')|^p \, ds' \, dt'$$

$$\lesssim \delta^p \cdot \operatorname*{ess\,sup}_{x \in X_1} \int_{\mathbb{R}^{d-1}} (1 + \|s - s'\|_2)^{-(d+1)p_1} \min(s - s')^{p_0}$$

$$\times \sum_{k=1}^{2^{d-1}} \int_{\mathbb{R}^d} \left(1 + \|V_k(s - s')S_{s'}^{-1}(t - t')\|_2\right)^{-Mp_0} dt' \, ds'$$

$$\lesssim \delta^p \cdot \operatorname*{ess\,sup}_{x \in X_1} \int_{\mathbb{R}^{d-1}} (1 + \|s - s'\|_2)^{-(d+1)p_1} \min(s - s')^{p_0 - 1} \sum_{k=1}^{2^{d-1}} \int_{\mathbb{R}^d} (1 + \|\tau\|_2)^{-Mp_0} d\tau \, ds'$$

$$= \delta^p \cdot \int_{\mathbb{R}^{d-1}} (1 + \|\sigma\|_2)^{-(d+1)p_1 + 1 - p_0} d\sigma \cdot \int_{\mathbb{R}^d} (1 + \|\tau\|_2)^{-Mp_0} d\tau \tag{5.5.110}$$

by substituting $\sigma = s - s'$ and $\tau = V_k(s - s')S_{s'}^{-1}(t - t')$, where $\det V_k(s - s')S_{s'}^{-1} = \min(s - s')$. Again, the integrals above are finite if and only if $Mp_0 > d$ and $(d + 1)p_1 + p_0 > d$, where $p_0 + p_1 = p > 1$. This can easily be obtained by simply choosing $p_1 = 1$ and $p_0 = p - 1 > 0$. Then, both inequalities are fulfilled for M sufficiently large, and we conclude

$$I_2 \leqslant C \cdot \delta^p. \tag{5.5.111}$$

By proving (5.5.96) and (5.5.111) we have thus shown, that the fourth integral in (5.5.82) is indeed finite for suficiently small $\delta > 0$.

Now we will show the boundedness of the first integral in (5.5.82) for any fixed $x = (a, s, t) \in X_0$. The second and third integral in (5.5.82) can be treated analogously, since the proofs rely on the same argument, namely the support properties given in Lemma 5.5.9. Similar to (5.5.83), we first write, using the abbreviation $w(a, a') = \max\left\{\frac{|a|}{|a'|}, \frac{|a'|}{|a|}\right\}$ as well as Lemma 5.5.3 and

Lemma 5.5.7,

$$\int_{\mathbb{R}^d}\int_{\mathbb{R}^{d-1}}\int_{-1}^{1}\sup_{z\in Q_y}|\langle\psi_{(a,s,t)},\psi_{(a',s',t')}-\psi_{(a'',s'',t'')}\rangle_{L_2}|^p w(a,a')^{pr}\frac{da'}{|a'|^{d+1}}\,ds'\,dt'$$

$$=\int_{\mathbb{R}^d}\int_{\mathbb{R}^{d-1}}\int_{-1}^{1}\sup_{z\in Q_y}\Big||\det A_{aa'^{-1}}|^{-\frac{1}{2}}\Big\langle\Psi,\Psi\big(A_{aa'^{-1}}^{-1}S^{-1}_{|a'|^{\frac{1}{d}-1}(s-s')}(\cdot-A_{a'}^{-1}S_{s'}^{-1}(t-t'))\big)\Big\rangle_{L_2}$$

$$-|\det A_{aa''^{-1}}|^{-\frac{1}{2}}\Big\langle\Psi,\Psi\big(A_{aa''^{-1}}^{-1}S^{-1}_{|a''|^{\frac{1}{d}-1}(s-s'')}(\cdot-A_{a''}^{-1}S_{s''}^{-1}(t-t''))\big)\Big\rangle_{L_2}\Big|^p w(a,a')^{pr}\frac{da'}{|a'|^{d+1}}\,ds'\,dt'$$

$$=\int_{\mathbb{R}^d}\int_{\mathbb{R}^{d-1}}\int_{-1}^{1}\sup_{z\in Q_y}\Big||\det A_{aa'^{-1}}|^{\frac{1}{2}}\mathcal{F}^{-1}\big(\hat{\Psi}\overline{\hat{\Psi}(A_{aa'^{-1}}S^T_{|a'|^{\frac{1}{d}-1}(s-s')}\cdot)}\big)(A_{a'}^{-1}S_{s'}^{-1}(t-t'))$$

$$-|\det A_{aa''^{-1}}|^{\frac{1}{2}}\mathcal{F}^{-1}\big(\hat{\Psi}\overline{\hat{\Psi}(A_{aa''^{-1}}S^T_{|a''|^{\frac{1}{d}-1}(s-s'')}\cdot)}\big)(A_{a''}^{-1}S_{s''}^{-1}(t-t''))\Big|^p W(a,a')^{pr}\frac{da'}{|a'|^{d+1}}\,ds'\,dt'$$

$$\leqslant\int_{\mathbb{R}^d}\int_{\mathbb{R}^{d-1}}\int_{-1}^{1}\sup_{z\in Q_y}\Big||\det A_{aa'^{-1}}|^{\frac{1}{2}}-|\det A_{aa''^{-1}}|^{\frac{1}{2}}\Big|^p$$

$$\times\Big|\mathcal{F}^{-1}\big(\hat{\Psi}\overline{\hat{\Psi}(A_{aa'^{-1}}S^T_{|a'|^{\frac{1}{d}-1}(s-s')}\cdot)}\big)(A_{a'}^{-1}S_{s'}^{-1}(t-t'))\Big|^p w(a,a')^{pr}\frac{da'}{|a'|^{d+1}}\,ds'\,dt'$$

$$+\int_{\mathbb{R}^d}\int_{\mathbb{R}^{d-1}}\int_{-1}^{1}\sup_{z\in Q_y}|\det A_{aa'^{-1}}|^{\frac{p}{2}}\Big|\mathcal{F}^{-1}\big(\hat{\Psi}\overline{\hat{\Psi}(A_{aa'^{-1}}S^T_{|a'|^{\frac{1}{d}-1}(s-s')}\cdot)}\big)(A_{a'}^{-1}S_{s'}^{-1}(t-t'))$$

$$-\mathcal{F}^{-1}\big(\hat{\Psi}\overline{\hat{\Psi}(A_{aa'^{-1}}S^T_{|a'|^{\frac{1}{d}-1}(s-s')}\cdot)}\big)(A_{a''}^{-1}S_{s''}^{-1}(t-t''))\Big|^p w(a,a')^{pr}\frac{da'}{|a'|^{d+1}}\,ds'\,dt'$$

$$+\int_{\mathbb{R}^d}\int_{\mathbb{R}^{d-1}}\int_{-1}^{1}\sup_{z\in Q_y}|\det A_{aa'^{-1}}|^{\frac{p}{2}}\Big|\mathcal{F}^{-1}\big(\hat{\Psi}\overline{\hat{\Psi}(A_{aa'^{-1}}S^T_{|a'|^{\frac{1}{d}-1}(s-s')}\cdot)}\big)(A_{a''}^{-1}S_{s''}^{-1}(t-t''))$$

$$-\mathcal{F}^{-1}\big(\hat{\Psi}\overline{\hat{\Psi}(A_{aa''^{-1}}S^T_{|a''|^{\frac{1}{d}-1}(s-s'')}\cdot)}\big)(A_{a''}^{-1}S_{s''}^{-1}(t-t''))\Big|^p w(a,a')^{pr}\frac{da'}{|a'|^{d+1}}\,ds'\,dt'$$

$$=:J_1+J_2+J_3.\tag{5.5.112}$$

Before we look at the three integrals above in detail, we know by virtue of Lemma 5.5.9 that $\hat{\Psi}\overline{\hat{\Psi}(A_aS_s\cdot)}\not\equiv0$ implies $|a|\in[\frac{a_0}{a_1},\frac{a_1}{a_0}]$ and $s\in Q_{d_1}$, where Q_{d_1} is a finite cuboid in $\mathbb{R}^{d-1}$ with $d_1=(a_0^{-1}+a_0^{-(1+\frac{1}{d})}a_1^{\frac{1}{d}})b$. This means that if both equations

$$\hat{\Psi}\overline{\hat{\Psi}(A_{aa'^{-1}}S^T_{|a'|^{\frac{1}{d}-1}(s-s')}\cdot)}\not\equiv0\quad\text{and}\quad\hat{\Psi}\overline{\hat{\Psi}(A_{aa''^{-1}}S^T_{|a''|^{\frac{1}{d}-1}(s-s'')}\cdot)}\not\equiv0$$

are fulfilled, then $|aa'^{-1}|,|aa''^{-1}|\in[\frac{a_0}{a_1},\frac{a_1}{a_0}]$. Since $|a'a''^{-1}|\in[1-\delta,1+\delta]$, see equation (5.5.73), it suffices to write $|aa'^{-1}|\in[\tilde{a}_0,\tilde{a}_1]=:\tilde{I}$, for some $0<\tilde{a}_0<\tilde{a}_1$. At the same time we have $|a'|^{\frac{1}{d}-1}(s-s'),|a''|^{\frac{1}{d}-1}(s-s'')\in Q_{d_1}$, which can be summerized by (5.5.75) to $|a'|^{\frac{1}{d}-1}(s-s')\in Q_{\tilde{d}_1}$ for some $\tilde{d}_1\in\mathbb{R}_+^{d-1}$. These support properties are vital for the rest of the proof and we briefly note that the following integral is finite:

$$\int_{Q_{\tilde{d}_1}}\int_{\tilde{I}}\frac{d\eta}{|\eta|}\,d\sigma<\infty.\tag{5.5.113}$$

Furthermore, by the definition of the weight w we have

$$\sup_{|aa'^{-1}|\in\tilde{I}}w(a,a')=\sup_{|aa'^{-1}|\in\tilde{I}}\max\Big\{\frac{|a|}{|a'|},\frac{|a'|}{|a|}\Big\}=\max\{\tilde{a}_1,\tilde{a}_0^{-1}\}<\infty,$$

so $w(a, a')$ can be estimated from above by a constant in all integrals in (5.5.112).

We start by estimating J_1 for some fixed $x = (a, s, t) \in X_0$. Invoking (5.5.73) yields

$$\left| |\det A_{aa'^{-1}}|^{\frac{1}{2}} - |\det A_{aa''^{-1}}|^{\frac{1}{2}} \right| = \left| |aa'^{-1}|^{1-\frac{1}{2d}} - |aa''^{-1}|^{1-\frac{1}{2d}} \right|$$

$$= |aa'^{-1}|^{1-\frac{1}{2d}} \cdot \left| 1 - |a'a''^{-1}|^{1-\frac{1}{2d}} \right|$$

$$\lesssim \delta^{1-\frac{1}{2d}} \tag{5.5.114}$$

for all $z \in Q_y$, since $|aa'^{-1}| \in \widetilde{I}$. Hence, by substituting $\eta = aa'^{-1}$, $\sigma = |a'|^{\frac{1}{d}-1}(s - s')$ and $\tau = A_{a'}^{-1} S_{s'}^{-1}(t - t')$, implying $|\eta| \in \widetilde{I}$ and $\sigma \in Q_{\tilde{d}_1}$, we obtain

$$J_1 \lesssim \delta^{p(1-\frac{1}{2d})} \cdot \int_{\mathbb{R}^d} \int_{\mathbb{R}^{d-1}} \int_{-1}^{1} \left| \mathcal{F}^{-1} \left(\widehat{\Psi} \overline{\widehat{\Psi}(A_{aa'^{-1}} S^T_{|a'|^{\frac{1}{d}-1}(s-s')} \cdot)} \right) (A_{a'}^{-1} S_{s'}^{-1}(t - t')) \right|^p \frac{da'}{|a'|^{d+1}} \, ds' \, dt'$$

$$= \delta^{p(1-\frac{1}{2d})} \cdot \int_{\mathbb{R}^d} \int_{Q_{\tilde{d}_1}} \int_{\widetilde{I}} \left| \mathcal{F}^{-1} \left(\widehat{\Psi} \overline{\widehat{\Psi}(A_\eta S^T_\sigma \cdot)} \right) (\tau) \right|^p \frac{d\eta}{|\eta|} \, d\sigma \, d\tau.$$

Since $\widehat{\Psi} \in \mathcal{S}(\mathbb{R}^d)$, for any $M \in \mathbb{N}$ the estimate

$$\left| \mathcal{F}^{-1} \left(\widehat{\Psi} \overline{\widehat{\Psi}(A_\eta S^T_\sigma \cdot)} \right)(x) \right| \leqslant C(M, \eta, \sigma) \cdot (1 + \|x\|_2)^{-M}$$

holds true with $C(M, \eta, \sigma) \leqslant C < \infty$ for all $\eta \in \widetilde{I}$ and $\sigma \in Q_{\widetilde{d}_1}$. Therefore, for M large enough we have

$$J_1 \lesssim \delta^{p(1-\frac{1}{2d})} \cdot \int_{Q_{\widetilde{d}_1}} d\sigma \cdot \int_{\widetilde{I}} \frac{d\eta}{|\eta|} \cdot \int_{\mathbb{R}^d} (1 + \|\tau\|_2)^{-Mp} \, d\tau \lesssim \delta^{p(1-\frac{1}{2d})}. \tag{5.5.115}$$

Next we estimate J_2 for some fixed $x \in X_0$ and apply a similar approach to (5.5.84). For fixed a, a' and s, s' let

$$H = H_{a,a',s,s'} := \mathcal{F}^{-1} \left(\widehat{\Psi} \overline{\widehat{\Psi}(A_{aa'^{-1}} S^T_{|a'|^{\frac{1}{d}-1}(s-s')} \cdot)} \right),$$

then $H \in \mathcal{S}(\mathbb{R}^d)$, because $\Psi \in \mathcal{S}(\mathbb{R}^d)$. Then, by virtue of the fundamental theorem of calculus, we write

$$\left| H(A_{a'}^{-1} S_{s'}^{-1}(t - t')) - H(A_{a''}^{-1} S_{s''}^{-1}(t - t'')) \right|$$

$$\leqslant \int_0^1 \left| \nabla H(\theta A_{a'}^{-1} S_{s'}^{-1}(t - t') + (1 - \theta) A_{a''}^{-1} S_{s''}^{-1}(t - t'')) \times (A_{a'}^{-1} S_{s'}^{-1}(t - t') - A_{a''}^{-1} S_{s''}^{-1}(t - t'')) \right| d\theta$$

$$\leqslant \int_0^1 \left\| \nabla H(\theta A_{a'}^{-1} S_{s'}^{-1}(t - t') + (1 - \theta) A_{a''}^{-1} S_{s''}^{-1}(t - t'')) \right\|_2 d\theta \tag{5.5.116}$$

$$\times \left\| A_{a'}^{-1} S_{s'}^{-1}(t - t') - A_{a''}^{-1} S_{s''}^{-1}(t - t'') \right\|_2.$$

Fix $1 \leqslant i \leqslant d$, then the partial derivative of H is given by

$$H_i := \frac{\partial}{\partial x_i} H = \mathcal{F}^{-1} \left(\overline{\frac{\partial}{\partial x_i} \Psi} \widehat{\Psi}(A_{aa'^{-1}} S^T_{|a'|^{\frac{1}{d}-1}(s-s')} \cdot) \right).$$

By the same arguments as above, we have $|H_i(x)| \lesssim (1 + \|x\|_2)^{-M}$ for any $M \in \mathbb{N}$. Plugging this into (5.5.116) yields

$$\left| H(A_{a'}^{-1} S_{s'}^{-1}(t - t')) - H(A_{a''}^{-1} S_{s''}^{-1}(t - t'')) \right|$$

$$\lesssim \int_0^1 \frac{\left\| A_{a'}^{-1} S_{s'}^{-1}(t - t') - A_{a''}^{-1} S_{s''}^{-1}(t - t'') \right\|_2}{\left(1 + \left\| \theta A_{a'}^{-1} S_{s'}^{-1}(t - t') + (1 - \theta) A_{a''}^{-1} S_{s''}^{-1}(t - t'') \right\|_2 \right)^M} \, d\theta. \tag{5.5.117}$$

To further estimate this we use the triangle inequality, (5.5.77) and (5.5.74) to obtain

$$\|A_{a'}^{-1}S_{s'}^{-1}(t-t')\|_2 - \|\theta A_{a'}^{-1}S_{s'}^{-1}(t-t') + (1-\theta)A_{a''}^{-1}S_{s''}^{-1}(t-t'')\|_2$$

$$\leqslant \|(1-\theta)(A_{a'}^{-1}S_{s'}^{-1}(t-t') - A_{a''}^{-1}S_{s''}^{-1}(t-t''))\|_2 +$$

$$\leqslant \|(I_d - A_{a'}^{-1}S_{s'}^{-1}S_{s''}A_{a''})A_{a'}^{-1}S_{s'}^{-1}(t-t') - A_{a''}^{-1}S_{s''}^{-1}(t'-t'')\|_2$$

$$\leqslant \|I_d - A_{a'a''^{-1}}^{-1}S_{|a'|^{\frac{1}{d}-1}(s'-s'')}^{-1}\|_{2\to2} \cdot \|A_{a'}^{-1}S_{s'}^{-1}(t-t')\|_2 + \|A_{a''}^{-1}S_{s''}^{-1}(t'-t'')\|_2$$

$$\leqslant \delta \cdot \|A_{a'}^{-1}S_{s'}^{-1}(t-t')\|_2 + \delta,$$

implying

$$1 + \|\theta A_{a'}^{-1}S_{s'}^{-1}(t-t') + (1-\theta)A_{a''}^{-1}S_{s''}^{-1}(t-t'')\|_2 \geqslant \frac{1}{2}(1 + \|A_{a'}^{-1}S_{s'}^{-1}(t-t')\|_2) \tag{5.5.118}$$

for sufficiently small $\delta > 0$. Plugging (5.5.76) and (5.5.118) into (5.5.117) gives us

$$\left|H(A_{a'}^{-1}S_{s'}^{-1}(t-t')) - H(A_{a''}^{-1}S_{s''}^{-1}(t-t''))\right| \lesssim \frac{\|A_{a'}^{-1}S_{s'}^{-1}(t-t') - A_{a''}^{-1}S_{s''}^{-1}(t-t'')\|_2}{\left(1 + \|A_{a'}^{-1}S_{s'}^{-1}(t-t')\|_2\right)^M}$$

$$\lesssim \delta \cdot \left(1 + \|A_{a'}^{-1}S_{s'}^{-1}(t-t')\|_2\right)^{-M+1}. \tag{5.5.119}$$

Therefore we can estimate J_2 in (5.5.112) by using (5.5.73) and substituting $\eta = aa'^{-1}$, $\sigma = |a'|^{\frac{1}{d}-1}(s-s')$ and $\tau = A_{a'}^{-1}S_{s'}^{-1}(t-t')$:

$$J_2 = \int_{\mathbb{R}^d}\int_{\mathbb{R}^{d-1}}\int_{-1}^{1}\sup_{z\in Q_y}|aa''^{-1}|^{p(1-\frac{1}{2d})}|H(A_{a'}^{-1}S_{s'}^{-1}(t-t')) - H(A_{a''}^{-1}S_{s''}^{-1}(t-t''))|^p \frac{da'}{|a'|^{d+1}}\,ds'\,dt'$$

$$\lesssim \delta^p \cdot \int_{\mathbb{R}^d}\int_{\sigma\in Q_{\widetilde{d_1}}}\int_{|\eta|\in\widetilde{I}}|aa'^{-1}|^{p(1-\frac{1}{2d})}\left(1 + \|A_{a'}^{-1}S_{s'}^{-1}(t-t')\|_2\right)^{-p(M-1)}\frac{da'}{|a'|^{d+1}}\,ds'\,dt'$$

$$= \delta^p \cdot \int_{\mathbb{R}^d}\int_{Q_{\widetilde{d_1}}}\int_{\widetilde{I}}(1 + \|\tau\|_2)^{-p(M-1)}\frac{d\eta}{|\eta|^{1-p(1-\frac{1}{2d})}}\,d\sigma\,d\tau$$

$$\lesssim \delta^p, \tag{5.5.120}$$

for sufficiently large M.

Finally, it remains to estimate J_3. We write

$$\mathcal{F}^{-1}\left(\overline{\widehat{\Psi}}\widehat{\Psi}(A_{aa'^{-1}}S_{|a'|^{\frac{1}{d}-1}(s-s')}^T\cdot)\right)(A_{a''}^{-1}S_{s''}^{-1}(t''-t))$$

$$- \mathcal{F}^{-1}\left(\overline{\widehat{\Psi}}\widehat{\Psi}(A_{aa''^{-1}}S_{|a''|^{\frac{1}{d}-1}(s-s'')}^T\cdot)\right)(A_{a''}^{-1}S_{s''}^{-1}(t''-t))$$

$$= \int_{\mathbb{R}^d}\overline{\widehat{\Psi}(\omega)}\left[\widehat{\Psi}\left(A_{aa'^{-1}}S_{|a'|^{\frac{1}{d}-1}(s-s')}^T\omega\right) - \widehat{\Psi}\left(A_{aa''^{-1}}S_{|a''|^{\frac{1}{d}-1}(s-s'')}^T\omega\right)\right]e^{-2\pi i\langle\omega,A_{a''}^{-1}S_{s''}^{-1}(t''-t)\rangle}\,d\omega$$

$$= G_{\widehat{\Psi}\left(A_{aa'^{-1}}S_{|a'|^{\frac{1}{d}-1}(s-s')}^T\cdot\right) - \widehat{\Psi}\left(A_{aa''^{-1}}S_{|a''|^{\frac{1}{d}-1}(s-s'')}^T\cdot\right)}\widehat{\Psi}(0, A_{a''}^{-1}S_{s''}^{-1}(t''-t)), \tag{5.5.121}$$

where G denotes the short time Fourier transform (STFT) given by

$$G_\psi f(x,\omega) = \int_{\mathbb{R}^d}f(t)\overline{\psi(t-x)}e^{-2\pi i\langle\omega,t\rangle}\,dt.$$

The STFT satisfies

$$G_\psi f(x,\omega) = e^{-2\pi i\langle\omega,x\rangle}G_\psi\widehat{f}(\omega,-x),$$

see [68, Lemma 3.1.1], as well as

$$|G_\psi f(x, \omega)| \lesssim \|f\|_{L_{\infty, m_\mu}} \|\psi\|_{L_{\infty, m_\mu}} (1 + \|x\|_2)^{-\mu}$$

for $\psi \in \mathcal{S}(\mathbb{R}^d)$ and $\mu > d$, where $m_\mu(x) = (1 + \|x\|_2)^\mu$ and $\|f\|_{L_{\infty, m_\mu}} = \operatorname{ess\,sup}_{x \in \mathbb{R}^d} |f(x)|(1 + \|x\|_2)^\mu$, see [68, page 232]. By (5.5.121) we thus obtain

$$\left| \mathcal{F}^{-1}\big(\overline{\widehat{\Psi}}\widehat{\Psi}(A_{aa'^{-1}} S^T_{|a'|^{\frac{1}{d}-1}(s-s')} \cdot)\big)(A_{a''}^{-1} S_{s''}^{-1}(t'' - t)) \right.$$

$$\left. - \mathcal{F}^{-1}\big(\overline{\widehat{\Psi}}\widehat{\Psi}(A_{aa''^{-1}} S^T_{|a''|^{\frac{1}{d}-1}(s-s'')} \cdot)\big)(A_{a''}^{-1} S_{s''}^{-1}(t'' - t)) \right|$$

$$= \left| G_{\widehat{\Psi}}\big(\widehat{\Psi}(A_{aa'^{-1}} S^T_{|a'|^{\frac{1}{d}}(s-s')} \cdot) - \widehat{\Psi}(A_{aa''^{-1}} S^T_{|a''|^{\frac{1}{d}}(s-s'')} \cdot)\big)\Psi(A_{a''}^{-1} S_{s''}^{-1}(t'' - t), 0) \right|$$

$$\lesssim \|\Psi\|_{L_{\infty, m_\mu}} \cdot \left\| \widehat{\Psi}(A_{aa'^{-1}} S^T_{|a'|^{\frac{1}{d}}(s-s')} \cdot) - \widehat{\Psi}(A_{aa''^{-1}} S^T_{|a''|^{\frac{1}{d}}(s-s'')} \cdot) \right\|_{L_{\infty, m_\mu}}$$

$$\cdot \big(1 + \|A_{a''}^{-1} S_{s''}^{-1}(t - t'')\|_2\big)^{-\mu}$$

$$\lesssim \|\Psi\|_{L_{\infty, m_\mu}} \cdot \left\| \widehat{\Psi}(A_{\eta_0} S^T_{\sigma_0} \cdot) - \widehat{\Psi}(A_{\eta_1} S^T_{\sigma_1} \cdot) \right\|_{L_{\infty, m_\mu}} \cdot \big(1 + \|A_{a'}^{-1} S_{s'}^{-1}(t - t')\|_2\big)^{-\mu} \qquad (5.5.122)$$

for some $\eta_0, \eta_1 \in \widetilde{I}$, $\sigma_0, \sigma_1 \in Q_{\widetilde{d_1}}$. Since $\Psi, \widehat{\Psi} \in \mathcal{S}(\mathbb{R}^d)$, the first two terms in (5.5.122) are finite. Plugging this into (5.5.112) and substituting $\eta = aa'^{-1}$, $\sigma = |a'|^{\frac{1}{d}-1}(s - s')$ as well as $\tau = A_{a'}^{-1} S_{s'}^{-1}(t - t')$ gives us

$$J_3 = \int_{\mathbb{R}^d} \int_{\mathbb{R}^{d-1}} \int_{-1}^{1} \sup_{z \in Q_y} |\det A_{aa''^{-1}}|^{\frac{p}{2}} \left| \mathcal{F}^{-1}\big(\overline{\widehat{\Psi}}\widehat{\Psi}(A_{aa'^{-1}} S^T_{|a'|^{\frac{1}{d}-1}(s-s')} \cdot)\big)(A_{a''}^{-1} S_{s''}^{-1}(t'' - t)) \right.$$

$$\left. - \mathcal{F}^{-1}\big(\overline{\widehat{\Psi}}\widehat{\Psi}(A_{aa''^{-1}} S^T_{|a''|^{\frac{1}{d}-1}(s-s'')} \cdot)\big)(A_{a''}^{-1} S_{s''}^{-1}(t'' - t)) \right|^p w(a, a')^{pr} \frac{\mathrm{d}a'}{|a'|^{d+1}} \, \mathrm{d}s' \, \mathrm{d}t'$$

$$\lesssim \int_{\mathbb{R}^d} \int_{\sigma \in Q_{\widetilde{d_1}}} \int_{|\eta| \in \widetilde{I}} |aa'^{-1}|^{p(1-\frac{1}{2d})} \big(1 + \|A_{a'}^{-1} S_{s'}^{-1}(t - t')\|_2\big)^{-p\mu} \frac{\mathrm{d}a'}{|a'|^{d+1}} \, \mathrm{d}s' \, \mathrm{d}t'$$

$$\lesssim \int_{\mathbb{R}^d} \int_{Q_{\widetilde{d_1}}} \int_{\widetilde{I}} (1 + \|\tau\|_2)^{-p\mu} \frac{\mathrm{d}\eta}{|\eta|^{1-p(1-\frac{1}{2d})}} \, \mathrm{d}\sigma \, \mathrm{d}\tau < \infty, \qquad (5.5.123)$$

since $p\mu > d$. By (5.5.115), (5.5.120) and (5.5.123) the expression $J_1 + J_2 + J_3$ is finite for any $x = (a, s, t) \in X_0$, therefore by (5.5.112) the first integral in (5.5.82) is finite. Using exactly the same arguments, which are solemnly based on the support properties developed in Lemma 5.5.9, the second and third integral in (5.5.82) are also finite. This completes the proof. $\qquad \square$

5.5.5 Discretization

Using the results of the preceding subsections, we are finally ready to give a discretization result for the inhomogeneous shearlet coorbit spaces $\mathcal{SC}^{\tau}_{\widetilde{s}, p}$. To be precise, we state a reconstruction result based on part (ii) of Theorem 5.2.17. Let us briefly recall the setting.

For an odd dimension $d \geqslant 3$ we suppose $\Psi \in \mathcal{S}(\mathbb{R}^d)$ is an admissible shearlet with $C_\Psi = 1$ satisfying the support condition

$$\widehat{\Psi} \subseteq ([-a_1, -a_0] \cup [a_0, a_1]) \times Q_b$$

for some $0 < a_0 < a_1$ and $b \in \mathbb{R}^{d-1}_+$. According to Remark 5.5.5, let $\Phi \in \mathcal{S}(\mathbb{R}^d)$ be chosen as in (5.5.19) such that condition (5.5.16) is fulfilled with $A = B = 1$ and the family $\mathfrak{F} = (\psi_x)_{x \in X}$ is a

Parseval frame for $L_2(\mathbb{R}^d)$ according to Theorem 5.5.4. The frame elements in $\mathfrak{F}$ are defined via

$$\psi_{(\infty,s,t)} := T_t D_s \Phi = \Phi(S_s^{-1}(\cdot - t)) \quad \text{and}$$

$$\psi_{(a,s,t)} := T_t D_s D_a \Psi = |\det A_a|^{-1/2} \Psi(A_a^{-1} S_s^{-1}(\cdot - t))$$

The kernel $K_{\mathfrak{F}}(x,y) = \langle \psi_y, \psi_x \rangle_{L_2}$ is contained in all $\mathcal{A}_{q,w_r}$ for $1 < q < \infty$, see Theorem 5.5.13. For $\alpha > 0$ and $\beta, \tau > 1$ we define the covering $\mathcal{U} = (U_\lambda)_{\lambda \in \Lambda}$ of X according to Definition 5.5.18, where

$$\Lambda := \Big(\{\infty\} \times \mathbb{Z}^{d-1} \times \mathbb{Z}^d \Big) \cup \Big(\{-1,1\} \times \mathbb{N}_0 \times \mathbb{Z}^{d-1} \times \mathbb{Z}^d \Big)$$

In Theorem 5.5.22 we then showed that the $\mathcal{U}$-oscillation of $K_{\mathfrak{F}}$,

$$\mathrm{osc}_{\mathcal{U}} K_{\mathfrak{F}}(x,y) = \sup_{z \in Q_y} |\langle \psi_y - \psi_z, \psi_x \rangle_{L_2}|, \quad x, y \in X,$$

is contained in all $\mathcal{A}_{p,w_r}$ for $r \in \mathbb{R}_{\geqslant 0}$ and $\frac{d}{d-1} < p < \infty$. We can therefore apply Theorem 5.2.17 and obtain the following.

Lemma 5.5.23. *Let $1 \leqslant p, q < \infty$ and $\frac{d}{d-1} < s < \infty$ such that $1/q + 1/s = 1 + 1/p$. Furthermore let $r \in \mathbb{R}_{\geqslant 0}$.*

Then for every $c = (c_\lambda)_{\lambda \in \Lambda} \in \ell_{q,m_r}(\Lambda)$ we have $T := \sum_{\lambda \in \Lambda} c_\lambda \psi_\lambda \in \mathcal{SC}_{\mathfrak{F},p}^r$ with

$$\|T\|_{\mathcal{SC}_{\mathfrak{F},p}^r} \leqslant C \cdot \|c\|_{\ell_{q,m_r}}, \tag{5.5.124}$$

where

$$C = d \cdot \alpha^r (\alpha^d - 1)^{-1} \beta^{-(d-1)} \tau^{-d} \cdot \Big(\|K_{\mathfrak{F}}\|_{\mathcal{A}_{s,w_r}} + \|\mathrm{osc}_{\mathcal{U}} K_{\mathfrak{F}}\|_{\mathcal{A}_{s,w_r}} \Big). \tag{5.5.125}$$

Proof. This is an application of Theorem 5.2.17 (ii), which shows (5.5.124) with

$$C = C_{\mathcal{U}}^{-1} \cdot \sup_{\lambda \in \Lambda} \sup_{x,y \in U_\lambda} w_r(x,y) \cdot N^{1/q'} \cdot \Big(\|\mathrm{osc}_{\mathcal{U}_n} K_{\mathfrak{F}}\|_{\mathcal{A}_{p,w_r}} + \|K_{\mathfrak{F}}\|_{\mathcal{A}_{p,w_r}} \Big).$$

By Lemma 5.5.19 we have $C_{\mathcal{U}} = \frac{1}{d}(\alpha^d - 1)\beta^{d-1}\tau^{-d}$—at least if $\alpha^d \leqslant d - 1$, which we reasonably assume—and Corollary 5.5.20 gives us $\sup_{x,y \in U_\lambda} w_r(x,y) \leqslant \alpha^r$ for all $\lambda \in \Lambda$. Finally, since the elements of $\mathcal{U}$ are pairwise disjoint we have $N = 1$, such that

$$C = d \cdot \alpha^r (\alpha^d - 1)^{-1} \beta^{-(d-1)} \tau^{-d} \cdot \Big(\|\mathrm{osc}_{\mathcal{U}_n} K_{\mathfrak{F}}\|_{\mathcal{A}_{p,w_r}} + \|K_{\mathfrak{F}}\|_{\mathcal{A}_{p,w_r}} \Big),$$

proving (5.5.125). $\qquad\square$

Remark 5.5.24. Of course we are also interested in the counterpart of Lemma 5.5.23, which is an application of part (i) of Theorem 5.2.17. However, for this we would need that Assumption 5.2.2 is fulfilled, i.e., the linear span of $\{K_{\mathfrak{F}}(\cdot,x)\}_{x \in X}$ is dense in the reproducing kernel space $\mathcal{M}_{p,m_r}$. As we have shown a sufficient condition for this assumption to be fulfilled is to propose that the integral operator associated to $K_{\mathfrak{F}}$ is bounded on $L_{p,m_r}(X,\mu)$. Showing this, however, is a very difficult task. And we frankly do not know whether this condition is fulfilled in the present setting, or maybe in a more restrictive setting.

As we have seen in Sections 5.2, 5.3 and 5.4 we have no hope of achieving a meaningful discretization of the inhomogeneous shearlet coorbit spaces $\mathcal{SC}_{\mathfrak{F},p}^r$ without proposing Assumption 5.2.2. A possible way to do this is layed out in the next chapter.

Chapter 6

Conclusion and Outlook

This very last chapter is dedicated to a short discussion on the results in this dissertation as well as an outlook on where the results might be improved or complemented. Many of the ideas presented are outside the scope of this thesis and maybe even impossible to achieve, but nonetheless the author considers them worthy of discussion.

Extension of the Generalized Coorbit Theory

The coorbit theory developed in Chapter 5 is a useful extension of both the generalized coorbit theory discussed in Chapter 4 as well as the coorbit theory with non-integrable kernel in Chapter 3. However, the only target spaces considered are weighted Lebesgue spaces, which is more restrictive than Definition 2.1.13 or Definition 4.1.6. It may be possible to also consider solid Banach spaces of functions Y on X where the continuous embeddings $\mathcal{A}_{q,w}(Y) \subseteq Y$ hold true for all $1 < q < \infty$, that is, for every $K \in \mathcal{A}_{q,w}$ and $f \in Y$ we have $\|K(f)\|_Y \leqslant \|K\|_{\mathcal{A}_{q,w}} \cdot \|f\|_Y$. Obviously, this is fulfilled for the setting of weighted Lebesgue spaces, but it is unclear if a suitable coorbit theory for target spaces as above is also possible in the setting of Chapter 5. This would especially affect Proposition 5.1.9. Similarly, an extension to quasi-Banach spaces of function as target spaces is also conceivable.

Since a main part of coorbit theory is the discretization, the discretization ideas of Chapters 3 and 5 would need to be adapted to the corresponding setting. Because some of the proofs are specifically designed for weighted Lebesgue spaces, possible changes would need to be examined carefully. However, considering that such discretizations are also possible for the classical setting this should extend for the generalizations as well.

In addition to different target spaces, one might ask if the choice of different kernel spaces is possible as well. Looking at the results of Section 3.1 we see that in the group setting we allowed the reproducing kernel to be an element of certain Fréchet spaces. As an example of such a Freéchet space we look at the intersection of all weighted Lebesgue spaces. Similarly, it might be possible to extend the idea of generalized coorbit spaces to cases, where the kernel is allowed to be an element of certain Fréchet spaces of kernels. The setting of Chapter 3, namely the intersection of all kernel spaces $\mathcal{A}_{p,w}$, $1 < p < \infty$, is a special example of this. This idea is probably a long shot but it would significantly generalize the approach.

Extension of the Inhomogeneous Shearlet Coorbit Spaces

Considering the inhomogeneous shearlet coorbit spaces developed in Section 5.5, there are still several open questions. First and foremost we have shown that the reproducing kernel is contained in all kernel spaces $\mathcal{A}_{p,w}$ for $1 < p < \infty$, but we have not shown that it is not an element of $\mathcal{A}_{1,w}$. This would show that we indeed have an exclusive example of the theory developed in

Chapter 5. Looking at the proof of Theorem 5.5.13, however, it already seems like the kernel is not integrable, just as we suspect.

The integrability of the oscillator of the reproducing kernel is not entirely satisfactory either, since we have proven in Theorem 5.5.22 that $\mathrm{osc}_\mathcal{U} K_\mathfrak{F} \in \mathcal{A}_{p,w}$ for all $1 + \frac{1}{d-1} < p < \infty$. Considering the integrability of the kernel $K_\mathfrak{F}$ itself, it is desirable to have the same integrability properties, i.e., it remains to show $\mathrm{osc}_\mathcal{U} K_\mathfrak{F} \in \mathcal{A}_{p,w}$ for $1 < p \leqslant 1 + \frac{1}{d-1}$ as well. It is not clear if this is possible or not, but proving this would improve the discretization results of Lemma 5.5.23.

Another interesting extension of inhomogeneous shearlet coorbit spaces would be to consider different shearlet transforms. In [23, 30] the idea of shearing has been extended to the so-called Toeplitz shearlet transform, which is also applicable in our setting. Or we might look at different scaling matrices, where the exponent of $|a|^\delta$ in (5.5.1) is modifiable for different $0 < \delta < 1$. It is highly likely that these different transforms yield meaningful coorbit spaces, but it would require some serious effort to alter the technical proofs to match the different transforms.

Additionally, in contrast to the homogeneous shearlet coorbit spaces in Subsection 2.3.2, we only considered band-limited shearlets Ψ. Then, following the construction in Remark 5.5.5, we were able to find some suitable function Φ such that $\mathfrak{F}$ constitutes an inhomogeneous shearlet frame. But it is unclear what happens for compactly supported shearlets Ψ and if they also allow us to find suitable functions Φ, which are maybe also compactly supported, such that $\mathfrak{F}$ constitutes a continuous frame. This question is very interesting since it could give rise to the possibility of achieving new embedding results involving inhomogeneous shearlet coorbit spaces. However, these new frames would require entirely new proofs for the integrability of the kernel and its oscillation since the existing proofs heavily rely on the support properties of the frame in the frequency domain.

Another tool to understand and define function spaces are the decomposition spaces [42, 43]. This approach allows to understand well-known function spaces in a different manner, such as homogeneous Besov spaces [43], inhomogeneous Besov spaces [105] and certain shearlet spaces [61]. This theory also gives rise to a variety of embedding theorems [59, 106]. If we could apply similar techniques to inhomogeneous shearlet coorbit spaces, this would pave the way for embedding results and a better understanding of these spaces.

Embeddings of Wavelet and Shearlet Spaces

But since embeddings between function spaces are a very useful tool for a better understanding of structural properties, naturally the question arises if embeddings of inhomogeneous shearlet coorbit spaces can also be obtained without the use of decompositions spaces. Looking at the homogeneous counterpart, we recall the following result from [22]. Let $d = 3$, then define a subset of $\mathcal{SC}_p^\sigma(\mathbb{R}^3)$ in the following manner. By the discretization results of Subsection 2.3.2, functions of the homogeneous shearlet coorbit space can be uniformly decomposed and reconstructed with suitable building blocks, see also (2.3.32) and (2.3.34). By restricting the shearing parameter to a bounded set depending on the dilation parameter, for $\eta \in \{0,1\}^2$ the space

$$\mathcal{SC}_p^{\sigma,\eta} = \left\{ f \in \mathcal{SC}_p^\sigma(\mathbb{R}^3) : \sum_{j \in \mathbb{Z}} \sum_{k \in \mathbb{Z}^2} \sum_{m \in \mathbb{Z}^3} c(j,k,m) \psi_{j,k,m}, \ c(j,k,m) = 0 \text{ for } |k_i| > \alpha^{\frac{2j}{3}} \text{ if } \eta_i = 1 \right\},$$

forms a subspace of $\mathcal{SC}_p^\sigma(\mathbb{R}^3)$, also called cone-adapted shearlet space. With these spaces at hand, [22, Theorem 5.1] shows that the embedding $\mathcal{SC}_p^{\sigma,\eta} \subseteq B_{p,p}^{\sigma_1}(\mathbb{R}^3) + B_{p,p}^{\sigma_2}(\mathbb{R}^3)$ holds true, where

$$\sigma_1 + 2\lfloor \sigma_1 \rfloor = 3r - \frac{21}{2} + \frac{9}{p} \quad \text{and} \quad \sigma_2 - \frac{2}{3}\lfloor \sigma_2 \rfloor = r + \frac{5}{3p} + \frac{7}{6}.$$

The proof is based on the following idea. Take a function $f \in \mathcal{SC}_p^{\sigma,\eta}$ and decompose it according to the definition of the space. Then, the building blocks $\psi_{j,k,m}$ can be rearranged such that they form atoms for the inhomogeneous Besov spaces in the sense of the definition in [54]. This means atoms are smooth functions, possibly having vanishing moments, that are supported on isotropic cubes. Obviously, this embedding therefore relies on the fact that we have compactly supported shearlets.

Naturally, one might ask if we can find embeddings in the opposite direction as well. The problem is, however, that in general the decomposition of function spaces is harder than reconstruction, which means that we cannot decompose Besov functions with arbitrary building blocks as above but are restricted in our choice. At the same time we are restricted in our choice of building blocks to reconstruct in the homogeneous shearlet coorbit space, which makes the embedding difficult. Moreover, both spaces have different geometric properties in the following sense. In the Besov space the dilation of the building blocks is isotropic, whereas the dilation in the shearlet setting is anisotropic. This makes it even harder to compare the two spaces.

But what about the inhomogeneous shearlet coorbit spaces, more precisely, can we embed subspaces of $\mathcal{SC}_{\mathfrak{Z},p}^r$ into (sums of) Besov spaces? The answer is unfortunately no. The problem here is that we can only reconstruct the inhomogeneous spaces; which is the easier part. But for an embedding into Besov spaces we would need decompositions of the inhomogeneous spaces and by the results of Section 5.2 this requires the boundedness of the kernel operator — something that we have not proven. Therefore, until a useful decomposition of the inhomogeneous shearlet coorbit spaces is available, there is no hope of embedding the spaces with the methods we know.

Boundedness of the Reproducing Kernel Operator

The following part will only be concerned with the setup of Chapter 5, but the same ideas hold for Chapter 3.

As we have seen in Theorem 5.3.1, any meaningful discretization requires the boundedness on weighted Lebesgue spaces of the operator associated with the reproducing kernel $K_{\mathfrak{Z}}$. Similarly this boundedness is sufficient for the decomposition of coorbit spaces in Theorem 5.2.17. But the only example we have seen so far where this is fulfilled is the Paley-Wiener spaces and the proof heavily relied on the Fourier transform—a very powerful machinery that only exists in the Euclidean case. In general though, the Fourier theory is not available and the question of the boundedness of the kernel operator remains unanswered. In the example of the inhomogeneous shearlet coorbit spaces, e.g., there is no hope of showing the boundedness ad hoc.

Let us briefly present a theory that might be helpful to tackle this problem, namely the theory of Calderón-Zygmund operators, or singular integral operators, on spaces of homogeneous type. Note that this notion of homogeneity is completely unrelated to the ones we have seen so far. We follow the lines of [13, Chapter 6] and [16] and first define a quasi-metric on X, i.e., a map $\rho\colon X \times X \to [0, \infty]$ satisfying

(i) $\rho(x,y) = 0$ if and only if $x = y$;

(ii) $\rho(x,y) = \rho(y,x)$ for all $x, y \in X$;

(iii) $\rho(x,y) \leqslant C_0 \cdot \rho(x,z) + C_0 \cdot \rho(z,y)$ for all $x, y, z \in X$ and some $C_0 > 0$.

The balls of radius $r > 0$ and center $x \in X$ associated to this quasi-metric are defined via

$$B(x,r) = \{y \in X \,:\, \rho(x,y) < r\}.$$

Then, we say the triple (X, μ, ρ) is a space of homogeneous type, if $\mu(B(x,r)) < \infty$ for all $x \in X$ and $r > 0$, and the doubling property is fulfilled, that is, there exists $C < \infty$ such that for all

$x \in X$ and $r > 0$:

$$\mu(B(x, 2r)) \leqslant C \cdot \mu(B(x, r)).$$

Examples for homogeneous spaces include connected Lie groups with a left-invariant Riemannian metric. But it is also possible to find a suitable quasi-metric for the inhomogeneous shearlet coorbit spaces.

Now fix a kernel $K \colon X \times X \to \mathbb{C}$ which fulfills the following properties. Assume there exists $\varepsilon > 0$ and $C < \infty$ such that for all $x \neq y \in X$ and $z \in X$ wth $\rho(x, z) < \varepsilon \cdot \rho(x, y)$ we have

$$|K(x, y)| \leqslant \frac{C}{\mu(B(x, r))}, \quad \text{where } r = \rho(x, y),$$

and

$$|K(x, y) - K(z, y)| + |K(y, x) - K(y, z)| \leqslant \left(\frac{\rho(x, z)}{\rho(x, y)}\right)^{\varepsilon} \cdot \frac{C}{\mu(B(x, r))}.$$

The first property can be seen as a decay property and the second property as a condition concerning the regularity, or the oscillation of K. While both properties may seem pretty straightforward, they are not necessarily easy to verify in practice. Especially considering that the kernels we are looking at are reproducing kernels, which are given by an inner product of two functions.

Let us also assume that the kernel operator associated with K is bounded on $L_2(X, \mu)$; which is always fulfilled for reproducing kernels by the frame properties. Then, the kernel operator fulfilling the two upper inequalities is also bounded on $L_p(X, \mu)$ for all $1 < p < \infty$ by [13, Theorem 9]. If we extend this result to the weighted setting we have show the intended boundedness. So the two conditions above give us a possibility to check if the kernel operator associated to a reproducing kernel is bounded on weighted Lebesgue spaces. Maybe in future work this can be used as a starting point for discretizations of generalized coorbit spaces.

The Second Kernel W

In Section 5.4 we have seen the existence of a second kernel W, which is contained in the kernel space $\mathcal{A}_{1,w}$ and fulfills either $W \circ K_{\mathfrak{F}} = K_{\mathfrak{F}}$ or $K_{\mathfrak{F}} \circ W = K_{\mathfrak{F}}$, ensures atomic decompositions and Banach frames for the generalized coorbit space with non-integrable reproducing kernel. The motivation behind this construction was the observation that exactly this setting is present for many Paley-Wiener spaces, see Chapter 3. Again, this is a consequence of the rich Fourier theory available in this particular setting. Obviously, in a more general setting this theory is not available and the question remains whether other examples exist. Unfortunately, to the best knowledge of the author no other examples are available, neither in the group setting nor in the more general setting. This should not mean, however, that the theory developed is void. Indeed, quite the opposite is true and our results can be understood as a limit of what is feasible.

Conclusion

The generalized coorbit theory presented in this dissertation provides a proper improvement to existing research and shows that a broader range of kernels can be used to define coorbit spaces. Moreover, different discretization techniques are laid out, starting with a reconstruction results as well as a basic decomposition result, provided the kernel operator fulfills certain boundedness conditions, followed by proper discretizations by means of atomic decompositions and Banach frames, under the additional assumption that there exists a second, well-behaving kernel W. Finally, the theory is applied to Paley-Wiener spaces and used to define new function spaces, namely inhomogeneous shearlet coorbit spaces.

We have seen above that it is still possible to extend, generalize or complement this theory. The main ideas and basic cases have been presented in this monograph and may serve as a starting point for further investigations. But at the same time, when saying goodbye to the integrability of the reproducing kernel, the discretization techniques are so much more comprehensive that the benefit for applications decreases. We have seen the Paley-Wiener spaces, which serve as great examples for the theory developed. But these spaces are the only application of the discretization techniques so far, not to mention applications without an underlying group structure. And when looking at inhomogeneous shearlet coorbit spaces, we see that showing a first discretization result is already very technically involved. All of this leads us to assume that the theory presented in this dissertation is, to a certain extend, the limit of what coorbit theory offers.

Zusammenfassung

In der vorliegenden Arbeit wird die bestehende Coorbit-Theorie auf zwei unterschiedlich Weisen verallgemeinert und auf Beispiele angewendet, unter anderen auf sogenannte Shearlets. Die Coorbit-Theorie erlaubt es, bestimmte Funktionenräume, sogenannte Coorbit-Räume, mittels Darstellungen von Gruppen auf Hilbert-Räumen zu definieren, und bietet eine einheitliche Strategie, um ebendiese Räume zu diskretisieren. Die grundsätzliche Idee dabei ist es, Coorbit-Räume als Glattheitsräume zu verstehen, wobei die Glattheit in dem asymptotischen Verhalten von Transformierten der Funktionen gemessen wird. Hierbei wird die Transformation über die Darstellung der Gruppe definiert.

Klassische Coorbit-Theorie

Die Coorbit-Theorie wurde in den 80er Jahren des vergangenen Jahrhunderts von Feichtinger und Gröchenig in mehreren Veröffentlichungen entwickelt [44–46], und sie wird in dieser Dissertation in Kapitel 2 zusammengefasst. Grundsätzlich wird dafür eine lokal-kompakte topologische Gruppe G mit unitärer Darstellung π auf einem Hilbert-Raum $\mathcal{H}$ verwendet. Zu einem zulässigen Vektor $\psi \in \mathcal{H}$ ist dann die Voice-Transformation V_ψ gegeben als die Abbildung

$$V_\psi \colon \mathcal{H} \to L_2(G), \quad V_\psi f(x) = \langle f, \pi(x)\psi \rangle_{\mathcal{H}}.$$

Die Zulässigkeit von ψ ist hierbei gerade die Wohldefiniertheit der obigen Abbildung, wir sagen auch die Abbildung ist quadrat-integrierbar. Eine Besonderheit der Voice-Transformation ist, dass die Faltung von rechts mit dem Element $V_\psi \psi$ gerade die Identität ist. Wenn wir mit $K_\psi = V_\psi \psi$ den sogenannten reproduzierenden Kern bezeichnen, dann bedeutet dies als Formel geschrieben $V_\psi f * K_\psi = V_\psi f$ für alle $f \in \mathcal{H}$. Zu einem fixierten Gewicht w auf G können wir nun die Räume $\mathcal{H}_{1,w} = \{f \in \mathcal{H} : V_\psi f \in L_{1,w}(G)\}$ definieren. Die Motivation dafür ist, dass wir an den Dual-Räumen von $\mathcal{H}_{1,w}$ interessiert sind. Diese Menge von Distributionen bildet das Reservoir, aus dem wir die Elemente der Coorbit-Räume wählen. Dazu erweitern wir zunächst die Voice-Transformation auf den Dualraum $\mathcal{H}'_{1,w}$ mittels eines Gelfand-Triples dichter Einbettungen und bezeichnen die Erweiterung mit $V_{e,\psi}$. Wir sind nun an den Distributionen interessiert, deren erweiterte Voice-Transformationen schnell genug asymptotisch fällt. Anders gesprochen sollen sie in einem gewissen Funktionenraum Y enthalten sein, die das asymptotische Verhalten von Funktionen charakterisiert:

$$\mathrm{Co}(Y) = \{T \in \mathcal{H}'_{1,w} : V_{e,\psi} T \in Y\}.$$

Dieser Raum ist nicht nur wohldefiniert sondern unter gewissen Annahmen auch ein Banach-Raum. Eine weiter wichtige und häufig genutzte Eigenschaft ist, dass $\mathrm{Co}(Y)$ isometrisch isomorph ist zu dem Raum $\{f \in Y : f * K_\psi = f\}$. Dies hat zur Folge, dass es für fast alle wichtigen Eigenschaften genügt zu zeigen, dass diese für den letzteren Raum gelten und sie sich direkt auf den Coorbit-Raum übertragen lassen. Wichtige Beispiele für Y sind die gewichteten Lebesgue-Räume $L_{p,m}(G)$ für gewisse positive Gewichte m auf G. Zusätzlich bietet die Coorbit-Theorie einen einheitlich Ansatz die Coorbit Räume zu diskretisieren. Genauer gesagt ist die Existenz von Zerlegungen in Atome und Banach-Frames gesichert. Dies ist ein sehr mächtiges Werkzeug, um die Struktureigenschaften der Räume analysieren zu können.

Coorbit-Theorie mit Nicht-Integrierbarem Kern

Allerdings basiert die Theorie nicht nur auf der grundsätzlichen Annahme, dass die Darstellung quadrat-integrierbar ist, sondern zusätzlich auch integrierbar ist, d.h. $K_\psi \in L_{1,w}(G)$. Um diese Einschränkung zu umgehen, haben Dahlke et al. [18] die Coorbit-Theorie weiterentwickelt, um auch solche Fälle zuzulassen, in denen der Kern in einem allgemeineren Fréchet-Raum enthalten ist. Dies beinhaltet den Fall, dass K_ψ nicht integrierbar ist, dafür aber in allen gewichteten Lebesgue-Räumen $L_{p,w}(G)$ für $1 < p < \infty$ enthalten ist. Wir sprechen daher auch von nicht-integrierbaren Kernen. Tatsächlich ist es auch in diesen Fällen möglich Coorbit-Räume zu definieren; die Theorie bedarf jedoch einiger Anpassung. So muss ein Ersatz für den Raum $\mathcal{H}_{1,w}$ gefunden werden und die Voice-Transformation folglich anders fortgesetzt werden.

Eine noch größere Herausforderung ist jedoch die Diskretisierung der Coorbit-Räume $\mathrm{Co}(L_{r,m})$ mit $1 < r < \infty$, denn die klassischen Methoden basieren stets auf Young'schen Faltungsungleichungen, welche die Integrierbarkeit des Kerns voraussetzen. Dies war bisher nicht möglich und wird in der vorliegenden Arbeit präsentiert; die Ergebnisse wurden in [19] veröffentlicht. In Abschnitt 3.2 wird dafür ein anderer Ansatz dargestellt, der nicht die Integrierbarkeit voraussetzt und stattdessen andere Faltungsungleichungen nutzt. Genauer beinhaltet Satz 3.2.17 die folgende Aussage zu Rekonstruktionen: zu abzählbaren Folgen $d \in \ell_{q,m}(I)$ in bestimmten gewichteten Folgenräumen gilt

$$T = \sum_{i \in I} d_i \pi(x_i)\psi \in \mathrm{Co}(L_{r,m}) \quad \text{für } q < r.$$

Hier sind $(x_i)_{i \in I}$ Punkte in G, die hinreichend nah beieinander sind. Man beachte, dass die Integrationsparameter q und r der diskreten Norm der Folgenräume und der Coorbit-Norm unterschiedlich sind.

Andersherum können Funktionen ebenfalls zerlegt werden, allerdings nur näherungsweise. Genauer gesagt existiert zu jedem $\varepsilon > 0$ eine endliche Folge $c \in \ell_{r,m}(J)$, sodass

$$\left\| T - \sum_{j \in J} c_j \pi(x_j)\psi \right\|_{\mathrm{Co}(L_{r,m})} \leqslant \varepsilon$$

gilt. Die Zerlegung ist hierbei zwar stetig aber nicht gleichmäßig.

Die Zerlegung basiert allerdings auf der Annahme, dass die Faltung mit dem reproduzierenden Kern K_ψ ein stetiger Operator auf $L_{r,m}(G)$ ist. Dies ist a priori nicht klar für nicht-integrierbare Kerne und ist daher eine zusätzliche Voraussetzung. Allerdings zeigt Satz 3.3.1, dass diese Annahme sogar notwendig ist, sollte man an atomaren Zerlegungen und Banach-Frames der Coorbit-Räume interessiert sein. Folglich scheint die Annahme unumgänglich.

Dennoch sind die obigen Diskretisierungen noch nicht optimal. Um stattdessen atomare Zerlegungen und Banach-Frames zu erlangen, wird in Abschnit 3.4 angenommen, dass ein zweiter Kern W auf G existiert, der unter anderem folgende Eigenschaften hat:

(i) $W \in L_{1,w}(G)$,

(ii) $W * K_\psi = K_\psi$, beziehungsweise $K_\psi * W = K_\psi$.

Diese Eigenschaften erlauben uns in Satz 3.4.8 zu zeigen, dass die Familie $(\pi(x_i)\psi)_{i \in I}$ für geeignete diskrete Punkte $(x_i)_{i \in I} \subset G$ einen Banach-Frame bildet. Analog zeigt Satz 3.4.15, dass dieselbe Familie unter ähnlichen Voraussetzungen eine atomare Zerlegung von $\text{Co}(L_{r,m})$ bildet. Folglich ist es möglich auch für Coorbit-Räume mit nicht-integrierbaren Kern dieselben Ergebnisse zu erzielen wie im klassischen Fall.

Verallgemeinerte Coorbit-Theorie

Allerdings ist es auch möglich die Coorbit-Theorie insofern zu verallgemeinern, als dass nicht nur Gruppen die Grundlage der Transformation bilden, sondern auch allgemeinere Maßräume (X, μ) zulässig sind. In Kapitel 4 wiederholen wir die Theorie von Fornasier und Rauhut [53], die genau dies besagt. Ohne Gruppenstruktur ergibt auch eine Darstellung von Gruppen auf Hilbert-Räumen keinen Sinn. Die Grundidee ist daher beides zu ersetzen durch sogenannte stetige Frames [1]. Angenommen wir haben eine Familie $\mathfrak{F} = (\psi_x)_{x \in X}$ indiziert durch den Raum X, welche einen stetigen Frame für einen Hilbert-Raum $\mathcal{H}$ bildet. Dann ersetzt dieser Frame die Elemente $\pi(x)\psi$ der klassischen Theorie, und folglich ist die Voice-Transformation, welche durch

$$V_{\mathfrak{F}} \colon \mathcal{H} \to L_2(X, \mu), \quad V_{\mathfrak{F}} f(x) = \langle f, \psi_x \rangle_{\mathcal{H}}$$

gegeben ist, wohldefiniert.

Nun ist aber, im Gegensatz zu dem obigen klassischen Fall, keine Faltung auf allgemeinen Maßräumen definiert. Stattdessen wird messbaren Funktionen $K \colon X \times X \to \mathbb{C}$ ein Integraloperator mittels

$$K f(x) = \int_X K(x, y) f(y) \, d\mu(y)$$

zugeordnet. Betrachten wir nun den Operator, der dem Kern $K_{\mathfrak{F}}(x, y) = V_{\mathfrak{F}} \psi_y(x)$ zugeordnet wird, dann erfüllt dieser ebenso wie oben die Gleichung $K_{\mathfrak{F}} V_{\mathfrak{F}} f = V_{\mathfrak{F}} f$ für alle $f \in \mathcal{H}$. In diesem Sinne ist der Kernoperator der Ersatz, oder die Verallgemeinerung, der Faltung. Um die Integrabilität des Kerns zu messen, können wir sogenannte Kernräume $\mathcal{A}_{p,w}$ für geeignete Gewichte w auf $X \times X$ und $1 \leqslant p \leqslant \infty$ einführen. Offensichtlich benötigen wir für diese Räume Analoga der Young'schen Faltungsungleichungen; diese sind in Abschnitt 1.8 enthalten. Wenn wir nun $K_{\mathfrak{F}} \in \mathcal{A}_{1,w}$ annehmen, und damit in gewisser Weise also die Integrabilität des Kerns annehmen, dann lassen sich analog zum klassischen Fall die gewichteten Räume $\mathcal{H}_{1,v} = \{f \in \mathcal{H} : V_{\mathfrak{F}} f \in L_{1,v}(X, \mu)\}$ definieren, deren Dualräume die Reservoirs der Coorbit-Räume bilden. Wieder lässt sich die Voice-Transformation auf $\mathcal{H}'_{1,v}$ durch $V_{e,\mathfrak{F}}$ erweitern und die Coorbit-Räume $\text{Co}(Y)$ für geeignete Funktionenräume Y sind nun definiert als

$$\text{Co}(Y) = \{T \in \mathcal{H}'_{1,v} : V_{e,\mathfrak{F}} T \in Y\}.$$

Diese Räume besitzen dieselben Eigenschaften wie oben, d.h. sie sind wohldefinierte Banach-Räume und isometrisch isomorph zu dem Raum $\{f \in Y : K_{\mathfrak{F}} f = f\}$. Ebenso können auch für diese Räume Bedingungen gestellt werden, unter denen sowohl atomare Zerlegungen als auch Banach-Frames existieren. Das heißt die Coorbit-Theorie ist auch in diesem allgemeineren Fall sinnvoll.

Verallgemeinerte Coorbit-Theorie mit Nicht-Integrierbarem Kern

Wie oben bemerkt basiert dieser Ansatz allerdings auf der grundlegenden Annahme, dass $K_{\mathfrak{F}} \in \mathcal{A}_{1,w}$ gilt. In Kapitel 5 entwickeln wir daher eine vollkommen neue Theorie, die auch nicht-integrierbare Kerne zulässt. Die Ergebnisse des Abschnitts wurden größtenteils in [50] veröffentlicht. Konkret fordern wir nun $K_{\mathfrak{F}} \in \mathcal{A}_{p,w}$ für alle $1 < p < \infty$, was eine schwächere

Voraussetzung als die Integrabilität ist. Durch diese geänderte Voraussetzung benutzen wir geänderte Ungleichungen vom Young'schen Typ und können damit Ersatz für die Reservoirs $\mathcal{H}'_{1,v}$ finden und auch Coorbit-Räume definieren. Dieser Fall ist ähnlich zu dem obigen Fall der Coorbit-Theorie mit nicht-integrierbaren Kernen für Gruppen.

Auch hier stehen wir wieder vor der Frage der Diskretisierung der Coorbit-Räume und können ähnliche Ideen wie in Kapitel 3 nutzen um vergleichbare Ergebnisse zu erzielen. In Abschnitt 5.2 zeigen wir Zerlegungs- und Rekonstruktionsergebnisse. Konkret beinhaltet Satz 5.2.17 die folgende Rekonstruktion: zu abzählbaren Folgen $d \in \ell_{q,m}(I)$ in gewissen gewichteten Folgenräumen gilt

$$T = \sum_{i \in I} d_i \psi_{x_i} \in \operatorname{Co}(L_{r,m}) \quad \text{für } q < r.$$

Die Integrationsparameter q und r der diskreten Norm der Folgenräume sowie der Coorbit-Norm unterscheiden sich auch hier.

Analog können Elemente $T \in \operatorname{Co}(L_{r,m})$ approximativ zerlegt werden, d.h. zu jedem $\varepsilon > 0$ existiert eine endlich Folge $d \in \ell_{r,m}(J)$ mit

$$\left\| T - \sum_{j \in J} c_j \psi_{x_j} \right\|_{\operatorname{Co}(L_{r,m})} \leqslant \varepsilon$$

Die diskreten Punkte $(x_i)_{i \in I}$ und $(x_j)_{j \in J}$ in X werden dabei hinreichend nah beieinander gewählt, was durch zulässige Überdeckungen von X sichergestellt wird. Auch hier ist die Zerlegung stetig aber nicht gleichmäßig.

Die Zerlegung fordert allerdings eine zusätzliche Bedingung and den Kern $K_{\mathfrak{F}}$. Genauer fordern wir, dass der zugehörige Operator als beschränkter Operator auf allen gewichteten Lebesgue-Räumen $L_{p,m}(X, \mu)$ verstanden werden kann. Auch diese Frage muss individuell geklärt werden. Nach den Ergebnissen aus Abschnitt 5.3 ist dies aber notwendig, um atomare Zerlegungen und Banach-Frames für die Coorbit-Räume zu erhalten.

Als hinreichende Bedingung formulieren wir dann in Abschnitt 5.4 die folgenden. Wir nehmen wieder als zusätzliche Forderung an, dass ein zweiter Kern W existiert, mit unter anderen den folgenden Eigenschaften:

(i) $W \in \mathcal{A}_{1,w}$,

(ii) $W \circ K_{\mathfrak{F}} = K_{\mathfrak{F}}$, beziehungsweise $K_{\mathfrak{F}} \circ W = K_{\mathfrak{F}}$.

Die Operation $\circ$ bezeichnet hier die Multiplikation von Kernen. Dieser zusätzliche Kern W ermöglicht uns schließlich in Satz 5.4.7 zu beweisen, dass die Familie $(\psi_{x_i})_{i \in I}$ für geeignet gewählte diskrete Punkte $(x_i)_{i \in I} \subset X$ einen Banach-Frame von $\operatorname{Co}(L_{r,m})$ bildet. In Satz 5.4.11 zeigen wir dann unter sehr ähnlichen Voraussetzungen, dass dieselbe Familie auch eine atomare Zerlegung von $\operatorname{Co}(L_{r,m})$ bildet.

Anwendungen

Die verschiedenen Theorien werden stets durch Anwendungen motiviert. Ein wichtige Anwendung der Coorbit-Theorie mit nicht-integrierbarem Kern sind die Paley-Wiener-Räume $B_\Omega^p = \{f \in L_p(\mathbb{R}) : \operatorname{supp}\widehat{f} \subseteq \Omega\}$ für $\Omega \subset \mathbb{R}$. In Kapitel 3 stellen diese Räume sowohl positive als auch negative Beispiele für Anwendungen der Theorie dar, je nach Wahl von Ω. Wählen wir aber z.B. für Ω ein symmetrisches Intervall, dann sind die Paley-Wiener-Räume nicht nur als Coorbit-Räume zu verstehen, sondern wir können sie sogar diskretisieren und Banach-Frames sowie atomare Zerlegungen finden. Damit erhalten wir unter anderem einen alternativen Beweis des Whittaker-Kotelnikov-Shannon-Abtasttheorems.

Am Schluss der Arbeit stellen wir zusätzlich neue Räume vor: die inhomogenen Shearlet-Coorbit-Räume. Die Coorbit-Theorie findet bekanntermaßen Anwendung um die homogenen und inhomogenen Besov-Räume zu beschreiben [45]. Dieses Prinzip lässt sich auf Shearlets verallgemeinern und es wurden bereits entsprechende Shearlet-Coorbit-Räume entwickelt [25]. Wir nehmen die Grundidee dieser Shearlet-Räume und verbinden sie mit der der inhomogenen Besov-Räume und definieren die inhomogenen Shearlet-Coorbit-Räume. Dazu betrachten wir die Indexmenge

$$X = \left(\{\infty\} \times \mathbb{R}^{d-1} \times \mathbb{R}^d\right) \cup \left([-1,1]\backslash\{0\} \times \mathbb{R}^{d-1} \times \mathbb{R}^d\right)$$

und den sogenannten inhomogenen Shearlet-Frame $\mathfrak{F} = (\psi_x)_{x \in X}$ definiert durch

$$\psi_{(\infty,s,t)} = \Phi(S_s^{-1}(\cdot - t)) \quad \text{und} \quad \psi_{(a,s,t)} = |\det A_a|^{-1/2}\Psi(A_a^{-1}S_s^{-1}(\cdot - t)).$$

Der reproduzierende Kern assoziiert zu diesem Frame ist tatsächlich nicht-integrierbar, aber in allen Räumen $\mathcal{A}_{p,w}$ für $p > 1$ enthalten, sodass wir die verallgemeinerte Coorbit-Theorie mit nicht-integrierbarem Kern anwenden können. Wir zeigen, dass der entsprechende Kern die nötigen technischen Voraussetzungen erfüllt, sodass die Räume nicht nur wohldefiniert sind, sondern auch Teile der Diskretisierungsergebnisse angewendet werden können.

Bibliography

[1] S.T. Ali, J.-P. Antoine, and J.-P. Gazeau, *Continuous frames in Hilbert spaces*, Ann. Physics **222** (1993), 1–37.

[2] ———, *Coherent States, Wavelets and their Generalizations*, Graduate Texts in Contemporary Physics, Springer-Verlag, New York, 2000.

[3] C.D. Aliprantis and K.C. Border, *Infinite Dimensinal Analysis, A Hitchhiker's Guide*, third ed., Springer-Verlag Berlin Heidelberg, 2005.

[4] H.W. Alt, *Linear Functional Analysis*, Universitext, Springer-Verlag London, Ltd., London, 2016.

[5] I. Babuška, *The finite element method with lagrangian multipliers*, Num. Math. **20** (1973), 179–192.

[6] D. Bernier and K.F. Taylor, *Wavelets from square-integrable representations*, SIAM J. Math. Anal. **27** (1996), no. 2, 594–608.

[7] A. Boettcher, *Private communications*.

[8] H. Brezis, *Functional Analysis, Sobolev Spaces and Partial Differential Equations*, Universitext, Springer, New York, 2011.

[9] F. Brezzi, *On the existence, uniqueness and approimation of saddle-point problems arising from lagrange multipliers*, R.A.I.R.O. Analyse Numérique 8, N° R2 (1974), 129–151.

[10] E.J. Candès and L. Demanet, *The Curvelet Representation of Wave Propagators Is Optimally Sparse*, Comm. Pure Appl. Math. **LVIII** (2005), 1472–1528.

[11] E.J. Candès and D.L. Donoho, *Ridgelets: a key to higher dimensional intermittency?*, Phil. Trans. R. Soc. Lond. A. **357** (1999), 2495–2509.

[12] ———, *New tight frames of curvelets and optimal representations of objects with piecewise C^2 singularities*, Comm. Pure Appl. Math. **57** (2004), no. 2, 219–266.

[13] M. Christ, *Lectures on singular integral operators*, CBMS Regional Conference Series in Mathematics, American Mathematical Society, 1991.

[14] J. G. Christensen and G. Ólafsson, *Coorbit spaces for dual pairs*, Appl. Comput. Harmon. Anal. **31** (2011), no. 2, 303–324.

[15] O. Christensen, *An Introduction to Frames and Riesz Bases*, second ed., Applied and Numerical Harmonic Analysis, Birkhauser, 2016.

[16] R. R. Coifman and G. Weiss, *Analyse harmonique non-commutative sur certain espaces homogènes*, Lecture Notes in Mathematics, Springer-Verlag Berlin Heidelberg, 1971.

[17] S. Dahlke, F. De Mari, E. De Vito, S. Häuser, G. Steidl, and G. Teschke, *Different Faces of the Shearlet Group*, J. Geometric Analysis **26** (2016), no. 3, 1693–1729.

[18] S. Dahlke, F. De Mari, E. De Vito, D. Labate, G. Steidl, G. Teschke, and S. Vigogna, *Coorbit spaces with voice in a Fréchet space*, J. Fourier Anal. Appl. **23** (2017), no. 1, 141–206.

[19] S. Dahlke, F. De Mari, E. De Vito, L. Sawatzki, G. Steidl, G. Teschke, and F. Voigtlaender, *On the atomic decomposition of coorbit spaces with non-integrable kernel*, Applied and Numerical Harmonic Analysis, ch. 4, pp. 75–144, Birkhäuser/Springer Basel AG, Basel, 2019.

[20] S. Dahlke, F. De Mari, P. Grohs, and D. Labate, *From Group Representations to Signal Analysis*, Appl. Numer. Harmon. Anal., Birkhäuser/Springer, Cham, 2015.

[21] S. Dahlke, M. Fornasier, H. Rauhut, G. Steidl, and G. Teschke, *Generalized coorbit theory, Banach frames, and the relation to alpha-modulation spaces*, Proc. London Math. Soc. **96** (2008), no. 2, 464–506.

[22] S. Dahlke, S. Häuser, G. Steidl, and G. Teschke, *Shearlet coorbit spaces: traces and embeddings in higher dimensions*, Monatsh. Math. **169** (2012), 15–32.

[23] S. Dahlke, S. Häuser, and G. Teschke, *Coorbit space theory for the Toeplitz shearlet transform*, Int. J. Wavelets Multiresolut. Inf. Process. **10** (2012), no. 4.

[24] S. Dahlke, G. Kutyniok, P. Maass, C. Sagiv, H. G. Stark, and G. Teschke, *The uncertainty principle associated with the continuous shearlet transform*, Int. J. Wavelets Multiresolut. Inf. Process. **6** (2008), 157–181.

[25] S. Dahlke, G. Kutyniok, G. Steidl, and G. Teschke, *Shearlet coorbit spaces and associated Banach frames*, Applied and Computational Harmonic Analysis **27** (2009), no. 2, 195–214.

[26] S. Dahlke, G. Steidl, and G. Teschke, *Coorbit spaces and Banach frames on homogeneous spaces with applications to the sphere*, Advances Comp. M. **21** (2004), 147–180.

[27] _____, *Weighted coorbit spaces and banach frames on homogeneous spaces*, J. Fourier. Anal. Appl. **10** (2004), no. 5, 507–539.

[28] _____, *The continuous shearlet transform in arbitrary space dimensions*, J. Fourier Anal. Appl. **16** (2010), no. 2, 340–354.

[29] _____, *Shearlet coorbit spaces: compactly supported analyzing shearlets, traces and embeddings*, J. Fourier Anal. Appl. **17** (2011), no. 6, 1232–1255.

[30] S. Dahlke and G. Teschke, *The continuous shearlet transform in arbitrary space dimensions: variations of a theme*, Group Theory: Classes, Representations and Connections, and Applications (2010), 167–175.

[31] I. Daubechies, *Ten Lectures on Wavelets*, SIAM Philadelphie, PA, USA, 1992.

[32] I. Daubechies, A. Grossmann, and Y. Meyer, *Painless nonorthogonal expansions*, J. Math. Phys. **27** (1986), 1271–1283.

[33] H.W. Davis, F.J. Murray, and J.K. Jr. Weber, *Families of L_p-spaces with inductive and projective topologies*, Pacific J. Math. **34** (1970), 619–638.

[34] M. Do and M. Vetterli, *Contourlets: a directional multiresolution image representation*, Proc. International Conference on Image Processing, IEEE, 2002.

[35] ———, *The contourlet transform: an efficient directional multiresolution image representation*, IEEE transactions on Image Processing **14** (2005), no. 12, 2091–2106.

[36] R.M. Dudley, *Real Analysis and Probability*, Cambridge Studies in Advanced Mathematics, vol. 74, Cambridge University Press, Cambridge, 2002.

[37] R.J. Duffin and A.C. Schaeffer, *A class of nonharmonic Fourier series*, Trans. Amer. Math. Soc. **72** (1952), 341–366.

[38] M. Duflo and C. Moore, *On the regular representation of a nonunimodular locally compact group*, J. Functional Analysis **21** (1976), no. 2, 209–243.

[39] J. Elstrodt, *Maß- und Integrationstheorie*, 7. ed., Grundwissen Mathematik, Springer-Verlag Berlin Heidelberg, 2010.

[40] H.G. Feichtinger, *A characterization of minimal homogeneous Banach spaces*, Proc. Amer. Math. Soc. **81** (1981), no. 1, pp. 55–61.

[41] ———, *Modulation spaces on locally compact abelian groups*, Tech. report, University Vienna, 1983.

[42] ———, *Banach spaces of distributions defined by decomposition methods II*, Math. Nachr. **132** (1987), 207–237.

[43] H.G. Feichtinger and P. Gröbner, *Banach spaces of distributions defined by decomposition methods I*, Math. Nachr. **123** (1985), 97–120.

[44] H.G. Feichtinger and K.H. Gröchenig, *A unified approach to atomic decompositions via integrable group representations*, Function spaces and applications (Lund, 1986), Lecture Notes in Math., vol. 1302, Springer, Berlin, 1988, pp. 52–73.

[45] ———, *Banach spaces related to integrable group representations and their atomic decompositions. I*, J. Funct. Anal. **86** (1989), no. 2, 307–340.

[46] ———, *Banach spaces related to integrable group representations and their atomic decompositions. II*, Monatsh. Math. **108** (1989), no. 2-3, 129–148.

[47] ———, *Multidimensional irregular sampling of band-limited functions in L_p-spaces*, Conf. Oberwolfach Feb. 1989 (1989), 135–142.

[48] ———, *Error analysis in regular and irregular sampling theory*, Applicable Analysis **50** (1993), no. 3-4, 167–189.

[49] H.G. Feichtinger and M. Pap, *Coorbit theory and Bergman spaces*, Harmonic and Complex Analysis and its Applications (2014), 231–259.

[50] F. Feise and L. Sawatzki, *Inhomogeneous shearlet coorbit spaces*, Int. J. Wavelets Multiresolut. Inf. Process. **16** (2018), no. 4.

[51] G.B. Folland, *A Course in Abstract Harmonic Analysis*, Studies in Advanced Mathematics, CRC Press, Boca Raton, FL, 1995.

[52] _____, *Real Analysis*, second ed., Pure and Applied Mathematics (New York), John Wiley & Sons Inc., New York, 1999.

[53] M. Fornasier and H. Rauhut, *Continuous frames, function spaces, and the discretization problem*, J. Fourier Anal. Appl. **11** (2005), no. 3, 245–287.

[54] J. Frazier and B. Jawerth, *Decomposition of Besov Spaces*, Indiana Univsersity Math. J. **34** (1985), no. 4.

[55] H. Führ, *Wavelet frames and admissibility in higher dimensions*, J. Math. Phys. **37** (1996), no. 12, 6353–6366.

[56] _____, *Coorbit spaces and wavelet coefficient decay over general dilation groups*, preprint (2012).

[57] _____, *Vanishing moment conditions for wavelet atoms in higher dimensions*, Advances Comp. M. **42** (2016), no. 1, 127–153.

[58] H. Führ and K.H. Gröchenig, *Sampling theorems on locally compact groups from oscillation estimates*, Math. Z **255** (2007), 177–194.

[59] H. Führ and R. Koch, *Embeddings of shearlet coorbit spaces into Sobolev spaces*, ArXiv e-prints (2019).

[60] H. Führ and R. Raisi Tousi, *Simplified vanishing moment criteria for wavelets over general dilation groups, with applications to abelian and shearlet dilation groups*, Appl. Comput. Harmon. Anal. **43** (2017), no. 3, 449–481.

[61] H. Führ and F. Voigtlaender, *Wavelet coorbit spaces viewed as decomposition spaces*, J. Funct. Anal. **269** (2015), no. 1, 80–154.

[62] J.-P. Gabardo and D. Han, *Frames associated with measurable spaces*, Adv. Comput. Math. **18** (2003), 127–147.

[63] F. Gensun, *Whittaker-Kotelnikov-Shannon Sampling Theorem and Aliasing Error*, J. Approx. Theory **85** (1996), 115–131.

[64] K. Goebel and S. Reich, *Uniform Convexity, Hyperbolic Geometry, and Nonexpansive Mappings*, Pure Appl. Math., CRC Press, 1984.

[65] L. Grafakos, *Classical Fourier Analysis*, Graduate Texts in Mathematics, Springer-Verlag, New York, 2004.

[66] _____, *Modern Fourier analysis*, third ed., Graduate Texts in Mathematics, vol. 250, Springer, 2014.

[67] K.H. Gröchenig, *Unconditional bases in translation and dilation invariant function spaces on $\mathbb{R}^n$*, Constructive Theory of Functions '87 (1988), 174–183.

[68] K.H. Gröchenig, *Describing functions: atomic decompositions versus frames*, Monatsh. Math. **112** (1991), no. 1, 1–42.

[69] _____, *Foundations of time-frequency analysis*, Applied and Numerical Harmonic Analysis, Birkhäuser/Springer Basel AG, Basel, 2001.

[70] A. Grossmann, J. Morlet, and T. Paul, *Transforms associated to square integrable group representations. I. General results*, J. Math. Phys. **26** (1985), no. 10, 2473–2479.

[71] K. Guo, G. Kutyniok, and D. Labate, *Sparse multidimensional representations using anisotropic dilation and shear operators*, Wavelets and Splines (2005).

[72] K. Guo and D. Labate, *Optimally sparse multidimensional representation using shearlets*, SIAM J. Math. Anal. **39** (2007), no. 1, 298–318.

[73] P.R. Halmos and V.S. Sunder, *Bounded Integral Operators on L^2 Spaces*, Ergebnisse der Mathematik und ihrer Grenzgebiete [Results in Mathematics and Related Areas], vol. 96, Springer-Verlag, Berlin-New York, 1978. MR 517709

[74] E. Hewitt and K. A. Ross, *Abstract Harmonic Analysis. Vol. I: Structure of Topological Groups. Integration Theory, Group Representations*, Die Grundlehren der mathematischen Wissenschaften, Bd. 115, Academic Press Inc., Publishers, New York, 1963.

[75] L. Hoermander, *The Analysis of Linear Partial Differential Operators I – Distribution Theory and Functional Analysis*, second ed., Classics in Mathematics, Springer-Verlag, Berlin, 2003.

[76] N. Holighaus, C. Wiesmeyr, and P. Balazs, *Continuous warped time-frequency representations - coorbit spaces and discretization*, Applied and Computational Harmonic Analysis (2018).

[77] L. Hörmander, *The Analysis of Linear Partial Differential Operators. I*, second ed., Springer Study Edition, Springer-Verlag, Berlin, 1990.

[78] G. Kaiser, *A friendly guide to wavelets*, Birkhäuser, Boston, 1994.

[79] H. Kempka, M. Schäfer, and T. Ullrich, *General coorbit space theory for quasi Banach spaces and inhomogeneous function spaces with variable smoothness and integrability*, J. Fourier Anal. Appl. **23** (2017), no. 6, 1348–1407.

[80] G. Kutyniok and D. Labate, *Resolution of the wavefront set using continuous shearlets*, Trans. Amer. Math. Soc. **361** (2009), no. 5, 2719–2754.

[81] G. Kutyniok and D. Labate (eds.), *Shearlets: Multiscale Analysis for Multivariate Data*, Appl. Numer. Harmon. Anal., Birkhäuser Basel, 2012.

[82] D. Labate, L. Mantovani, and P. Negi, *Shearlet Smoothness Spaces*, J. Fourier Anal. Appl. **19** (2013), no. 3, 577–611.

[83] V. Lebedev and A. Olevskiĭ, *Idempotents of Fourier multiplier algebra*, Geom. Funct. Anal. **4** (1994), no. 5, 539–544.

[84] R. Meise and D. Vogt, *Introduction to Functional Analysis*, Oxford Graduate Texts in Mathematics, Clarendon Press, 1997.

[85] T.J. Morrison, *Functional Analysis: An Introduction to Banach Space Theory*, John Wiley and Son, Inc., Hoboken, NJ, USA, 2000.

[86] R. Murenzi, *Wavelet transforms associated to the n-dimensional Eucdledian group with dilations: signal in more than one dimension*, Wavelets, Time-Frequency Methods and Phase Space Proceedings of the International Conference (1987), 239–246.

[87] M. Plancherel and G. Pólya, *Fonctions entières et intégrales de fourier multiples*, Comment. Math. Helv. **10** (1937), no. 1.

[88] H. Rauhut, *Banach frames in coorbit spaces consisting of elements which are invariant under symmetry groups*, Appl. Comput. Harmon. Anal. **18** (2005), no. 1, 94–122.

[89] _____, *Radial time-frequency analysis and embeddings of radial modulation spaces*, Sampl. Theory Signal Image Process. **5** (2006), no. 2, 201–224.

[90] _____, *Coorbit space theory for quasi-Banach spaces*, Studia Mathematica **180** (2007), no. 3, 237–253.

[91] _____, *Wiener amalgam spaces with respect to quasi-Banach spaces*, Colloquium Math. **109** (2007), 345–361.

[92] W. Rudin, *Functional Analysis*, International Series in Pure and Applied Mathematics, McGraw Hill, 1991.

[93] C. Schneider, *Besov spaces with positive smoothness*, Ph.D. thesis, Universiät Leipzig, 2009.

[94] L. Schumaker et al. (eds.), *Curvelets - a surprisingly effective nonadaptive representation for objects with edges*, Vanderbilt University Press, Nashville, 2000.

[95] L. Schwartz, *Théorie des Distributions.*, Hermann & Cie., Paris, 1966.

[96] C. Shannon, *A Mathematical Theoryof Communication*, Bell System Technical Journal **27** (1948), no. 4, 623–656.

[97] H. Triebel, *Characterization of Besov-Hardy-Sobolev spaces: a unified approach*, J. Approx. Theory **52** (1988), 162–203.

[98] _____, *Theory of Function Spaces II*, Monographs in Mathematics 84, Birkhäuser Basel, 1992.

[99] _____, *Theory of Function Spaces III*, Monographs in Mathematics 100, Birkhäuser Basel, 2006.

[100] _____, *Theory of Function Spaces*, Modern Birkhäuser Classics, Birkhäuser/Springer Basel AG, Basel, 2010, Reprint of 1983 edition.

[101] T. Ullrich, *Continuous Characterizations of Besov-Lizorkin-Triebel Spaces and New Interpretations as Coorbits*, J. Function Spaces App. **2012** (2010).

[102] T. Ullrich and H. Rauhut, *Generalized coorbit space theory and inhomogeneous function spaces of Besov-Lizorkin-Triebel type*, J. Funct. Anal. **11** (2011), 3299–3362.

[103] D. Vera, *Triebel Lizorkin spaces and shearlets on the cone in $\mathbb{R}^2$*, Appl. Comput. Harmon. Anal. **35** (2013), 130–150.

[104] _____, *Shear anisotropic inhomogeneous Besov spaces in $\mathbb{R}^d$*, Int. J. Wavelets Multiresolut. Inf. Process. **12** (2014), no. 1.

[105] F. Voigtlaender, *Embedding Theorems for Decomposition Spaces with applications to Wavelet Coorbit Spaces*, Ph.D. thesis, RWTH Aachen University, 2015.

[106] _____, *Embeddings of decomposition spaces into sobolev and bv spaces*, ArXiv e-prints (2016).

[107] P. Wojtaszczyk, *A Mathematical Introduction to Wavelets*, Cambridge University Press, Cambridge, 1997.

[108] K. Yosida, *Functional analysis*, Springer, 1974.